2006 Distinguished Instructor Short Course

Distinguished Instructor Series, No. 9

sponsored by the
Society of Exploration Geophysicists
European Association of Geoscientists & Engineers

Seismic Attribute Mapping of Structure and Stratigraphy

presented by

Kurt J. Marfurt

written by

Satinder Chopra and Kurt J. Marfurt

SEG

EAGE
EUROPEAN ASSOCIATION OF GEOSCIENTISTS & ENGINEERS

ISBN 1-56080-086-0 (Series)
ISBN 1-56080-135-2 (Volume)

Society of Exploration Geophysicists
P.O. Box 702740
Tulsa, OK USA 74170-2740

Printed in the U.S.A.

Library of Congress Control Number: 2006921277

SEG and EAGE

wish to thank the following
for their generous contributions

SEG Foundation

UNIVERSITY OF HOUSTON
Learning. Leading.

About the Authors

Kurt J. Marfurt began his geophysical career as an assistant professor teaching mining geophysics at Columbia University's Henry Krumb School of Mines in New York. After five years, he joined Amoco at its Tulsa Research Center. Through successive reorganizations at Amoco, Marfurt has obtained diverse experience in seismic modeling, migration, signal analysis, basin analysis, seismic-attribute analysis, reflection tomography, seismic inversion, and multicomponent data analysis. Through Amoco, he won five patents, two in seismic coherence technology.

Marfurt joined the University of Houston in 1999 as a professor in the Department of Geosciences and as director of the Center for Applied Geosciences and Energy (CAGE), where his primary emphasis is on development and calibration of new seismic-attribute technology. He is the author of several dozen articles and coeditor of two books published by SEG. He has received the SEG Best Presentation Award for work on seismic modeling (1989) and SEG Best Paper Award for work on seismic coherence (1999) and has served 16 years as an associate or assistant editor for GEOPHYSICS. Marfurt is a member of GSH, SEG, EAGE, AAPG, and AGU.

Satinder Chopra is manager of reservoir services at Arcis Corporation, Calgary, Alberta, Canada. He joined the Oil and Natural Gas Corporation Limited (ONGC) of India in 1984 and worked there until 1997. In 1998, he joined CTC Pulsonic in Calgary, which later became Scott Pickford and Core Laboratories Reservoir Technologies. He has worked in regular seismic processing and interactive interpretation but has spent more time in special processing of seismic data involving seismic attributes, including coherence and textures, seismic inversion, AVO, VSP processing, and frequency enhancement of seismic data. Chopra's research interests focus on techniques aimed at characterization of reservoirs. He has written three books and more than 80 papers and abstracts and has given many presentations.

Chopra received M.S. and M.Phil. degrees in physics from Himachal Pradesh University, Shimla, India. He is chairman of the SEG Publications Committee and is editor of the *CSEG Recorder*. He received several awards at ONGC, and he and his colleagues received the CSEG Best Poster Awards from 2002 through 2005. He is a member of SEG, CSEG, EAGE, AAPG, the Association of Professional Engineers, Geologists and Geophysicists of Alberta, and the Texas Board of Professional Geoscientists. Chopra lives in Calgary with his wife and two children.

SEG

EAGE

EUROPEAN
ASSOCIATION OF
GEOSCIENTISTS &
ENGINEERS

Dear DISC Participant:

It is a great pleasure to welcome you to the ninth annual SEG/EAGE Distinguished Instructor Short Course (DISC), "Seismic Attribute Mapping of Structure and Stratigraphy," by Dr. Kurt J. Marfurt. SEG, EAGE, and your local society are proud to provide this premier course in geophysics education.

With rapidly changing technologies, geoscientists around the world have an increasing need to acquire expert technical knowledge. SEG's Professional Development program — the SEG/EAGE DISC, the SEG Distinguished Lecturer, the SEG/AAPG Distinguished Lecturer, and live and online short courses — aids in the promotion of technologies that will have a significant impact on geophysics and geology. Likewise, EAGE supports geoscience education through its Distinguished Lectures and short courses and its Learning Geoscience e-learning initiative.

Previous DISC programs were:

- "Time-Lapse Seismic in Reservoir Management," by Ian Jack, in 1998
- "The Seismic Velocity Model as an Interpretation Asset," by Phil Schultz, in 1999
- "Shear Waves from Acquisition to Interpretation," by Robert Garotta, in 2000
- "Seismic Amplitude Interpretation," by Fred J. Hilterman, in 2001
- "Understanding Seismic Anisotropy in Exploration and Exploitation," by Leon Thomsen, in 2002
- "Geostatistics for Seismic Data Integration in Earth Models," by Olivier Dubrule, in 2003
- "Petroleum Systems of Deepwater Settings," by Paul Weimer, in 2004
- "Insights and Methods for 4D Reservoir Monitoring and Characterization," by Rodney Calvert, in 2005

Education in geophysics and geoscience is one of the top priorities for both SEG and EAGE. The DISC affords important opportunities for local organizations to provide first-rate geophysical education at modest cost. The program is truly a cooperative effort of many people dedicated to the promotion and advancement of geophysics. Your participation is key to the continued success of this annually renewed program.

We are honored to have Kurt Marfurt for our 2006 DISC program. This is a great opportunity to learn from one of our profession's recognized experts in the use of seismic attributes to improve the quality of interpretations. We encourage you to take full advantage of this opportunity to broaden your perspectives through participation in the 2006 DISC.

Sincerely,

Terence K. Young
SEG President

Gareth Williams
EAGE President

Table of Contents

Acknowledgments

The SEG/EAGE 2006 Distinguished Instructor Short Course, "Seismic Attribute Mapping of Structure and Stratigraphy," is based on Kurt J. Marfurt's SEG short course, "3D Seismic Attributes for Prospect Identification and Reservoir Characterization." This book, which accompanies the 2006 DISC, is drawn from a larger volume initiated by Satinder Chopra, cowritten by Marfurt, and scheduled to be published by SEG later in 2006.

Marfurt is particularly indebted to his students and colleagues at Allied Geophysical Laboratories, University of Houston, who helped to write, debug, calibrate, proofread and, of course, reject many of the concepts that became part of the original book and the DISC lecture. Specifically, he wishes to recognize the contributions of Charlotte Sullivan, Chuck Blumentritt, and Peter Bartok, who "kept Marfurt honest," he says, by grounding him in geology, and Saleh al-Dossary, Jianlei Liu, and Cory Hoelting, who helped with algorithm development.

Many of the examples in this course come from thesis work by University of Houston students, including Yomi Oyedele, Claudio d'Agosto-Palladino, Prasad Jyosula, Ivana Lazarevic, Yves Simon, Marija Djordjevic, and Joel Famini. Financial support for work done at the University of Houston came from the U. S. Department of Energy, the American Chemical Society Petroleum Research Fund, the Texas Applied Technology Program, and the industrial sponsors of the Allied Geophysical Laboratories.

In addition, after 31 years of marriage, Marfurt sincerely thanks his wife, Stephanie, for her longtime support and understanding. Although he says he may have rocks in his head, it is Stephanie who is always in his heart.

Satinder Chopra began the book that is the foundation for this course while he was working at Coherence Technology Company (CTC), which later became part of Core Lab. He is particularly indebted to Vasudhaven Sudhakar (Sudha), who first introduced him to coherence technology. Sudhaker's belief in Chopra's abilities sustained him in his work, wherein the intriguing and fascinating aspects of an emerging technology are revealed.

Chopra also is indebted to his former CTC/Core Lab colleagues Bob Stevens, Laura Evins, and Vladimir Alexeev, who were always willing to help. Many examples included in the text have been borrowed from the work of other colleagues as well, and Chopra and Marfurt gratefully acknowledge those contributions.

Chopra thanks his parents, who nurtured in him a thirst for knowledge that continues still. He thanks the love of his life, his wife, Amrita, for her love, encouragement, and personal sacrifice, without which he would not have been able to accomplish this work. He also extends special thanks to his two children. A lot of precious time, which was rightfully theirs, was spent in the writing of this book, he says.

Together, Chopra and Marfurt acknowledge and thank their coauthors of previously published works who were or are affiliated with Amoco, Coherence Technology Company, and the University of Houston. Those authors will recognize many of their detailed interpretations incorporated into Units 4 and 5 of this DISC 2006 book. Chopra and Marfurt are deeply indebted to Steve Hill, who took on the Herculean task of reviewing the "big book" from which this DISC book is extracted. They also are indebted to the SEG publications team. In particular, they thank staff members Rowena Mills, manager of GEOPHYSICS and books, and William D. Underwood, manager of professional development; copy editor Anne Thomas of Thomas, Ink; and Pat Dock of ProType Inc., in Tulsa, Oklahoma.

Finally, Chopra and Marfurt thank the following companies for permission to use their data to illustrate this work:

Alberta Energy Company
Amoco Production Company
Anadarko
Arcis
BP
Burlington Resources
CanOxy
CGAS
CGG
ChevronTexaco
ConocoPhillips
Continental Resources of Illinois, Inc.
Core Lab
Crestar Energy Development Co. Ltd.
Devon Energy
EnCana Corporation
ExxonMobil
Geco-Prakla
Geospectrum
Geotexture
Kicking Horse Resources
Marathon Oil Co.
Mull Energy Company
Olympic Seismic
Output Exploration LLC
OXY Permian
Paradigm Geophysical
PdVSA
Pemex
Petrovera Resources
Phillips Petroleum
Pohle Oil and Gas
Pure Resources
Saudi Aramco
Seitel
Shell Hibernia Management
Talisman
Texaco
Unocal Resources
WesternGeco

Satinder Chopra
Calgary, Alberta, Canada

Kurt J. Marfurt
Houston, Texas, U.S.A.

Preface

This book, *Seismic Attribute Mapping of Structure and Stratigraphy*, by Satinder Chopra and Kurt J. Marfurt, is designed specifically as a companion to Marfurt's 2006 DISC presentation of the same title and is composed primarily of figures from a forthcoming SEG book by the same authors, *Seismic Attributes for Prospect Identification and Reservoir Characterization.*

Seismic data are incredibly rich in information and provide interpreters with measurements of amplitude, frequency, and the configuration or morphology of reflection events. Seismic attribute analysis is a rapidly evolving subject because new attributes are added every year. Attributes such as volumetric estimates of dip and azimuth, coherence, curvature, amplitude texture, and spectral decomposition can greatly accelerate interpretation of newly acquired 3D surveys and can provide new insights into old 3D surveys. Successful use of seismic attributes requires an understanding of not only their physical basis but also of the underlying seismic data quality and sedimentary and tectonic processes.

Both books discuss basic concepts of seismic attributes, their physical significance, and descriptions of the algorithms that generate those attributes. Examples illustrate how seismic attributes prove useful in interpretation and in understanding reservoir properties.

Unit 1 Introduction

Overview

Good seismic attributes and effective attribute-analysis tools mimic a good interpreter. Seismic attributes are specific measures of geometric, kinematic, dynamic, and statistical features derived from seismic data. General attributes include measures of reflector amplitude, reflector time, formation thickness, energy between formation top and bottom, reflector dip and azimuth, complex amplitude and frequency, illumination, coherence, amplitude versus offset, and spectral decomposition. If we have a working geologic model, often we can relate general attributes directly to stratigraphic deposition, tectonic deformation, and reservoir properties.

In contrast, hundreds of sometimes statistically valid specific attributes also exist. An attempt to correlate hundreds of general or specific attributes in an exploratory data-analysis work flow seeking to predict reservoir properties risks generating a false-positive correlation. Our analogy would be to predict a week of weather (sunshine or rain) by flipping coins from 100 countries. Although we all agree that it would be foolish to plan tomorrow's activities on the basis of a coin toss, we are statistically assured that one of those coins will correlate strongly with the seven days of weather.

One way to avoid a false-positive correlation is to use only attributes that have a physical reason to correlate to a given lithologic property or geologic process. In this course, we focus on a suite of recently developed volumetric attributes that form the building blocks for statistical analysis, 3D seismic stratigraphy, and seismic geomorphology. We emphasize the theoretical basis for modern attributes that measure seismic textures, and we explore those attributes' algorithmic implementation, data sensitivity, and interpretational use.

Volumetric estimates of dip and azimuth, coherence, curvature, and lateral changes in amplitude or energy commonly are called geometric attributes. By including spectral decomposition, texture analysis, and certain measures of seismic reflectivity within azimuth- and offset-limited volumes, we broaden our scope considerably beyond that of attributes that clearly map changes in 3D reflector geometry.

We have two major interpretational objectives in using seismic attributes. The first is to intelligently extrapolate sparse well measurements, such as reservoir thickness, porosity, and hydrocarbon saturation, onto the much denser seismic grid. Prediction tools include multivariable statistics, geostatistics, and neural networks applied to attributes sensitive to amplitude and thickness. Practitioners are typically petrophysicists, geophysicists, and engineers.

The second interpretational objective in using seismic attributes is to reconstruct the tectonic, depositional, and diagenetic history — which will allow us to infer lithology, porosity, seal, fracture density, and fracture orientation. Prediction tools include geologic models and modern and paleoanalogs. Typically, practitioners are structural geologists and stratigraphers. In this course, we will see that geometric attributes can be applied readily to this second objective. In contrast, we will see that quantitatively predicting fractures and permeability requires further refinement of our statistical analysis and validation of our work flows.

Third, seismic attributes can provide quality control in data processing. On uninterpreted time slices, processors easily can identify geologic features, such as faults and channels, and processing artifacts, such as false structure and acquisition footprint, and thereby can optimize their choice of processing parameters.

We also will summarize the methods we will use for attribute display in this course. We review the red-green-blue (RGB), cyan-magenta-yellow-black (CMYK), and hue-lightness-saturation (HLS) color models. Then we discuss conventional but effective 1D color bars for displaying seismic data and attributes. We evaluate combining displays of attributes either with seismic data or with other attributes, using overlays, composite displays, 2D color bars, and 3D color bars. We identify good and bad color display practices, use 2D and 3D color bars to display multiple attributes in a single image, and apply color schemes that allow interpreters to communicate these features to others.

Summary

Seismic attributes usually are displayed in color to capture as much detail as possible and to help them resemble a natural analog familiar to team members or investors who may not be geoscientists. The best color scales have analogs to everyday human perception and/or experience. For example, complex-trace-envelope displays often use a rainbow color scheme with hot colors representing high values and cool colors indicating low values.

Time-structure maps also are displayed using a rainbow color scheme, but usually with shallow depths (low values) plotted against red and deeper values against blue, to draw the interpreter's attention to shallow structural features that may trap hydrocarbons. In dip-magnitude and coherence images, we wish to draw attention to anomalous zones. Thus, we plot low values of dip magnitude and high values of coherence as white, and we plot high values of dip magnitude and low values of coherence as black.

Because hue is cyclic, it is a natural choice for attributes that exhibit cyclic behavior, such as phase and azimuth. Although the choice of a discontinuous color bar can highlight features of interest for a presentation, such color bars prevent the data from speaking for themselves. Instead, interpreters should use single or dual gradational color bars. For seismic data or attribute displays exhibiting both positive and negative values, one should always use a neutral background color (such as white, gray, or black) for attribute values close to zero.

In this unit, we have shown how we combine the information from more than one attribute within a single display. The commonest techniques include seismic overlays and blended or mixed maps. Composite color maps, which are less common, map a first attribute with its color bar if its amplitude exceeds a given threshold, and a second attribute with a distinctive color bar if it does not. Color tables for 2D and3D displays are extremely powerful in mapping vector data, such as dip magnitude and dip azimuth, and energy-weighted coherent-amplitude gradients. Color bars for 2D and 3D displays also allow interpreters to modulate attributes by a measure of confidence. For example, estimates of reflector dip and azimuth are valid only if the reflector is coherent locally. Meaningful estimates of reflector shape are valid only if the reflector curvedness is not zero (i.e., is not planar). Meaningful estimates of a peak spectral frequency require the amplitude of that frequency component to be significantly above the average spectral amplitude.

Through meaningful color-coding schemes, multiattribute seismic visualization tools allow interpreters to link selected attributes visually, thereby making data analysis, communication, and presentation more efficient.

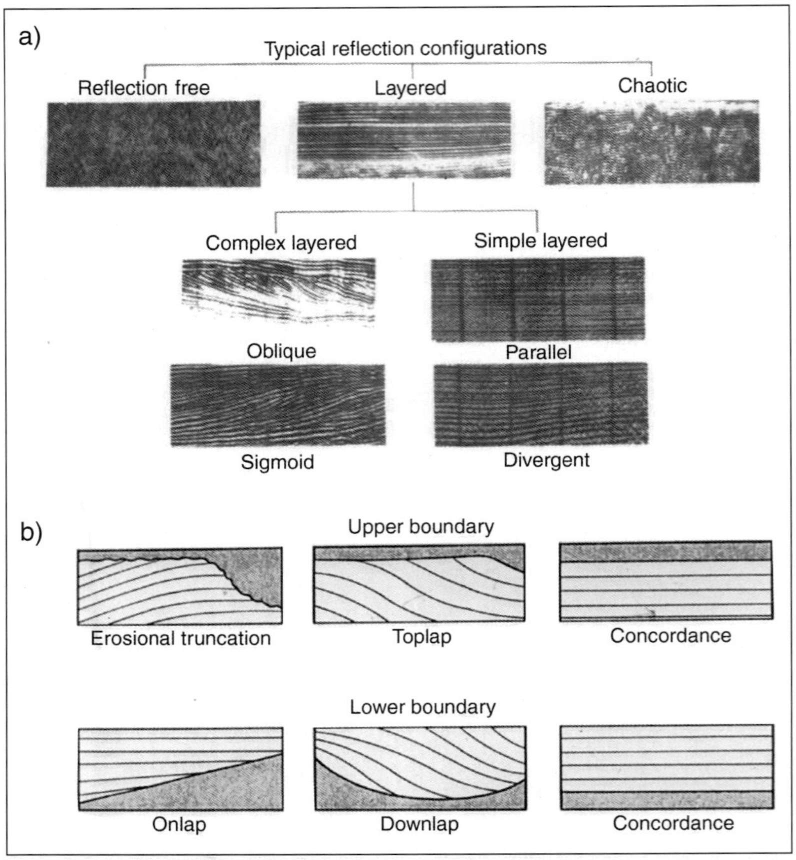

Figure 1.1. Scanned copies of two slides used by Tury Taner in the AAPG-sponsored school on seismic stratigraphy during the middle 1970s and the 1980s. (a) Representative reflection characters seen on 2D seismic lines. (b) Idealized characters used in seismic stratigraphy interpretation. These early interpretation-workflow concepts inspired later developments in geometric attributes (including the study of volumetric dip and azimuth, reflector parallelism, continuity, and unconformity indicators). These images are courtesy of Tury Taner, Rock Solid Images.

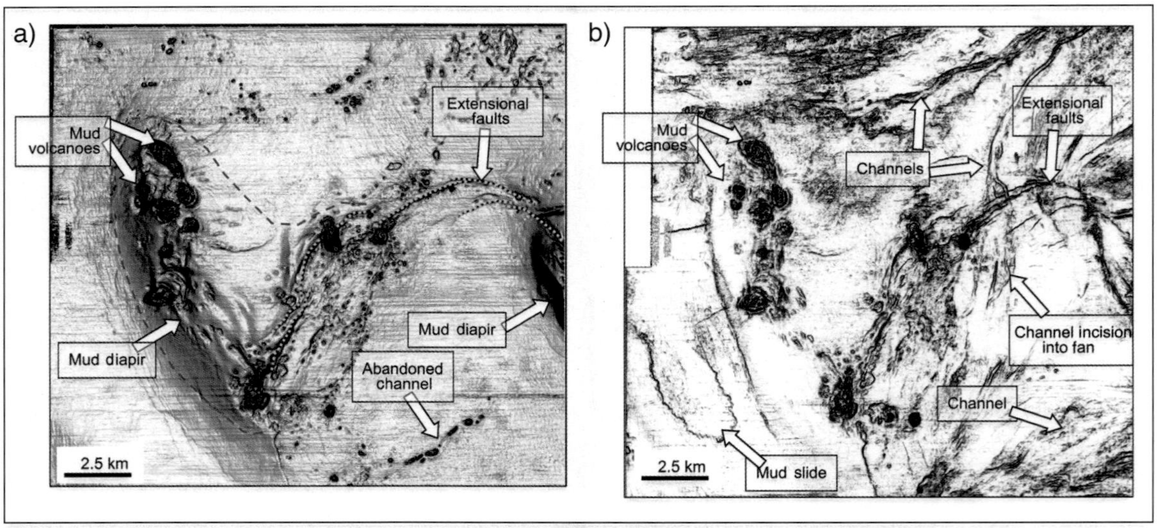

Figure 1.2. (a) A conventional dip-magnitude map generated from water-bottom picks. (b) A horizon slice, 250 ms below the water bottom, through an "edge" cube generated using a variation of the Sobel filter. After Adeogba et al. (2005).

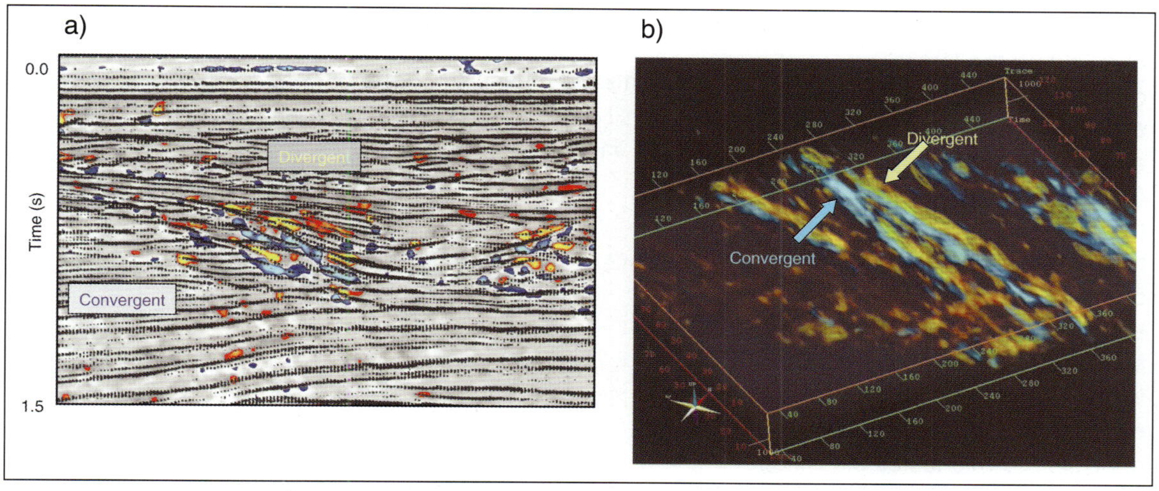

Figure 1.3. Calculation of reflector convergence and divergence is the first step toward 3D computer-aided seismic stratigraphy. In these images, positive convergence is shown as blue, and negative convergence (divergence) is shown as yellow. Like dip and azimuth, reflector convergence is a vector. Here, we display the component of convergence/divergence in the inline direction. After Barnes (2000b).

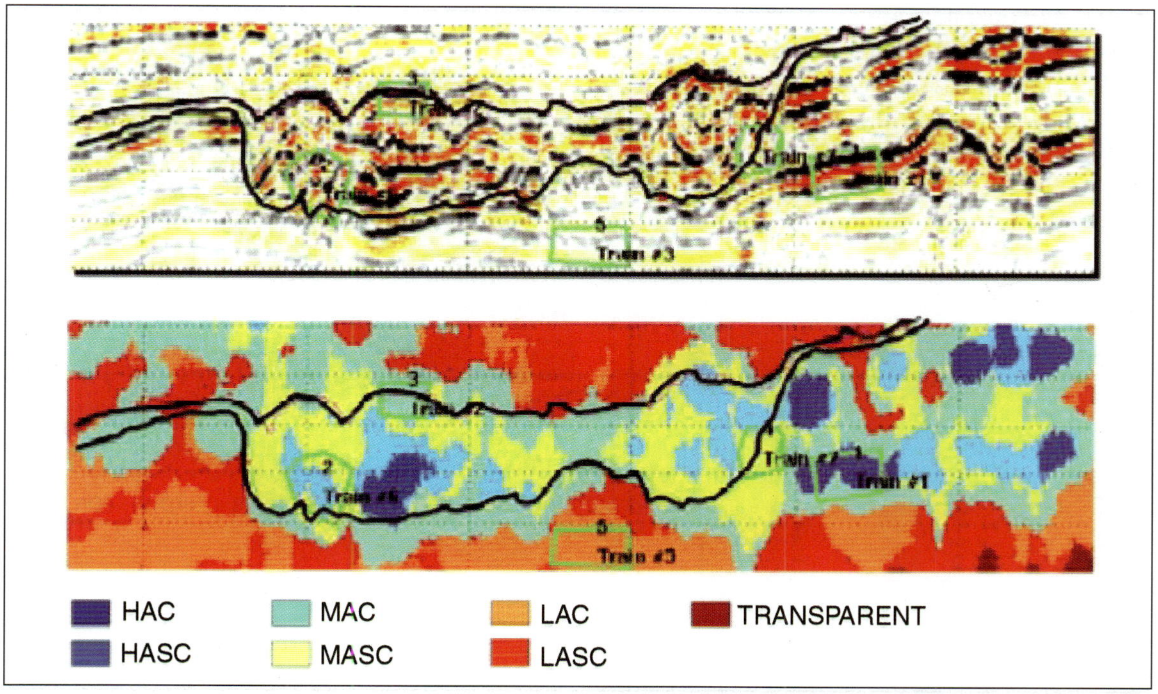

Figure 1.4. A typical seismic facies classification using an interpreter-trained probabilistic neural network, in which multiple facies classes have been identified. The seismic classification scheme on the right consists of high-amplitude (HA), moderate-amplitude (MA), low-amplitude (LA), continuous (C), and semicontinuous (SC) seismic facies. After West et al. (2002).

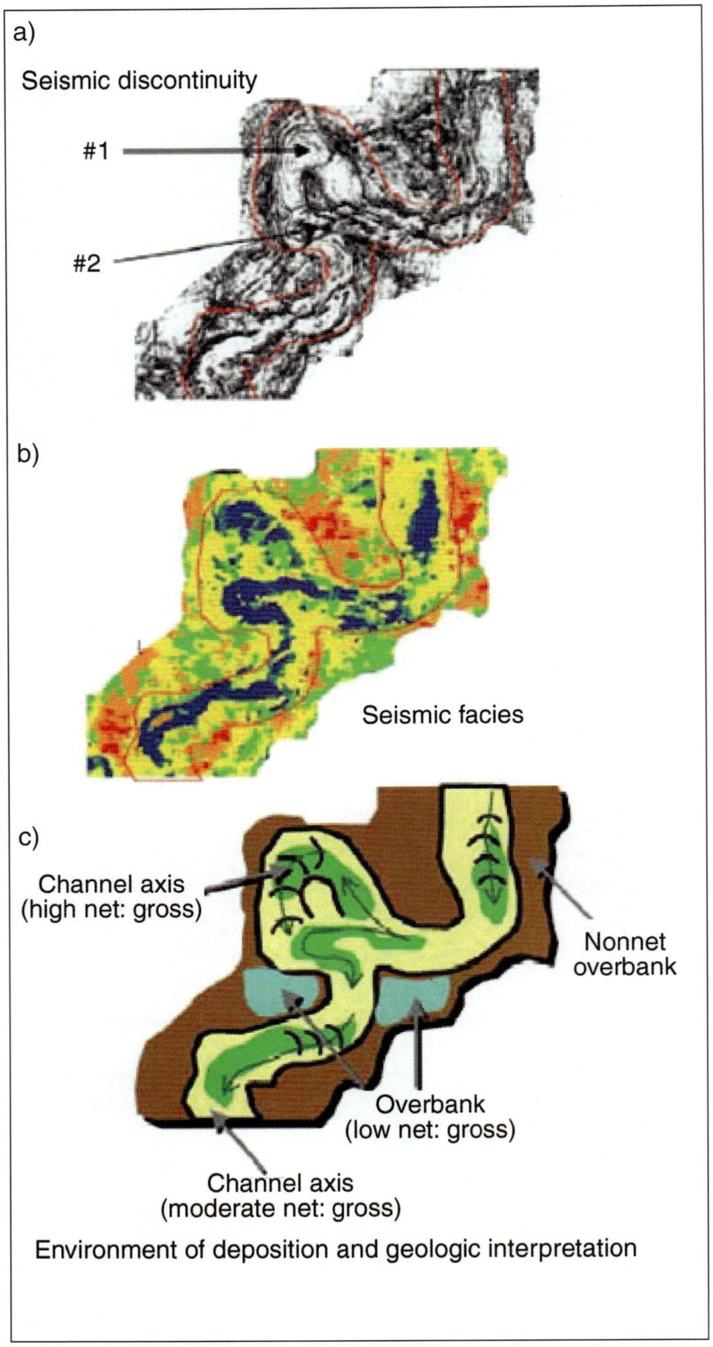

Figure 1.5. (a) A coherence slice highlighting the lateral edges of a channel (red lines); a broad, older sinuous element (1); and a narrower, younger sinuous element (2). (b) A seismic facies slice, showing that sinuous element 1 is composed of high-amplitude-continuous (HAC) to moderate-amplitude semicontinuous (MASC) seismic facies, whereas sinuous element 2 is composed primarily of HAC seismic facies. Because the texture-analysis seismic-facies classification is a volume, this type of analysis can be applied at different stratigraphic levels within an interval of interest, whether or not the interval is bounded by mapped horizons. (c) Considering the conceptual relationships between seismic facies and their potential associated geologic fill, the net-to-gross environment can be understood and 3D regions can delineate differing depositional and geologic properties. After West et al. (2002).

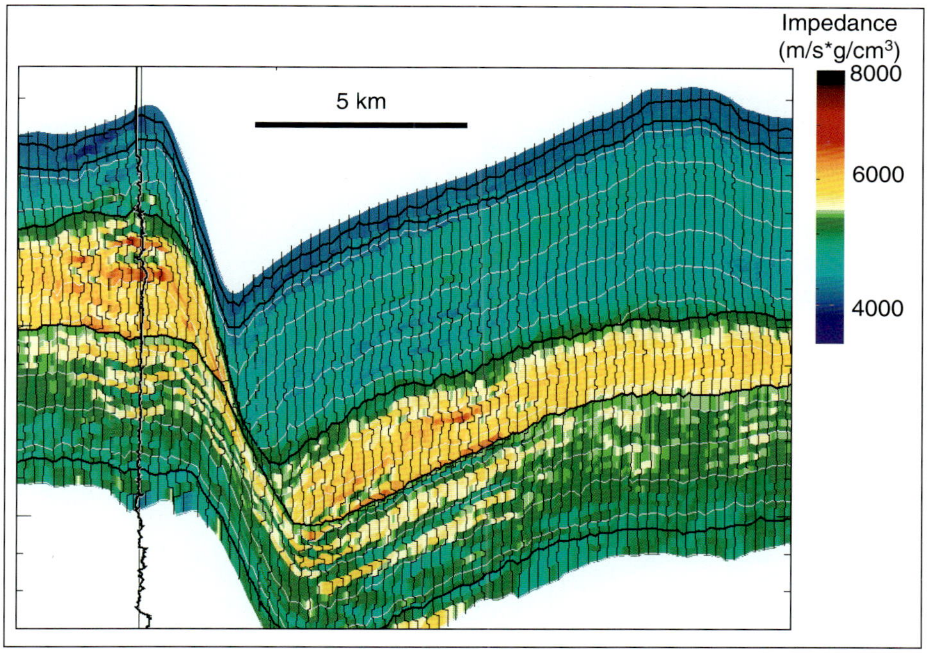

Figure 1.6. There are two major uses of seismic attributes. The first is to intelligently extrapolate sparse well measurements of reservoir thickness, porosity, and hydrocarbon saturation onto the much denser seismic grid. Prediction tools include multivariable statistics, geostatistics, and neural networks applied to attributes that are sensitive to amplitude and thickness. Practitioners typically are petrophysicists, geophysicists, and engineers. This example shows prediction of reservoir porosity from an acoustic-impedance inversion attribute computed for a survey in Mexico. After Veekan et al. (2002).

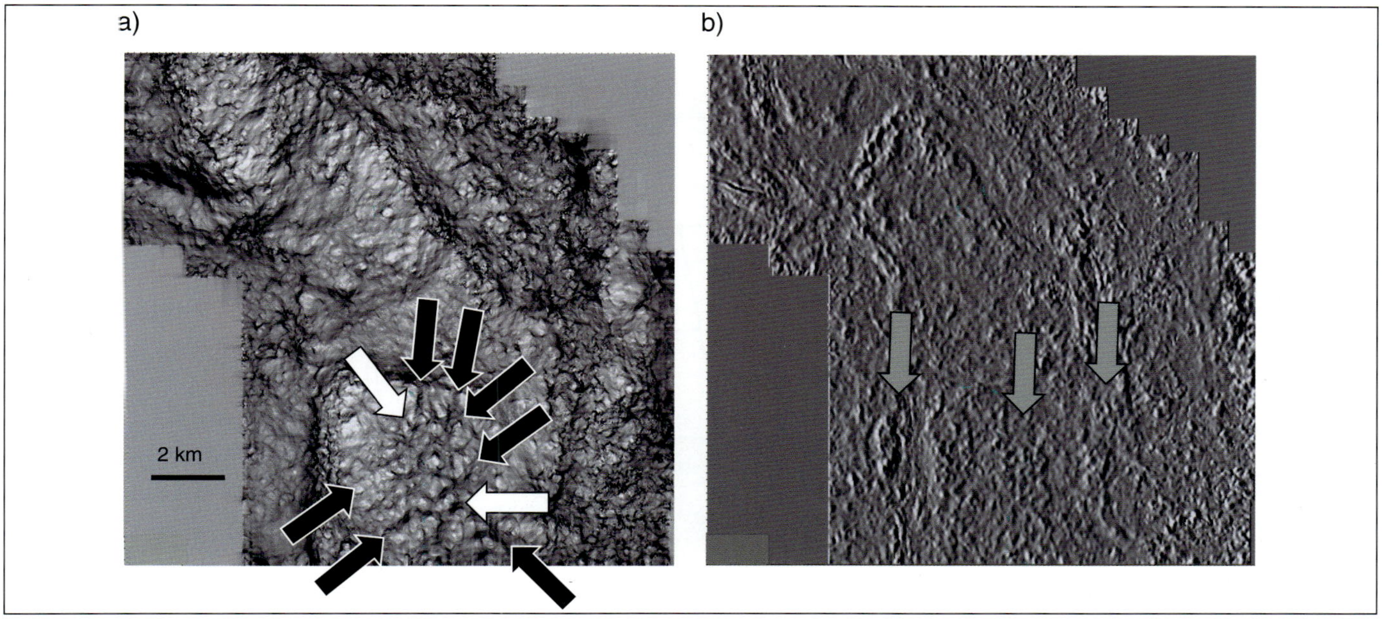

Figure 1.7. The second use of seismic attributes is to reconstruct the tectonic, depositional, and diagenetic history, thereby allowing us to *infer* lithology, porosity, seal, fracture density, and fracture orientation. Prediction tools include geologic models, as well as modern analogs and paleoanalogs. Practitioners typically are structural geologists and stratigraphers. In (a), we display a time slice, at $t = 1.440$ s, through a most-negative-curvature volume generated for a survey acquired over the Central Basin Platform, Texas, U.S.A. Black arrows indicate lineaments that we interpret to be fractures. White arrows point to bowl-shaped areas that suggest karstification of the Ellenburger formation at this level. In (b), we display a time slice, at $t = 1.440$ s, through the north-south component of the energy-weighted coherent-amplitude gradient. Arrows indicate channels that flowed from north to south. These channels help define the depositional environment at the time, and with well control, they can help infer lithology.

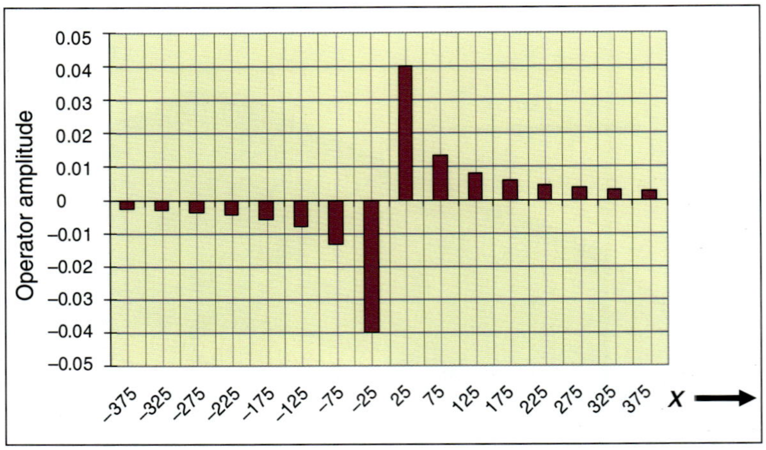

Figure 1.8. The Hilbert transform displayed as a linear combination of the difference operators displayed in Figure 3.77.

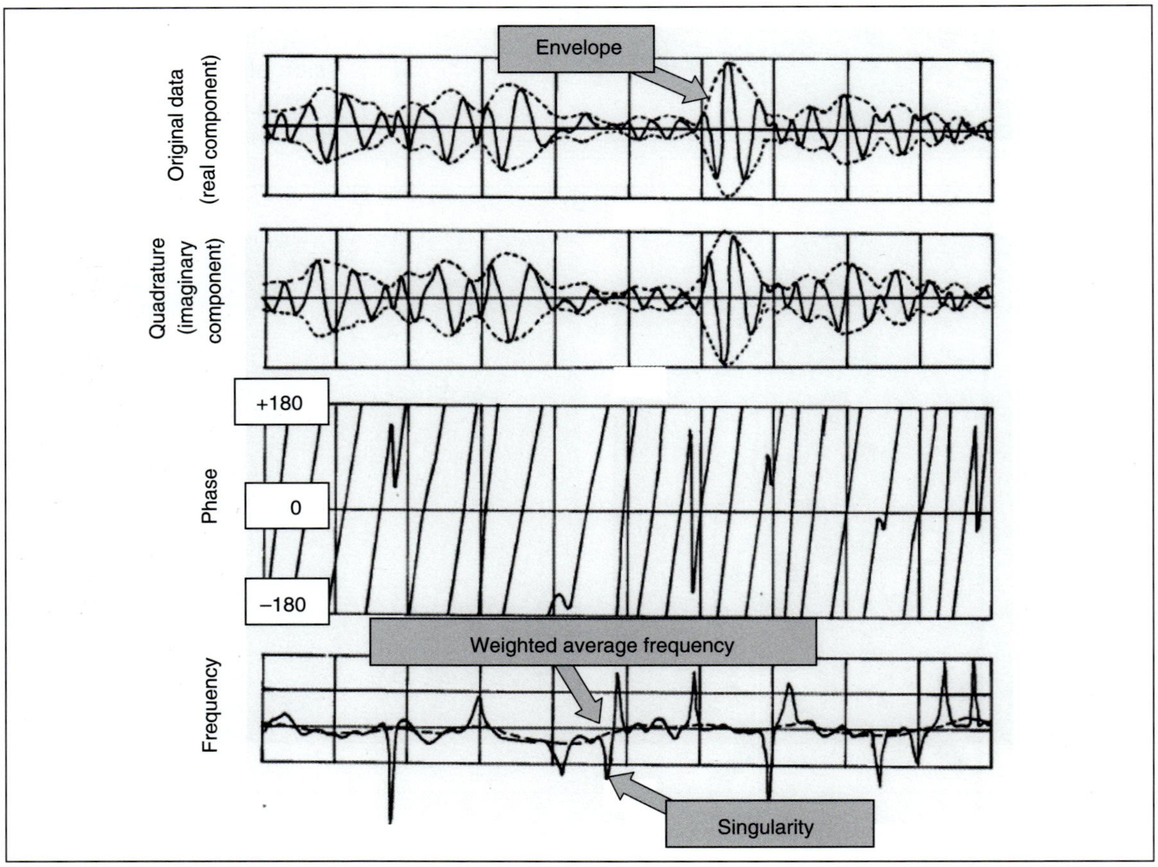

Figure 1.9. Complex trace attributes, including the real trace, its quadrature, its envelope, and its instantaneous phase and instantaneous frequency. Waveform interference results in singularities in the instantaneous frequency, such as the negative excursion indicated by the lowermost arrow. Such singularities can be suppressed by calculating a weighted-average frequency obtained by weighting the instantaneous frequency by the envelope and smoothing over a short time window. After Taner et al. (1979).

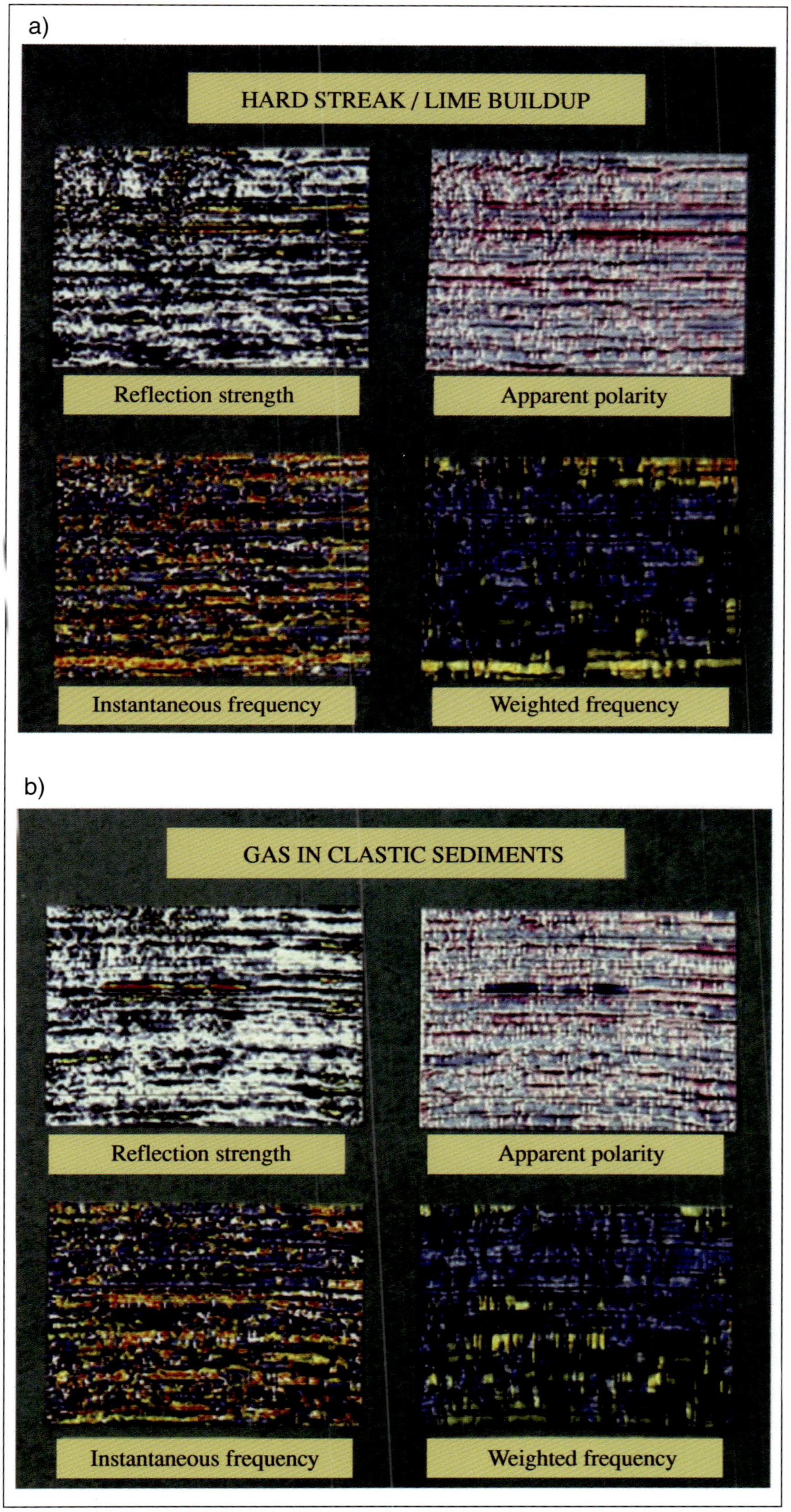

Figure 1.10. Scanned copies of two slides used by Tury Taner in the AAPG-sponsored school on seismic stratigraphy during the middle 1970s and the 1980s, to demonstrate the value of complex trace attributes. (a) An example of hard lime buildup and (b) an example of gas in clastic sediments, including reflection strength (also called envelope), apparent polarity, instantaneous frequency, and weighted frequency. Note that although the reflection strength of both examples is strong (appearing as red and yellow), the apparent polarity of the hard lime buildup appears as red, whereas that of the gas embedded in clastic sediments appears as blue. These images are courtesy of Tury Taner, Rock Solid Images.

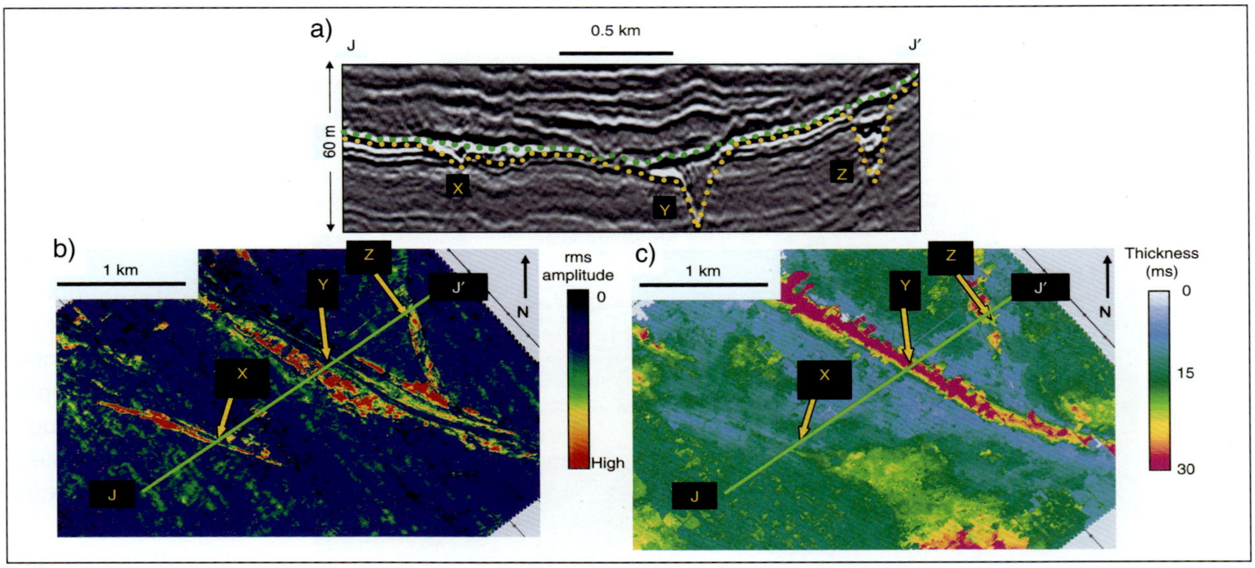

Figure 1.11. (a) An east-west seismic section cutting across three channel-like features, which are labeled X, Y, and Z. (b) A root-mean-square (rms) amplitude map calculated over a 20-ms window centered about the green pick shown in (a). High amplitudes indicate levee complexes. (c) A time-thickness map between the green and yellow picks. After Oyedele (2005).

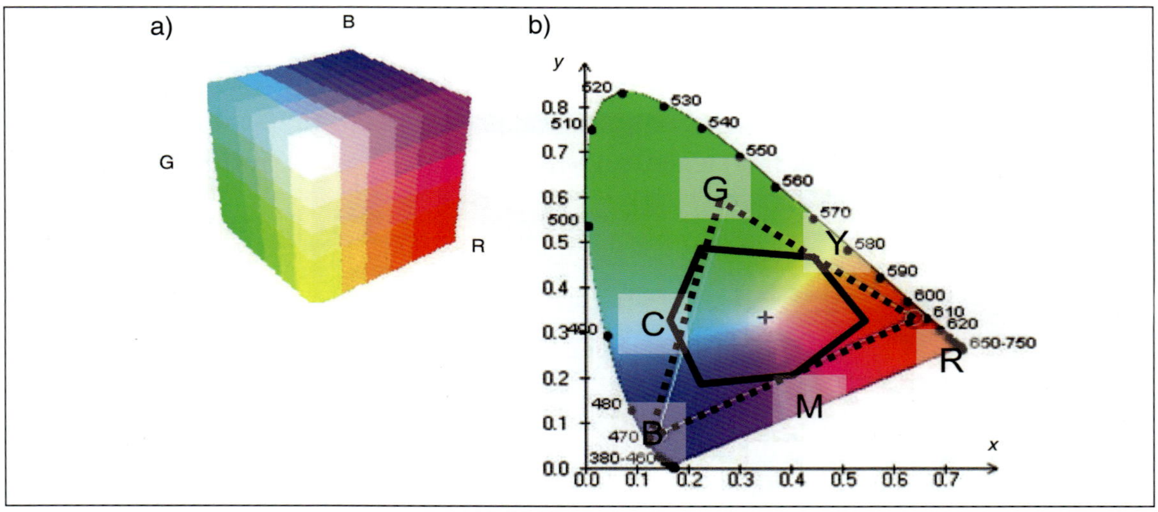

Figure 1.12. (a) The RGB color model (using the colors red, green, and blue, the colors commonly used for visual displays, such as computer monitors). (b) An RGB model and a CMYK model (CMYK refers to cyan, magenta, yellow, and black, the pigment colors used in the printing process), mapped here to the human visual system defined by the CIE (Commission Internationale de l'Éclairage) standard (the area between the RGB triangle and the perimeter of the CIE color model). Not all colors perceivable by humans can be created using positive amounts of red, green, and blue. In addition, many colors displayed on a monitor cannot be generated with printing inks using the CMYK model (the area between the CMYK polygon and the RGB triangle in b). Modified from http://www.hf.faa.gov.

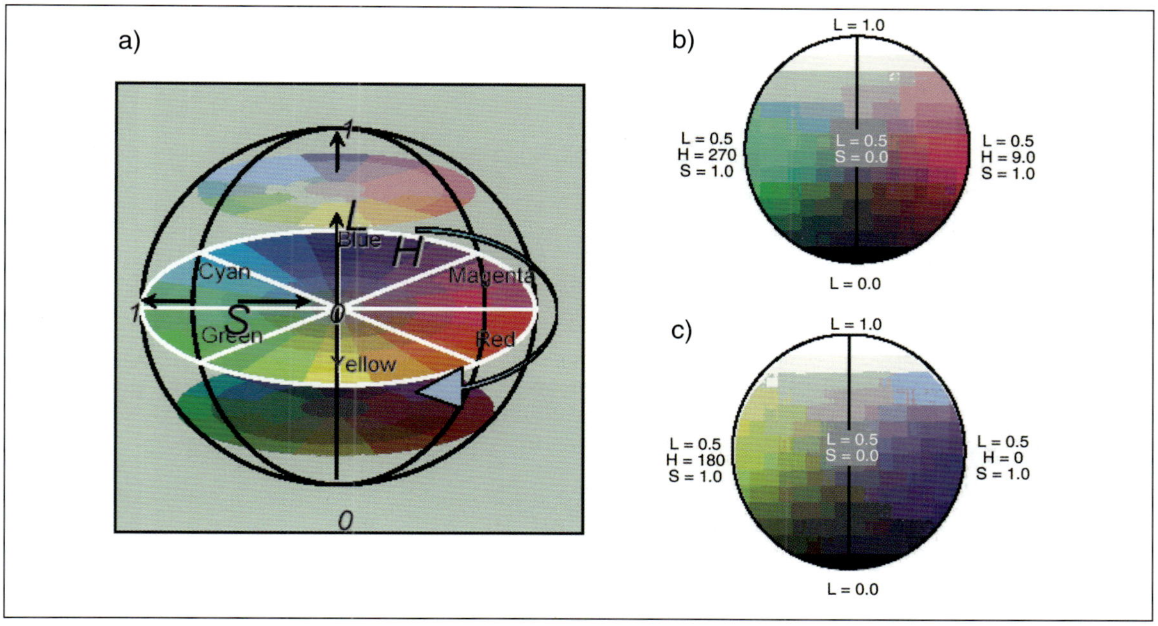

Figure 1.13. (a) The HLS (hue-lightness-saturation) color model mapped to spherical coordinates. Here we link the hue, 0° < H < 360°, to azimuth; the lightness, 0.0 < L < 1.0, to the vertical axis; and saturation, 0.0 < S < 1.0, to radius. (b) An east-west vertical slice through the HLS color model and (c) a north-south vertical slice through the HLS color model. We will use the horizontal plane of this image for dip and azimuth images. By mapping coherence to the vertical axis, we obtain a three-attribute display that is simple to interpret. After Lin et al. (2003).

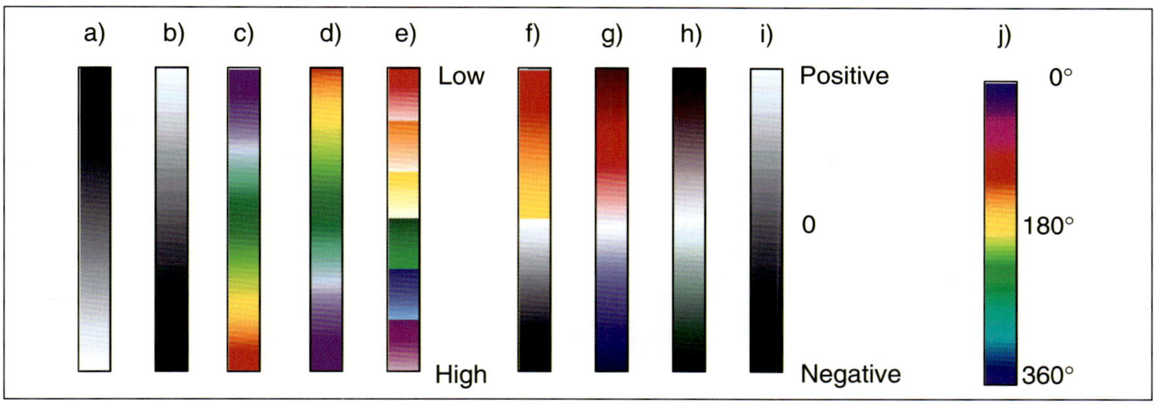

Figure 1.14. The 1D color bars used to display (a)-(d) single-polarity attributes, such as time-structure, horizon slices along seismic amplitude peaks or troughs, envelope, rms (root-mean-square) amplitude, coherence, curvedness, and dip magnitude. (e) A specialized color bar used to provide a "contoured" appearance. (f)-(h) Dual-polarity color bars used to plot attributes such as seismic data and attributes that have both negative and positive polarity, including seismic time and vertical slices, most positive and most negative curvature, coherent amplitude gradients, and apparent reflector dip. (i) A single gradational 1D color bar that also is used commonly to plot dual-polarity-attribute data, particularly when the interpreter wishes to plot interpreted faults and horizons. (j) A 1D cyclical color bar that is useful for displaying phase and azimuth.

Figure 1.15. Examples of good and bad color bars used in displaying seismic data. (a)-(b). Single-gradational color bars. (c)-(d) Dual-gradational color bars. (a) An excellent choice for displaying single- or dual-polarity seismic data or attributes, with colored faults or other interpretation superposed. The background color is a neutral gray. (b) A good choice for displaying single-polarity data, such as a horizon slice through picked seismic peaks. However, this is a poor choice for displaying dual-polarity seismic data or attributes because the image is dominated by green. A good color bar should assign a neutral color (white, gray, or black) to data that have little informational content. (c)-(d) Excellent color bars for displaying dual-polarity seismic data, or attributes, such as curvature, that have both positive and negative values.

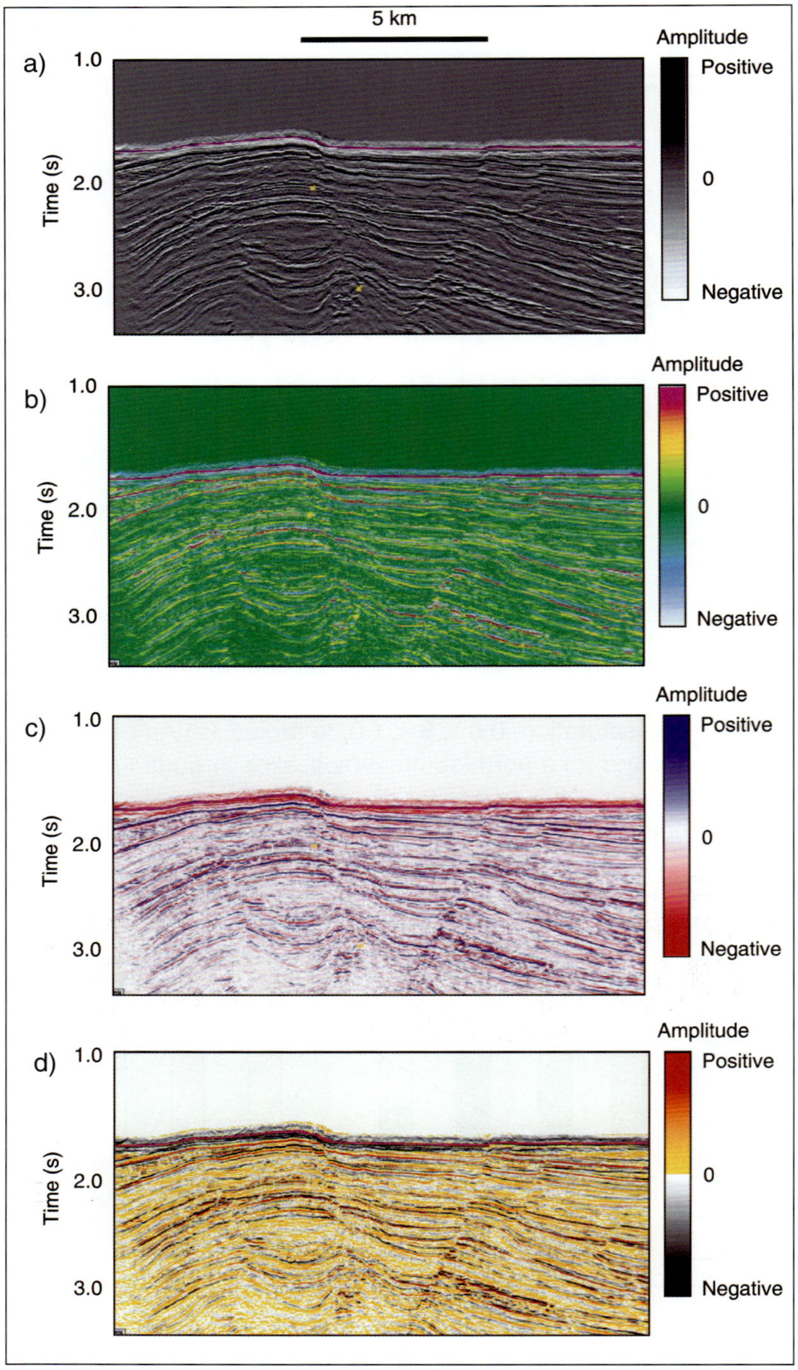

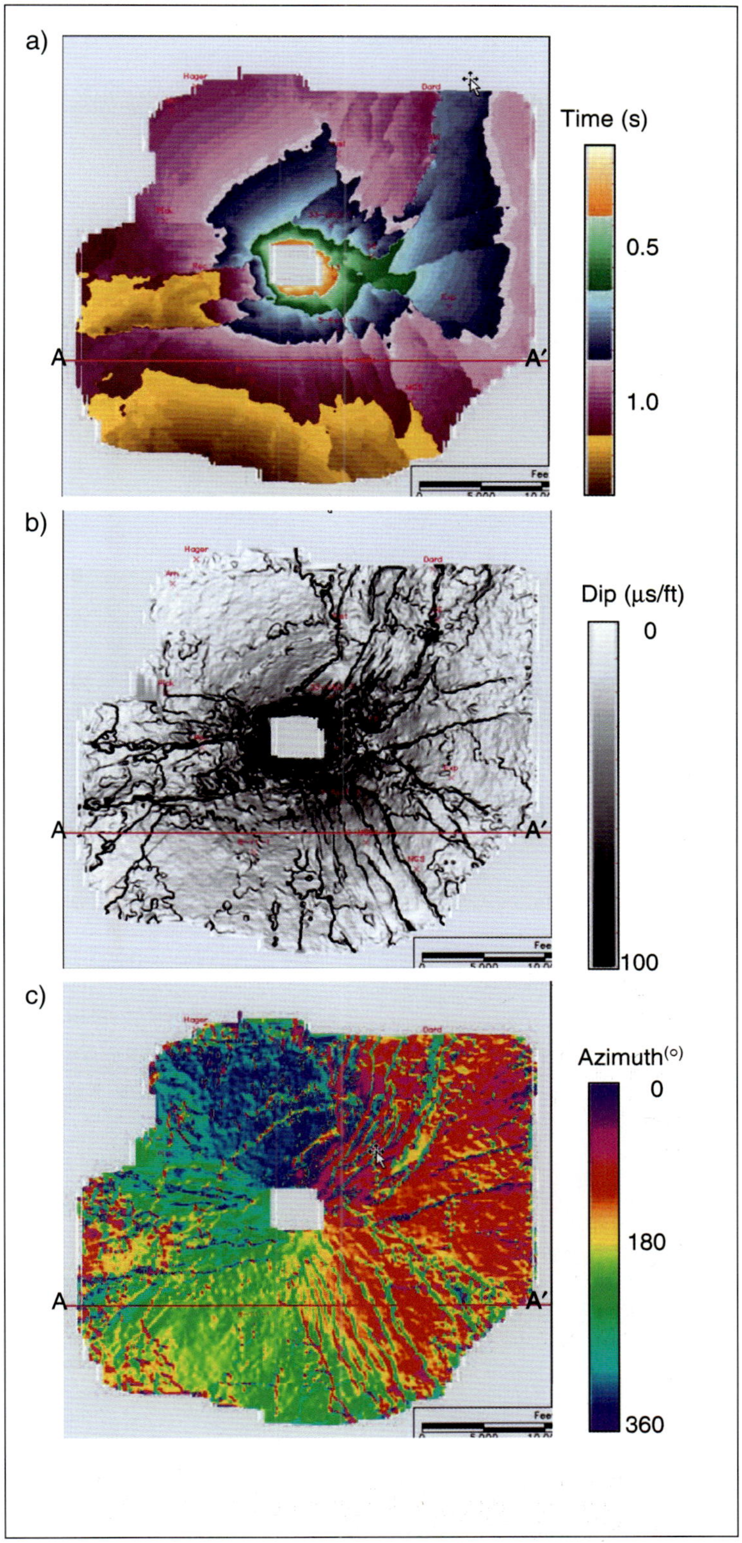

Figure 1.16. Conventional workstation displays of horizon attributes extracted along the middle Miocene strata of Vinton Dome, Louisiana, U.S.A.: (a) time, (b) time/dip magnitude, and (c) time/dip azimuth. Here, we have mapped time against a discontinuous color scale, providing a contour effect at each discontinuity. We have mapped dip magnitude against lightness, which ranges from L = 0.0 for strong dip to L = 1.0 for zero dip. Finally, we have mapped dip azimuth against hue (H), with the values of H ranging from 0 to 360 and with H = 0 and H = 360 (blue) being north and H = 180 (yellow) being south. Because this horizon lies over a salt dome, the azimuth follows the color wheel. Data are courtesy of OPEX.

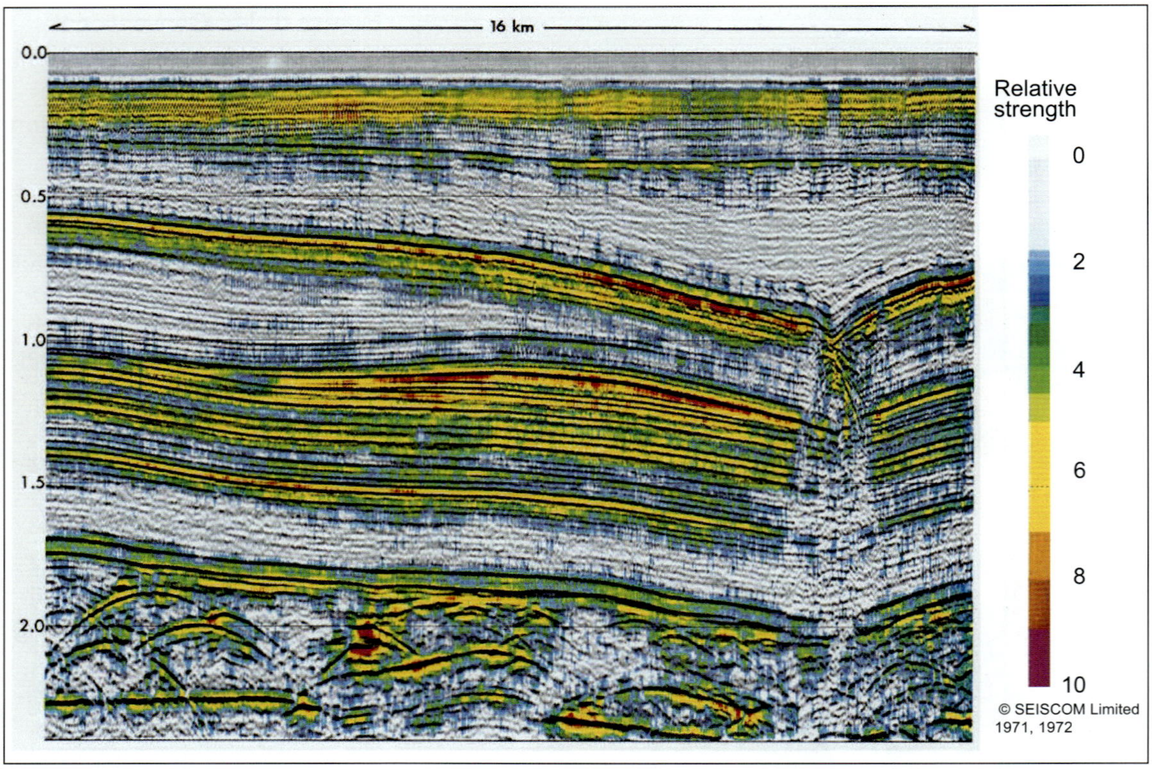

Figure 1.17. A composite attribute image from the early 1970s, showing reflection strength (which in early days was the most popular attribute) superimposed on a structural seismic section. After Anstey (2005).

Figure 1.18. A "blended" or "mixed" attribute image. Here, the two seismic attributes to be blended are the seismic data (plotted against a single gradational gray scale) and the green fault plane. Each voxel of the seismic display has its own R_1, G_1, B_1 color components. The green fault plane has the color components $R_2 = 0.0$, $G_2 = 1.0$, $B_2 = 0.0$. The blended image is simply the weighted average of the two input images: $R = aR_1 + (1 - a)R_2$, $G = aG_1 + (1 - a)G_2$, $B = aB_1 + (1 - a)B_2$. In this example, the blending factor $a = 0.5$. After Meyer et al. (2001).

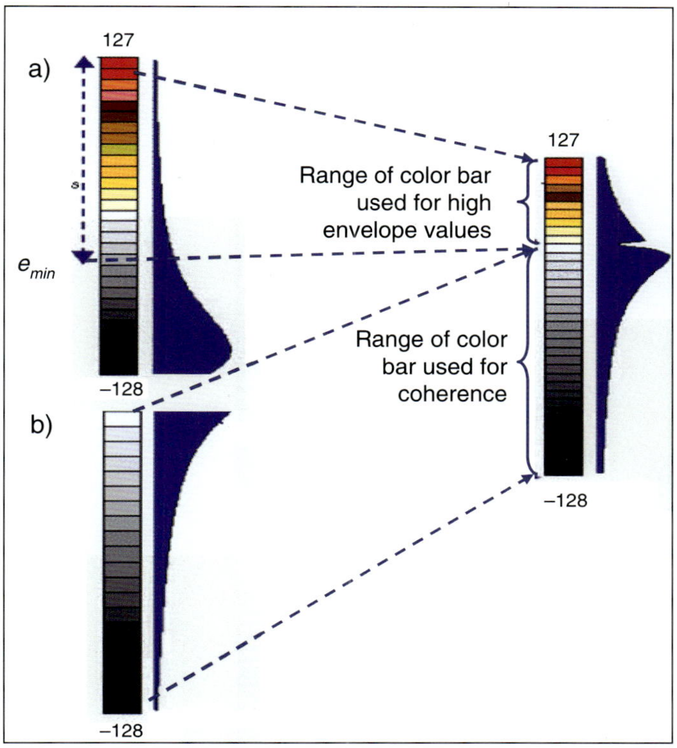

Figure 1.19. Integrating seismic attributes and seismic data using a composite display — (a) an amplitude envelope color bar, (b) a coherence gray-scale bar, and (c) a composite multiattribute color bar. We generate our composite attribute displays in two steps. First, assuming our graphics software is limited to 256 colors, we construct the color bar shown. Second, we perform Boolean arithmetic on the two input attributes and generate a third volume that has values ranging from 0 to 255. In this example, if our envelope value is greater than some threshold ($e > e_{min}$), we assign a color index that uses one of the higher (64) colors corresponding to the envelope portion of the composite color bar. If $e < e_{min}$, we assign a color index that uses one of the lower (192) colors corresponding to the coherence attribute. After Chopra (2001b).

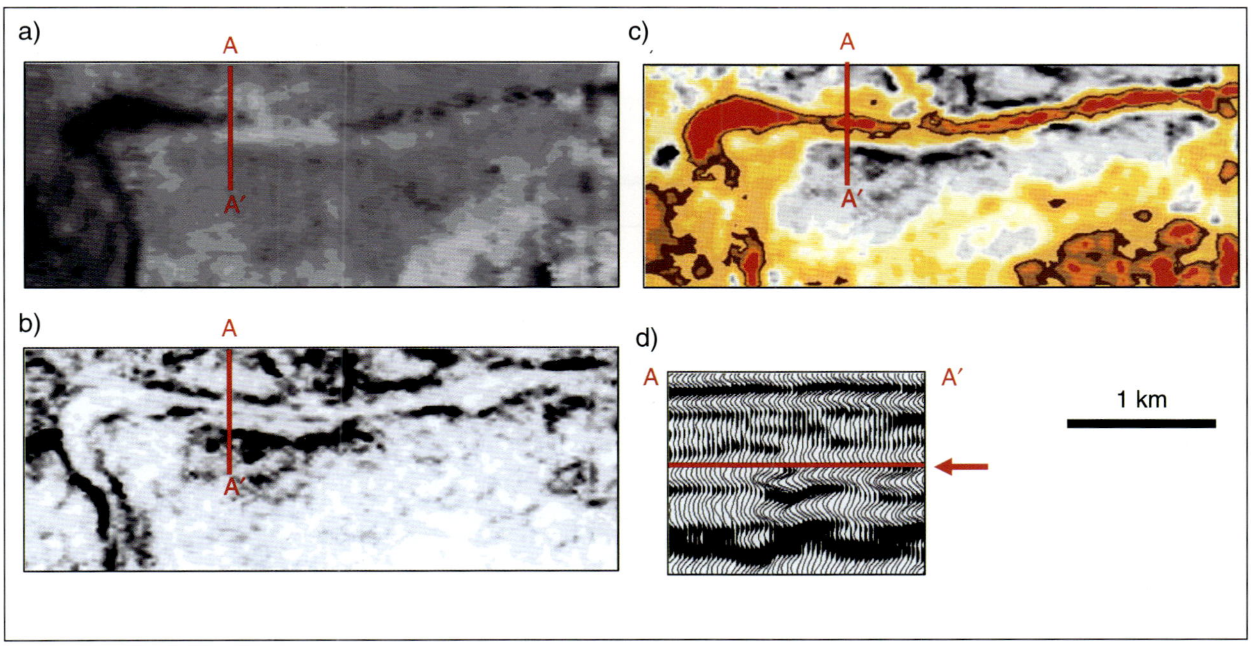

Figure 1.20. A channel from Alberta, Canada, seen on time slices through (a) seismic data, (b) coherence data, and (c) a composite coherence and reflector envelope, using the color bar described in the previous figure. (d) A vertical slice through the channel corresponding to line AA′. The arrow indicates the level of time slices. After Chopra (2001b).

Figure 1.21. A channel system appearing on time slices through (a) a coherence volume, (b) an acoustic-impedance volume, and (c) a composite acoustic-impedance and coherence volume. The respective color bars for (a), (b), and (c) are indicated.

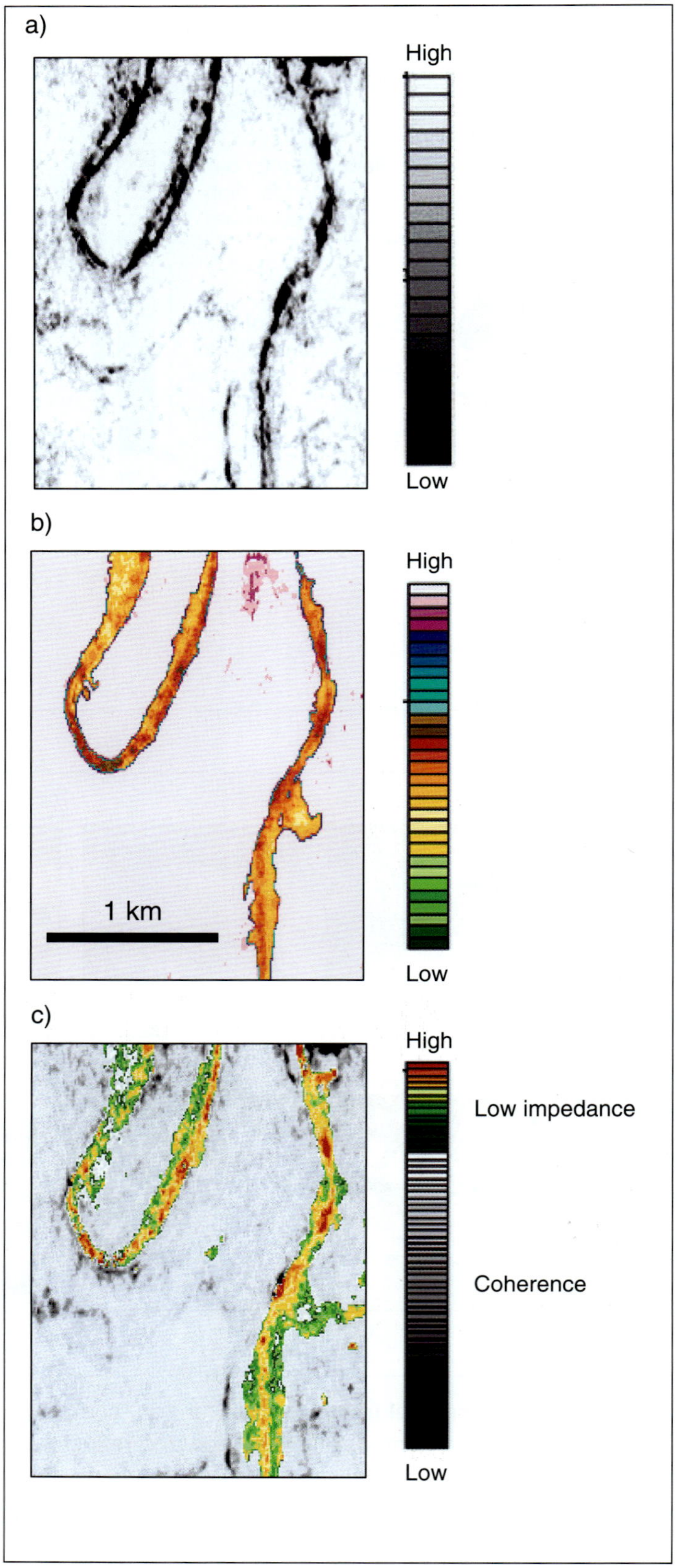

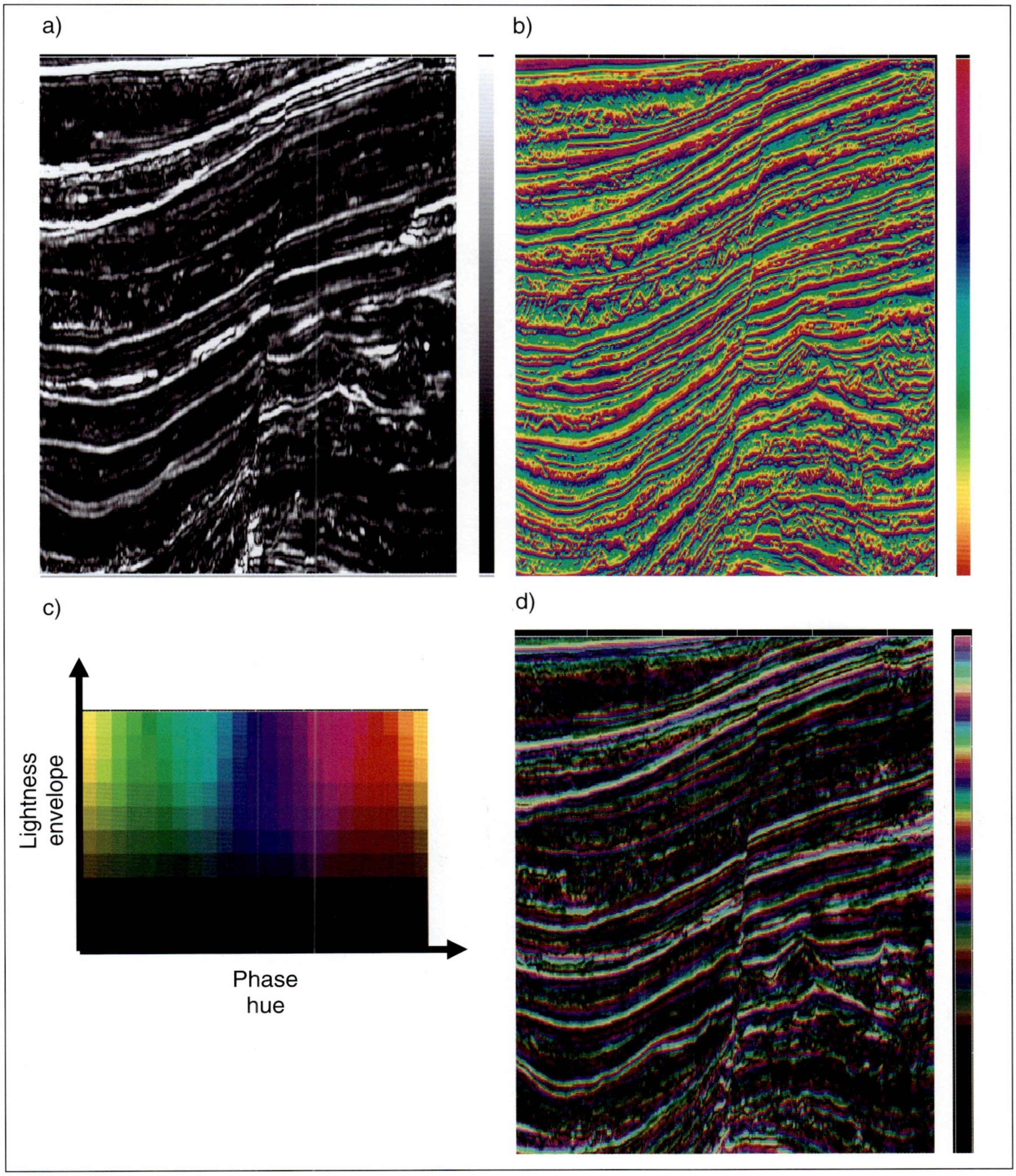

Figure 1.22. Combining (a) reflector envelope against (b) reflector phase, using (c) a composite color scale to form (d) a composite image. Conventional 256 1D color bars used by the plotting software are displayed on the right of (a), (b), and (d). This technique, originally presented by Knobloch (1982), emphasizes the phase of the stronger reflection events and provides an effective tool for tracking waveforms across faults. Note that the 2D color legend in (c) has been mapped to a more conventional 1D color legend in (d), to use conventional plotting software.

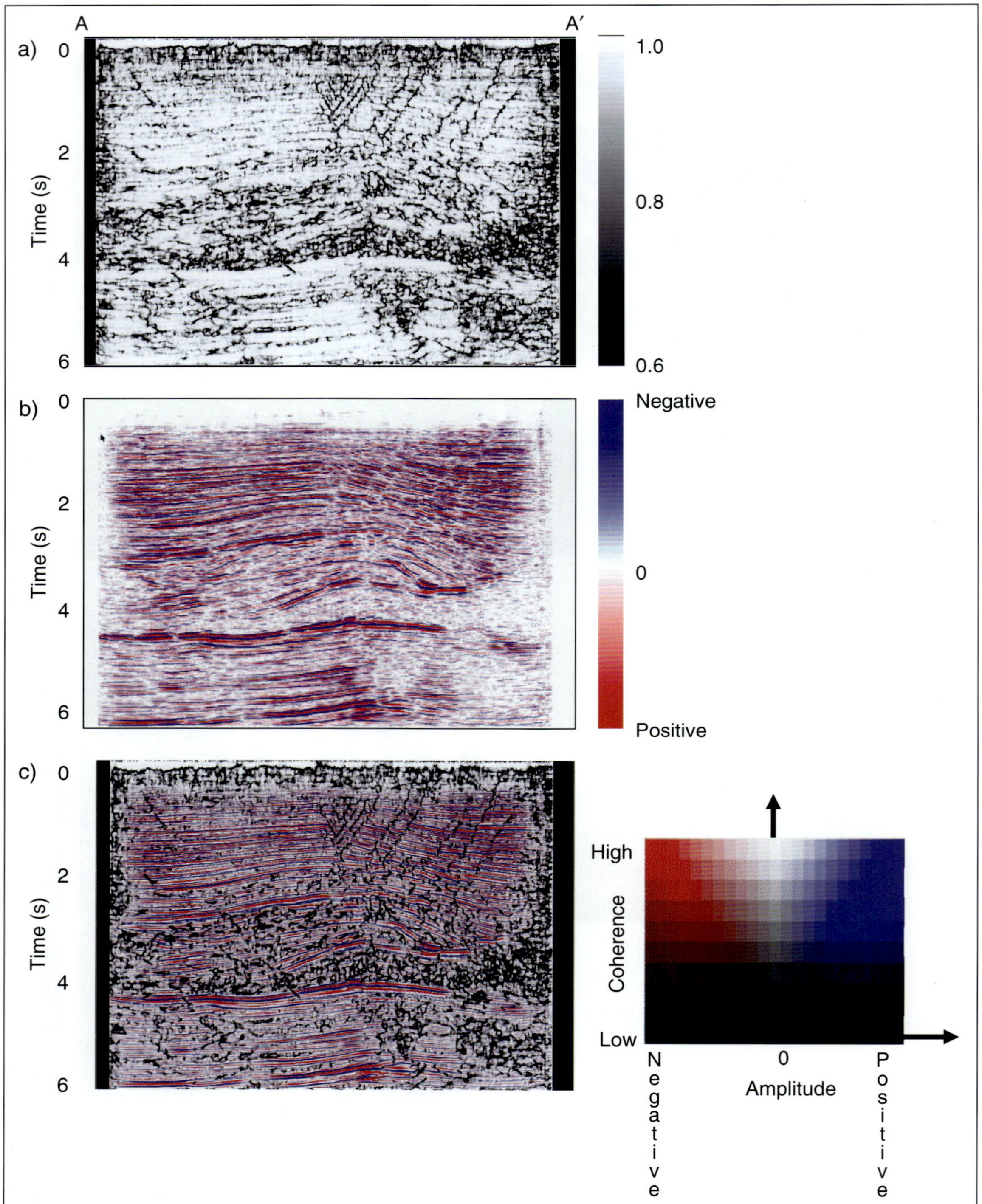

Figure 1.23. (a) Coherence plotted using the lightness, L, axis. (b) Seismic amplitude using a conventional dual-gradational red-white-blue color bar. (c) Composite coherence and seismic amplitude obtained by using the 2D color bar shown to the right. After Lin et al. (2003).

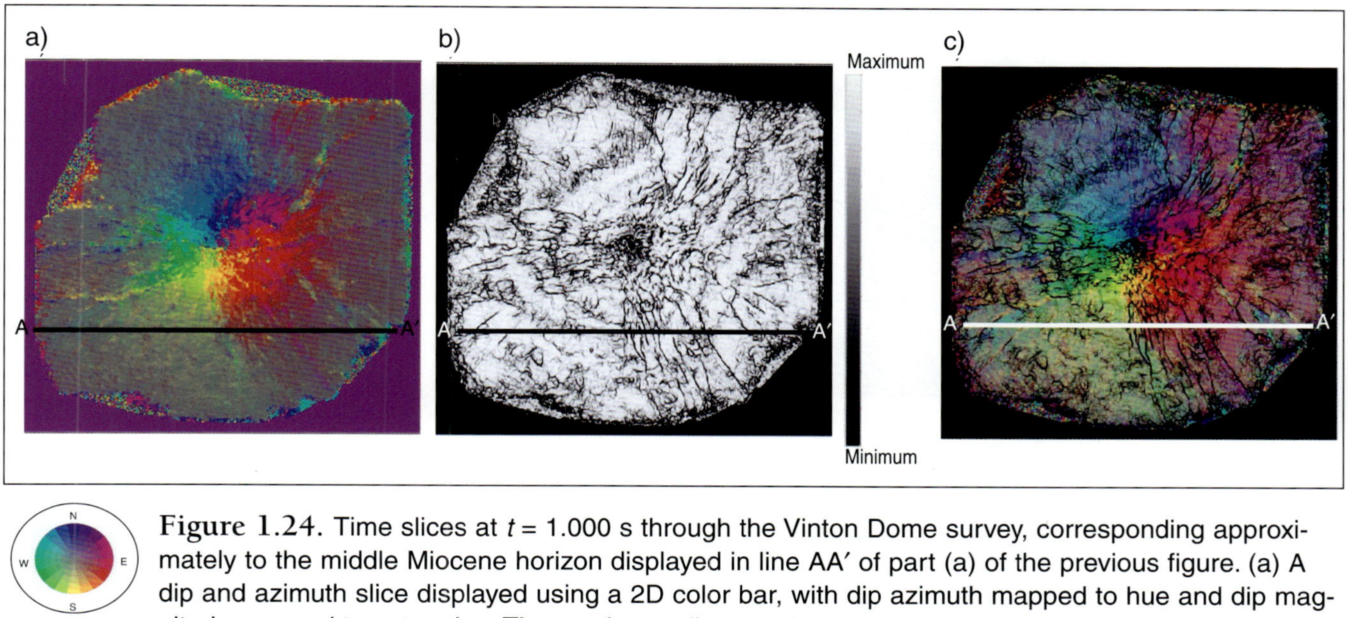

Figure 1.24. Time slices at t = 1.000 s through the Vinton Dome survey, corresponding approximately to the middle Miocene horizon displayed in line AA′ of part (a) of the previous figure. (a) A dip and azimuth slice displayed using a 2D color bar, with dip azimuth mapped to hue and dip magnitude mapped to saturation. The maximum dip magnitude of 50° is plotted at full saturation. The minimum dip magnitude of 0° is plotted as zero saturation, or gray. Lightness is fixed at L = 0.5. (b) A coherence slice displayed using a single gradational color bar along the lightness axis. (c) Dip and azimuth and coherence displayed using a 3D color bar. Here, dip and azimuth are mapped as they were in (a), and coherence is mapped against lightness, where $0 < L < 0.6$. Data are courtesy of OPEX.

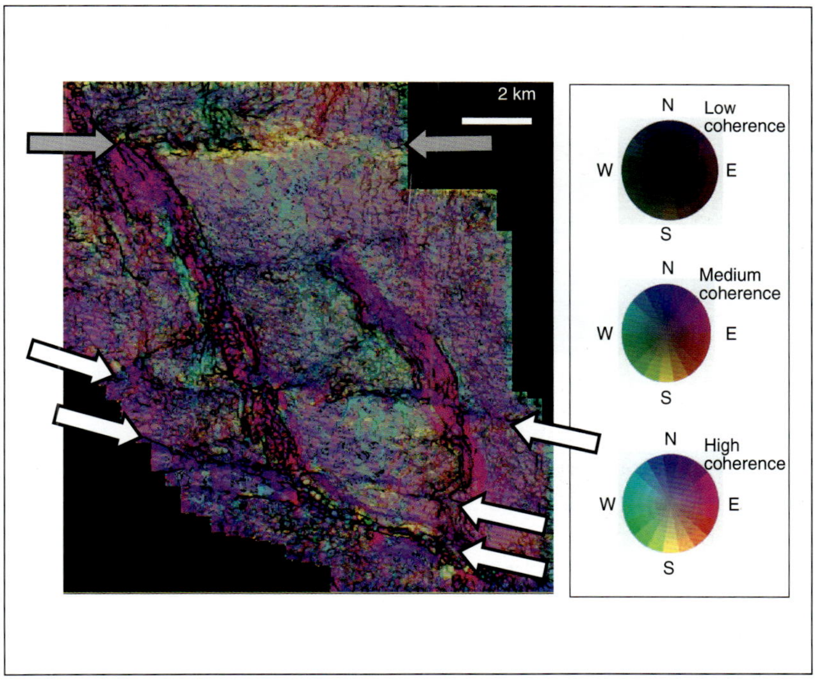

Figure 1.25. A time slice through a survey acquired over the Central Basin Platform, Texas, U.S.A., using a 3D color legend. Here, we have plotted dip azimuth against hue, dip magnitude against saturation, and coherence against lightness. Note how the low-coherence fault indicated on the northwest by a gray arrow becomes in the northeast a more coherent flexure dipping to the south. The two faults farther south, indicated by white arrows, are seen either as a band of low coherence or as a flexure dipping to the north. The juxtaposition of magenta (northeast dip) and green (southwest dip) in the center of the image indicates a local anticlinal feature associated with northwest-southeast-trending reverse faults. Data are courtesy of OXY.

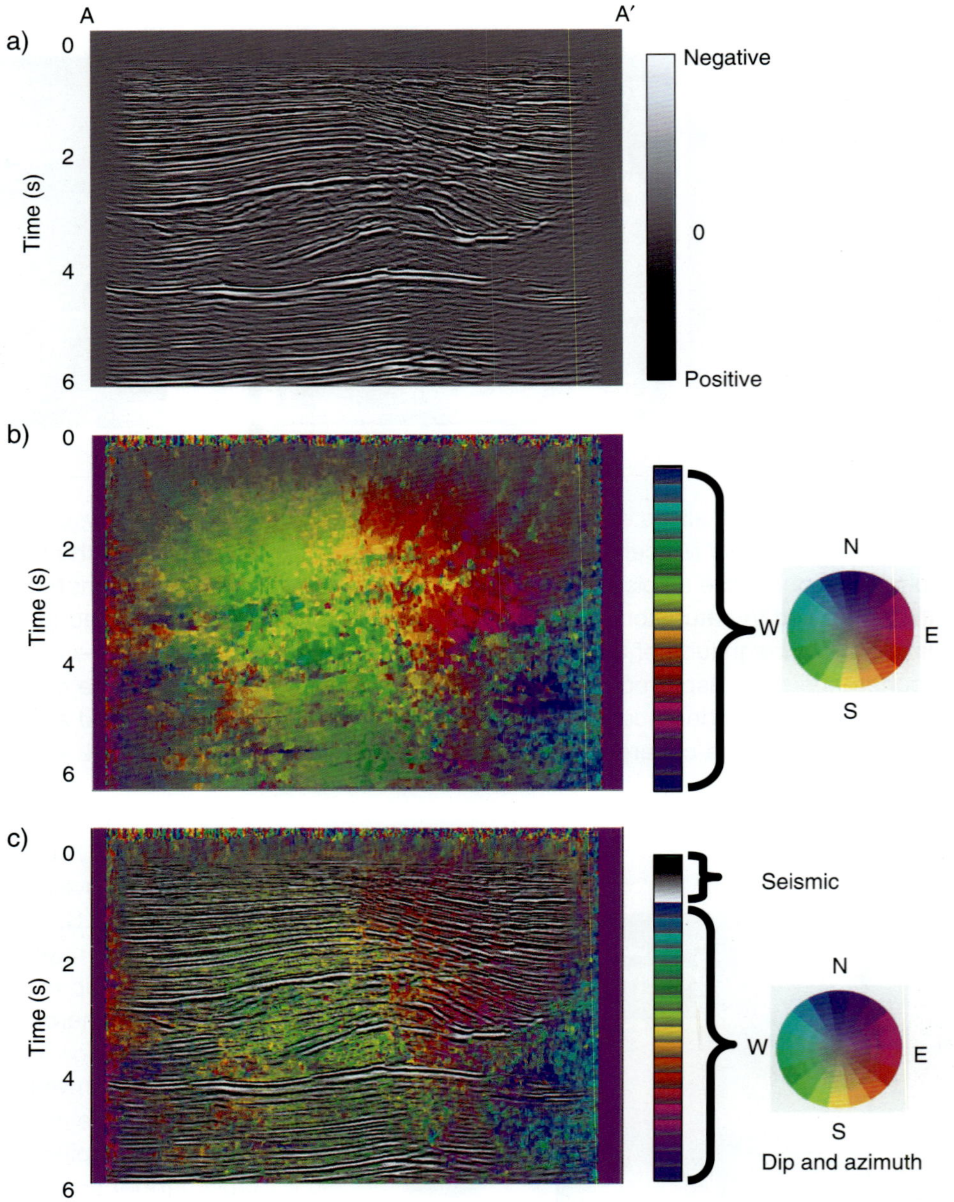

Figure 1.26. A vertical section corresponding to line AA′ in Figures 1.23, 1.16, and 1.24 from Vinton Dome, Louisiana, U.S.A. (a) Seismic data plotted using a conventional single-gradational gray-scale color bar. (b) Dip and azimuth plotted using a 2D hue-saturation color map with dip magnitude plotted against saturation, S, and dip azimuth plotted against hue, H. (c) A composite image obtained by plotting the seismic data at voxels where $|a| > a_{min}$ and plotting the dip and azimuth where $|a| < a_{min}$. After Lin et al. (2003). Data are courtesy of OPEX.

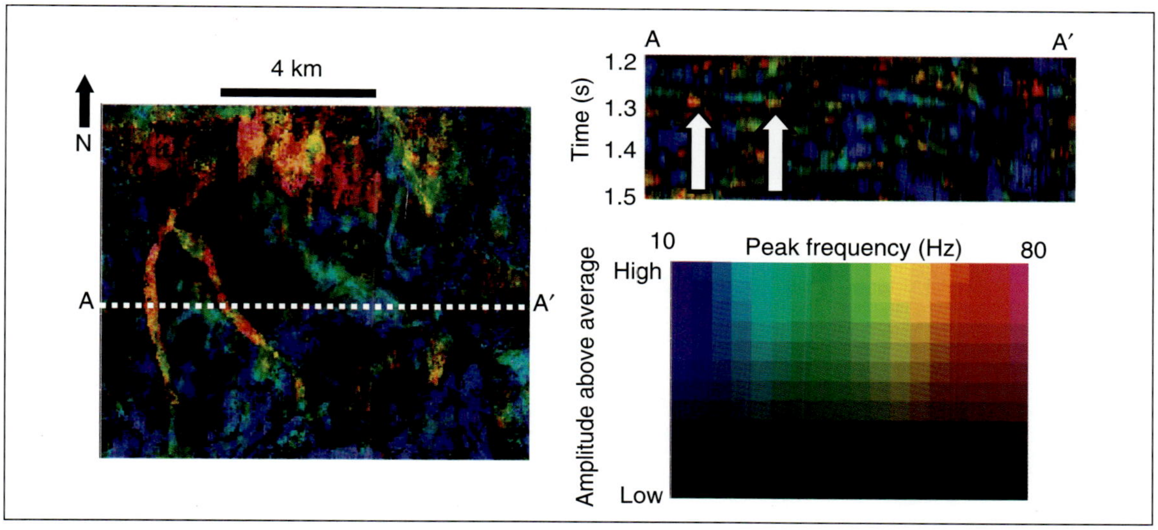

Figure 1.27. A composite image summarizing the information shown in Figure 4.48, here using a 2D color bar and plotting peak frequency against the amplitude of that peak frequency above the average normalized amplitude for all frequencies at each sample. Note that the channel tunes in as orange, implying a peak frequency of about 65 Hz. Similarly, the bright blue area in the southern part of the image tunes in at 25 Hz. Such composite images can summarize much of the information contained in the individual components displayed in Figure 4.48.

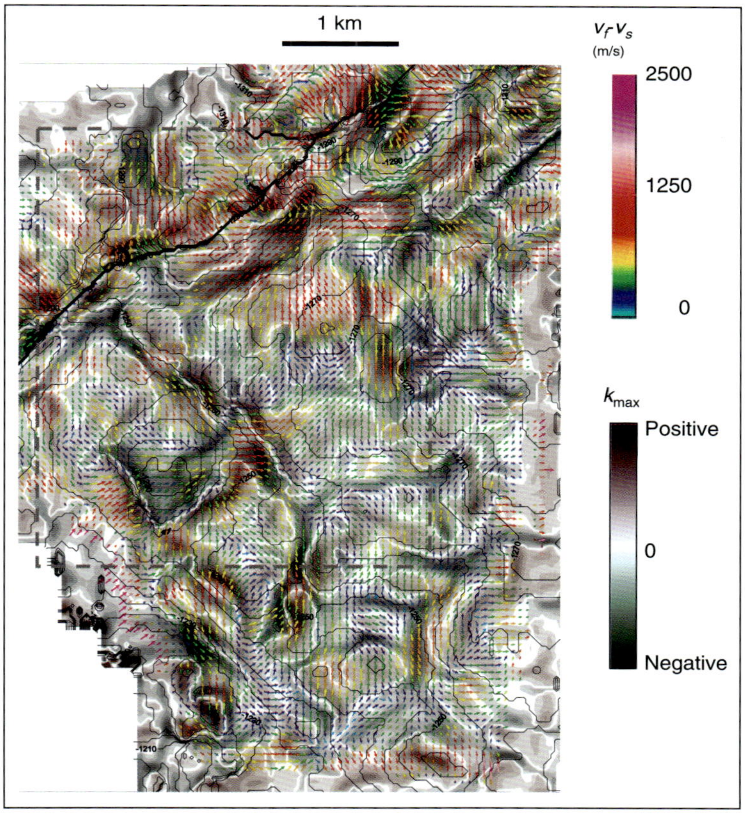

Figure 1.28. Use of colored icons (arrows) to display vector data. The color of the icon represents the magnitude of the P-wave velocity anisotropy. The length and orientation of the arrow represent the magnitude and azimuth of the faster velocity, $\mathbf{v}_f$. These icons are overlaid on a horizon slice through a maximum-curvature volume displayed using a dual-gradational red-white-green color bar. After Simon (2005).

Unit 2 Spectral Decomposition and Wavelet Transforms

Overview

Since the beginning of digital recording, geophysical data processors have decomposed the measured seismic signal into its Fourier (or spectral) components, to attenuate low-frequency ground roll, 50- or 60-Hz cultural noise, and high-frequency random noise. In addition, data processors routinely have balanced the source spectrum through seismic-deconvolution and wavelet-shaping techniques, to account for the input source signature, spectral changes resulting from short-period multiples, and attenuation of the overburden.

Running-window spectral-filtering and spectral-balancing techniques have been applied to seismic data since the 1970s. In such applications, each seismic trace is broken into a suite of short, overlapping traces centered about the output sample. Longer windows (typically 500 or 1000 ms long) provide more robust statistics, whereas 250-ms windows yield poorer results. These poor results stem from an assumption that the underlying reflectivity has a white spectrum. If the underlying reflectivity were not white, it would be distorted during the spectral-balancing step. Because of this focus on spectral balancing, spectral analysis of shorter windows was overlooked until the 1990s — and the impetus came from seismic interpretation rather than from seismic processing. Generally, seismic interpreters are content with relative spectral measurements and are more interested in knowing at which frequency tuning occurs than in knowing the absolute value of each frequency component (the measure that data processors desire).

Because we wish only to *interpret* spectral components rather than to filter components and efficiently reconstruct the data, we are not constrained by orthogonal transforms. For example, if we wish to analyze seismic data within a 100-ms window, Nyquist's sampling criterion states that we need only to decompose the data at 10-Hz increments — say, at 10, 20, 30, 40, 50, and 60 Hz for band-limited seismic data, with a highest-frequency contribution between 60 and 70 Hz and no amplitude at 0 Hz. Knowledge of the spectral components at these six frequencies completely and uniquely describes this time series. Fast Fourier transforms (FFT) efficiently generate these components. Unfortunately, the efficiency of the FFT is so ingrained in processors' heads that they forget they could use a simple slow Fourier transform – that is, they could simply crosscorrelate a sine and cosine with the data for spectral analysis. We will use slow Fourier transforms (more properly called discrete Fourier transforms) to decompose the data into sinusoidal components separated by 1 or 2 Hz. That allows the interpreter to inspect a more finely sampled spectrum for features of interest.

We begin by reviewing basic concepts of seismic resolution and thin-bed tuning. Next, we review the properties of Fourier transforms, emphasizing the impact of the shape and size of the seismic analysis window. Thus we gain a quantitative understanding of the similarities and differences between two methods — short-window discrete Fourier transforms (SWDFT, also commonly called spectral decomposition) and wavelet transforms. We find it more useful to use nonorthogonal Fourier transforms for our analysis, and we also find it more useful to use nonorthogonal wavelet transforms rather than those commonly used in seismic data compression. We show that by using a matched-pursuit technique, we can fit the seismic data and also estimate their complex spectrum for an entire volume.

After defining our terms and theoretical basis, we study a suite of examples that exhibits work flows and features amenable to such an analysis. A recent innovation is to estimate fluid mobility (a function of fluid viscosity and reservoir permeability) by analyzing the behavior near the low end of the complex spectrum. Another recent innovation is to map singularities in the complex spectrum, thereby mapping unconformities in the seismic data, which is the basis of the SPICE (spectral imaging of correlative events) algorithm. We close by showing the relation between SPICE and earlier work on a thin-bed-indicator attribute based on the analytic trace.

Summary

Short-window discrete Fourier transforms and wavelet transforms differ from each other in the choice of their analysis window. Spectral decomposition uses a fixed-length analysis window for all frequencies, thereby allowing us to compare which frequency component is dominant in a geologic formation. In contrast, continuous-wavelet transforms use windows that are inversely proportional to the frequency being analyzed, thereby providing maximum temporal resolution at the expense of analyzing different regions of a geologic formation. Matched-pursuit wavelet transforms (or other types of seismic data fitting) are preferred for 3D volumetric analysis, with each seismic sample being represented accurately by overlapping analysis windows.

In general, spectral components produce images with higher lateral resolution than a simple horizon slice through the seismic amplitude data provides. Such image improvements have statistical, model, and interpretation bases. First, we statistically combine 25–50 samples in our spectral analysis, thereby improving our signal-to-noise ratio over a single amplitude slice. Second, we relate the geologically useful thin-bed tuning effect directly to our spectral components. Third, the interpreter winnows through a multitude of spectral components, compares them to mental images of anticipated geologic features, and chooses only frequency components that make geologic sense.

Constructive and destructive interferences from the top and bottom of a thin bed generate changes in the seismic amplitude and phase spectra. Components of these spectra can be used to detect lateral changes in layer thickness and heterogeneity, well below the limits of classic one-quarter-wavelength seismic resolution. SWDFT and wavelet-transform spectral components are sensitive to thin-bed tuning and to attenuation resulting from overlying scatterers and hydrocarbons and also are sensitive to hydrocarbon products.

Because spectral decomposition operates on a fixed, user-defined, temporal window within which the spectral content is balanced, we conclude that the SWDFT is the method of choice when formation-based attributes are appropriate — such as estimating tuning thickness or mapping stratigraphic variability. In contrast, because the matched-pursuit wavelet transformation reconstructs the data and minimally mixes temporally separate events, we feel that wavelet-transform decomposition is the method of choice for volumetrically analyzing direct hydrocarbon indicators, for studying seismic attenuation, and for enhancing seismic bandwidth.

Spectral decomposition has the disadvantage of generating a high volume of output. Although an interpreter easily can animate through multiple frequency slices computed along a horizon or vertical section of interest, animation through multiple volumes of attributes is intractable for large surveys. The peak spectral frequency (the mode of the spectrum) and the amplitude of this peak spectral component above the mean spectral amplitude together are a good zero-order representation of the seismic spectral response.

Recently, workers have focused on analyzing the low end of the seismic spectrum. Several reasons exist for low-frequency anomalies in properly processed seismic data, including waveform healing below an overlying gas reservoir, fault whispers, and reservoir permeability and the viscous response of oil. Finally, spectral decomposition forms the basis of the recently introduced SPICE algorithm, which maps discontinuities between reflector wavelets and promises to be a powerful tool for mapping unconformities and subtle sequence boundaries.

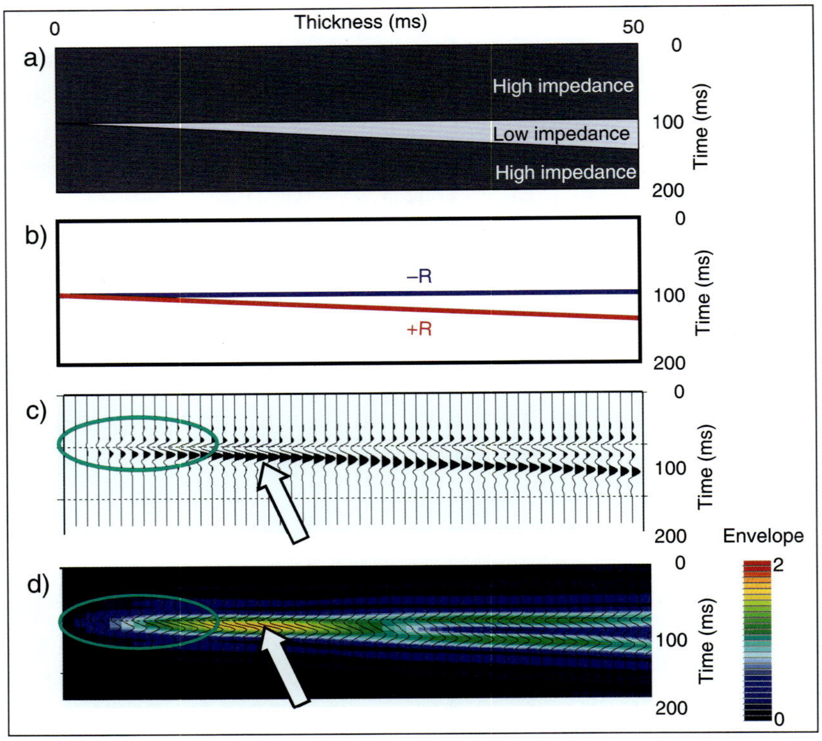

Figure 2.1. (a) Thin-bed resolution and the classic wedge model of a low-impedance layer encased in a higher-impedance matrix. (b) Reflectivity and thickness: The wedge increases in thickness from 0 to 50 ms from left to right, with a negative reflection on the top and a positive reflection on the bottom. (c) Seismic synthetic data generated by convolving (b) with an 8-10-40-50-Hz Ormsby filter. (d) Instantaneous envelope (also called reflection strength) computed from the seismic data in (c). The peak energy occurs at the one-quarter-wavelength (one-half-period two-way traveltime) tuning thickness indicated by the arrow. Below the tuning thickness, which is indicated by the green ellipse, only the amplitude changes, whereas the phase and frequency remain fixed. Such subtle changes in amplitude allow us to map lateral changes in bed thickness using lateral changes in amplitude variability and spectral decomposition. After Partyka (2001).

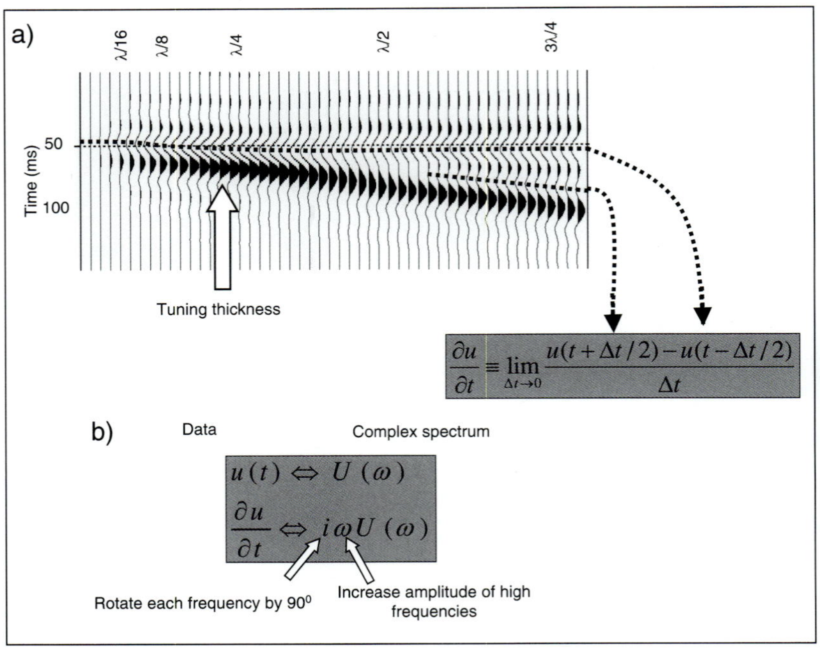

Figure 2.2. A review of basic math to help readers understand the effects of thin-bed tuning discussed in Figure 2.1. We define the thickness of the wedge at any point along the horizontal axis as Δt. Then we can interpret the reflector from the top of the wedge as having a negative polarity and a phase advance of $\Delta t/2$. We also can interpret the reflector from the bottom of the wedge as having a positive polarity and a phase delay of $\Delta t/2$. If we divide the composite reflection response by Δt, we obtain the finite difference approximation to the derivative shown in (a). Thus, for small values of Δt, the composite response will be the derivative of the original seismic wavelet. In (b), we review some of the spectral properties of the derivative operator so that we can gain insight into the spectral response of a thin bed. If the Fourier transform (complex spectrum) of a seismic waveform, $u(t)$, is $U(\omega)$, where ω is the temporal frequency in radians/s, then the Fourier transform (complex spectrum) of the derivative du/dt is $i\omega\, U(\omega)$, such that for a very thin bed, the phase is rotated by 90° and the amplitude spectrum is increased proportionally to the frequency, ω.

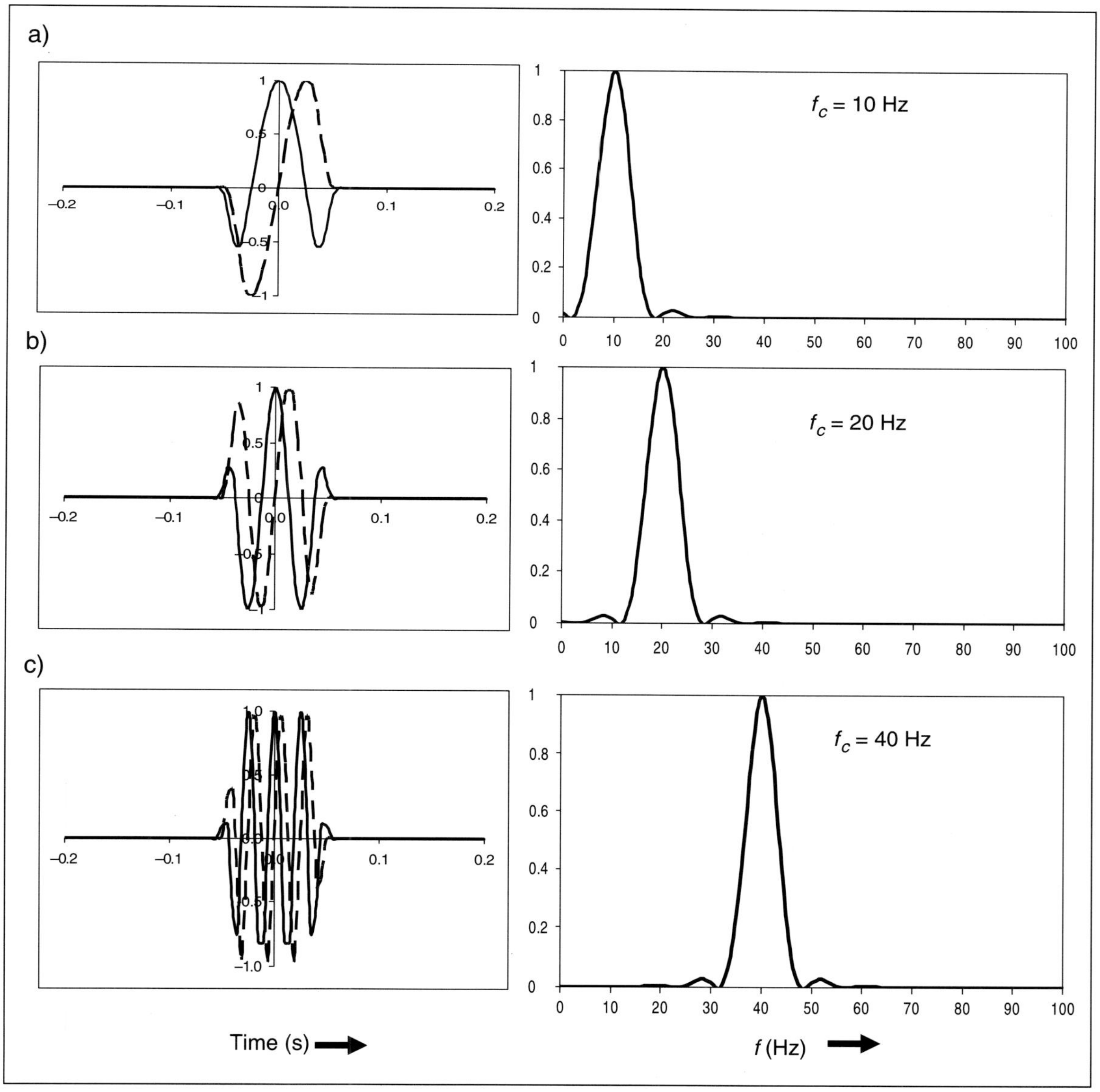

Figure 2.3. Representative seismic wavelets and their Fourier spectra for carrier frequencies at (a) 10, (b) 20, and (c) 40 Hz, commonly used in short-window discrete Fourier transform (SWDFT) spectral-decomposition analysis. The window width is a constant 0.100 s; the window taper is 0.020 s. Solid and dashed lines indicate the cosine and sine wavelets, respectively. Note the side lobes of the spectra. Also note that the spectrum of the 10-Hz wavelet extends into negative frequencies.

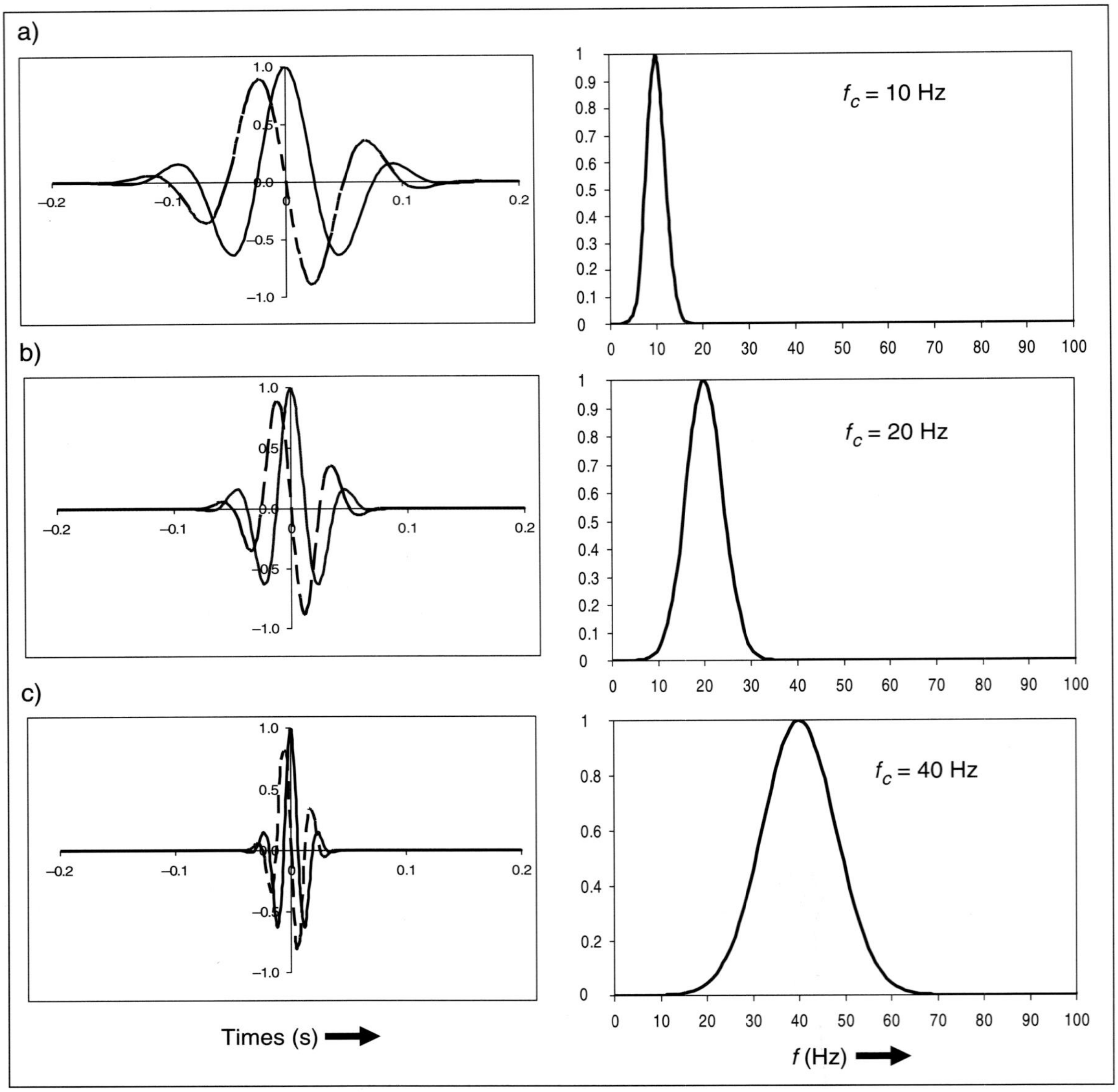

Figure 2.4. Representative seismic wavelets and their Fourier spectra for carrier frequencies at (a) 10, (b) 20, and (c) 40 Hz, commonly used in the continuous-wavelet transform (CWT). Morlet wavelets commonly are used in wavelet compression and in instantaneous spectral analysis, using the window defined by $w(k\Delta t) \equiv w_k = \frac{1}{\sqrt{\sigma}} \exp\left[\frac{-(k\Delta t)^2}{2\sigma^2}\right]$, where k is the sample number, Δt is the sample increment, and $\sigma = 1/f_c$. Solid and dashed lines indicate the cosine and sine wavelets, respectively. Note that the bandwidth increases with carrier frequency.

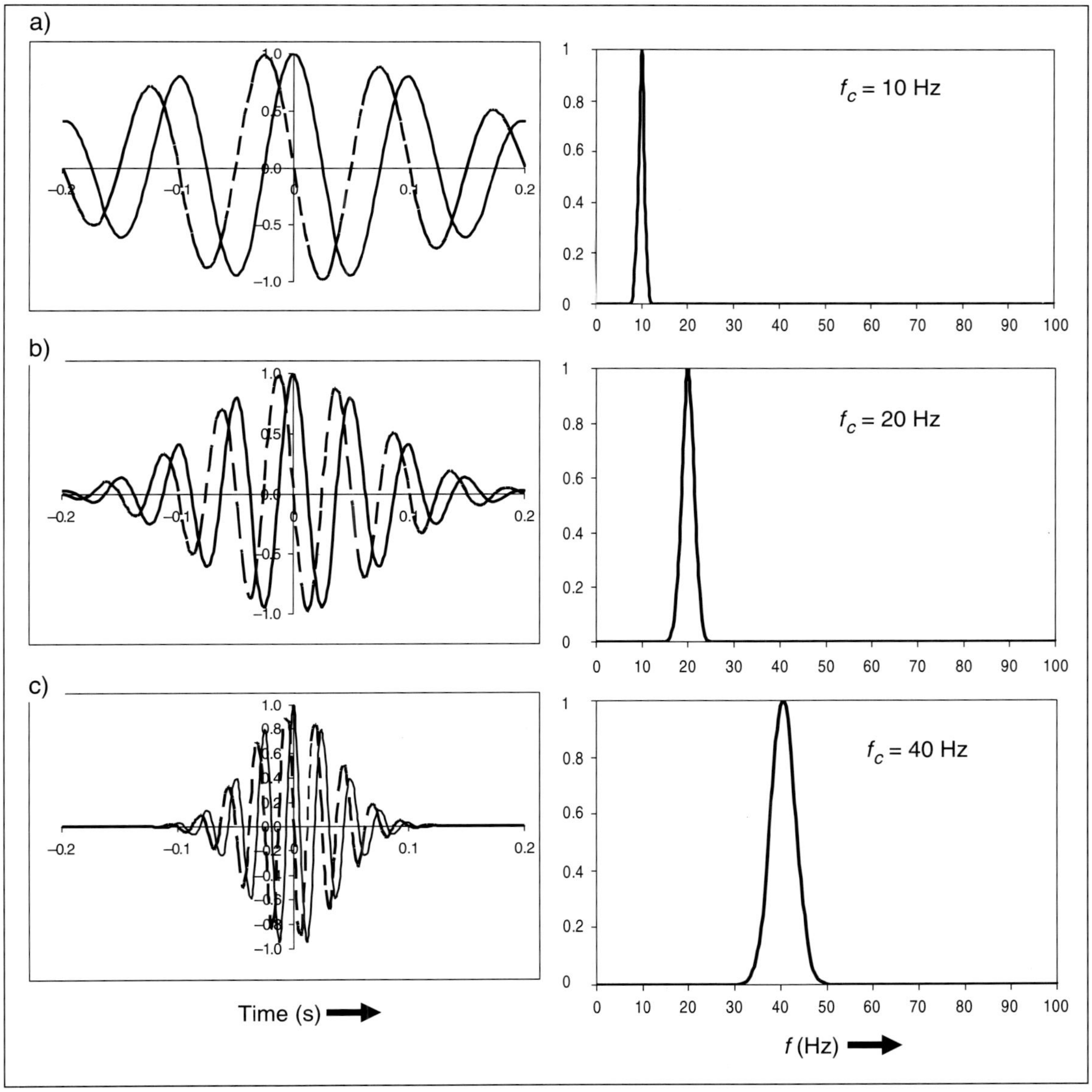

Figure 2.5. Representative seismic wavelets and their Fourier spectra for carrier frequencies at (a) 10, (b) 20, and (c) 40 Hz, commonly used in the CWT for the SPICE (spectral imaging of correlative events) algorithm, using the window defined by $w(k\Delta t) \equiv w_k = \frac{1}{\sqrt{\sigma}} \exp\left[\frac{-(k\Delta t)^2}{2\sigma^2}\right]$, where k is the sample number, Δt is the sample increment, and $\sigma = 3/f_c$. Solid and dashed lines indicate the cosine and sine wavelets, respectively. The seismic wavelet for 10 Hz goes beyond the scale of this plot.

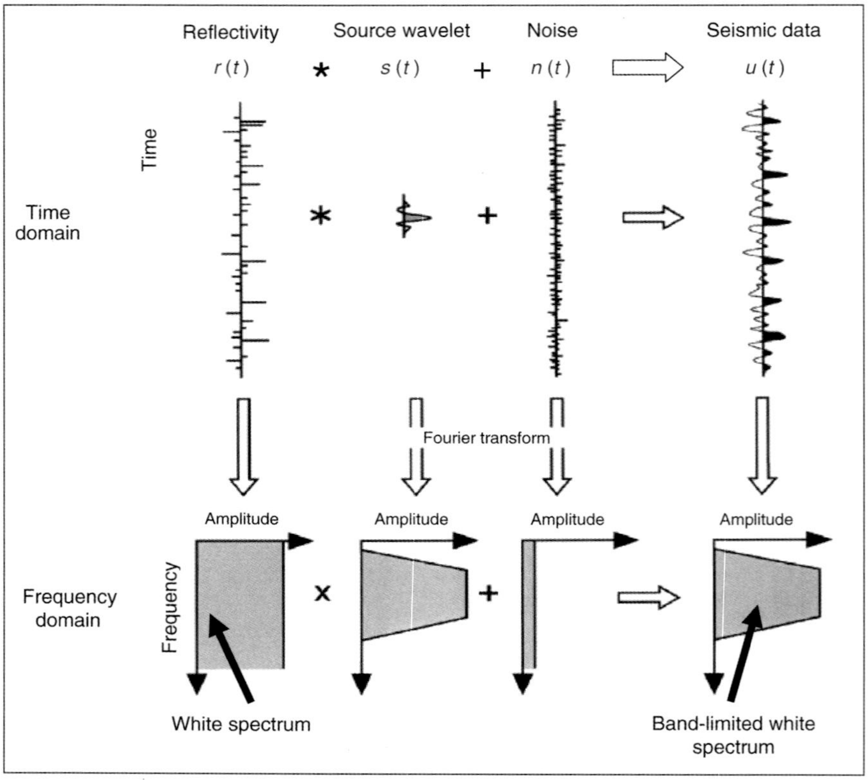

Figure 2.6. Long-window spectral decomposition and the convolutional model. Typically, we assume that the multiple-free reflectivity spectrum has very little structure (in this case, we display a white spectrum). Next, we calculate the Fourier spectrum of the entire (unwindowed) trace. Finally, we explicitly flatten its spectrum, which, under the assumption of white additive noise, will flatten the spectrum of the source wavelet. Similar assumptions are made in the deconvolution algorithms routinely used in seismic processing. After Partyka et al. (1999).

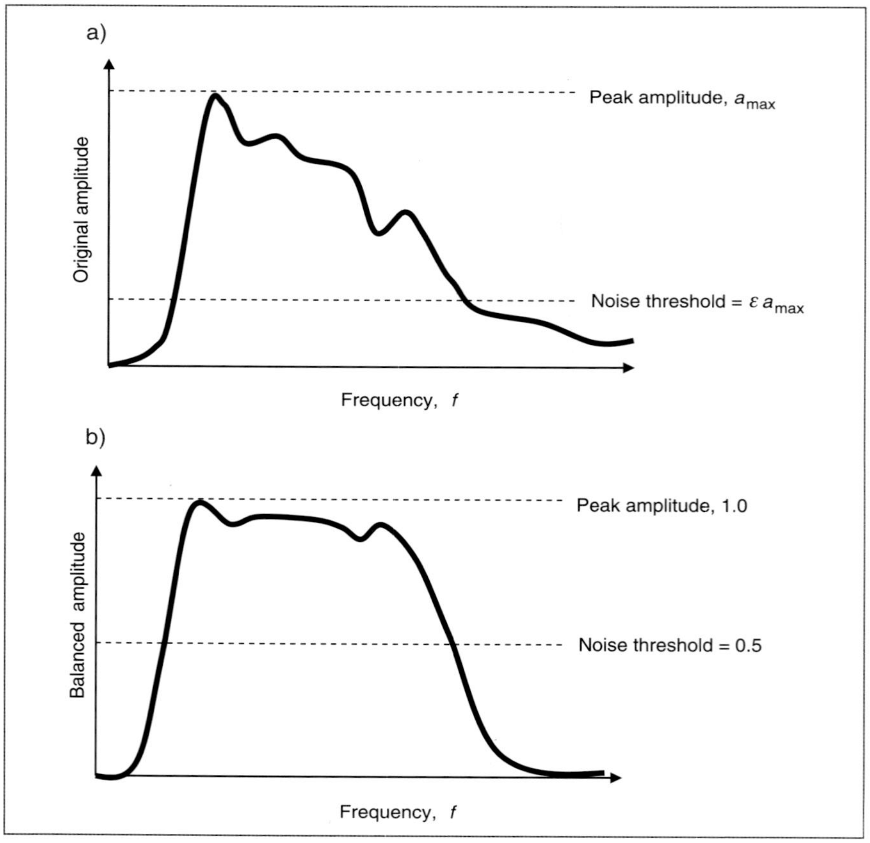

Figure 2.7. The concept of spectral balancing to achieve a band-limited white spectrum in the presence of noise. (a) First we calculate the magnitude of each spectral component, $a(f)$, and its maximum, a_{max}. Next, we estimate a noise level as a fraction, ε, of the peak spectral amplitude. (b) Finally, we rescale each spectral component by $1./[a(f) + \varepsilon a_{max}]$, such that the new peak spectrum is approximately 1.0.

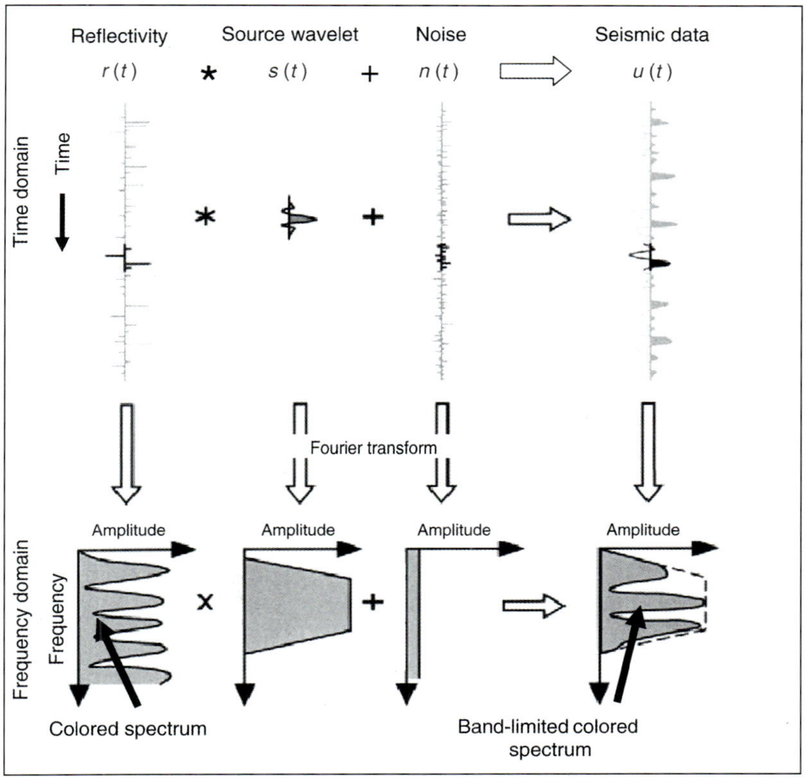

Figure 2.8. Short-window spectral decomposition and the convolutional model. Although we may wish to assume that the reflectivity has a more or less white spectrum, any short-windowed realization of this white distribution will have only a few discrete reflection spikes and thus invariably will have a colored spectrum. If we process the data to provide a band-limited white source spectrum and assume we have white noise, the colored spectrum of the windowed seismic data will be a band-limited representation of the colored spectrum of the reflectivity within the window. After Partyka et al. (1999).

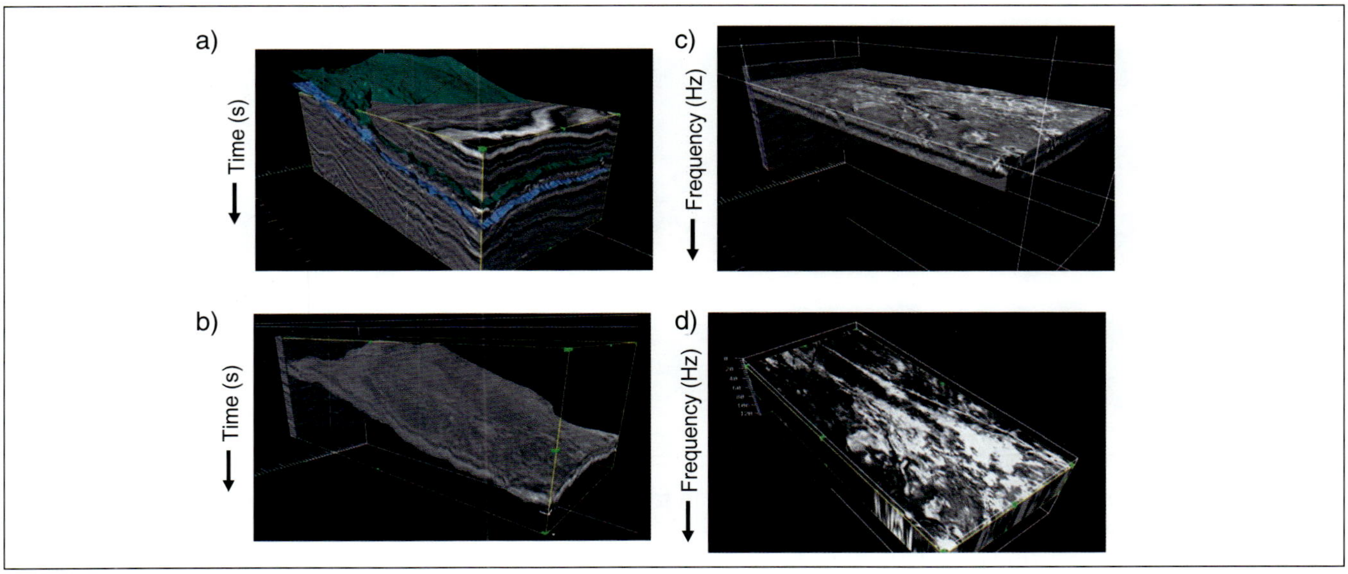

Figure 2.9. Basic steps in short-window discrete Fourier transform spectral decomposition. (a) First, we define a short analysis window parallel to an interpreted horizon. (b) Next, we extract and then flatten seismic data falling within the analysis window. (c) Then we calculate the discrete Fourier transform (DFT) of the window for each frequency by crosscorrelating with sines and cosines. (Steps [b] and [c] are equivalent to directly applying the windowed DFT shown in Figure 2.3.) (d) Finally, we spectrally balance each frequency slice, after which we interpret spectral components through animation, composite displays, 3D visualization, or statistical analysis (as in Figures 2.16 and 2.18). After Johann et al. (2003).

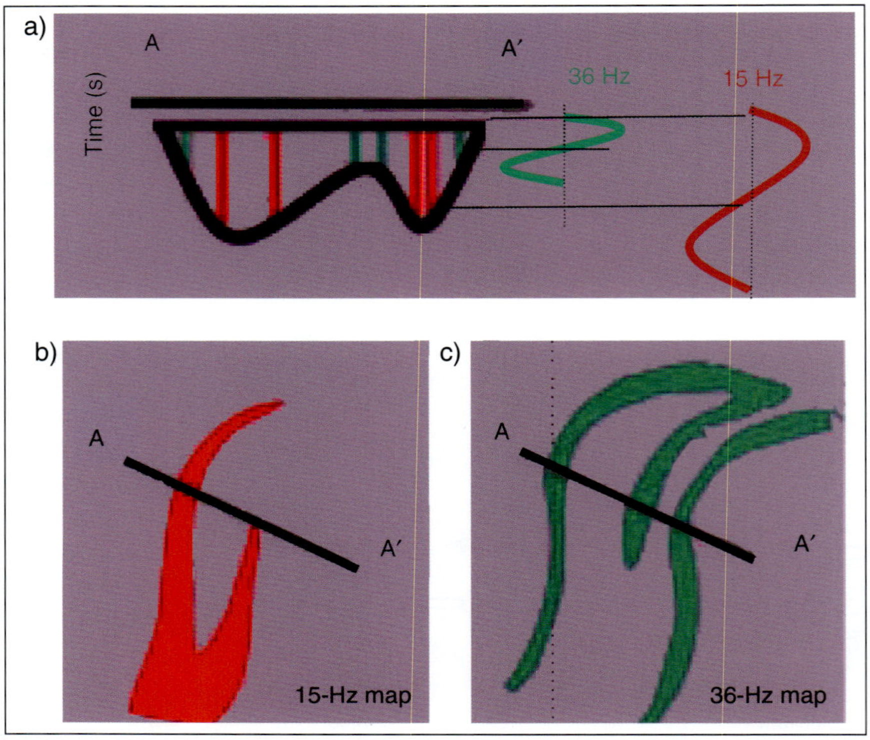

Figure 2.10. Schematic diagrams showing the interrelationship between thin-bed tuning and the amplitude of spectral components through an idealized channel. (a) A vertical cross section and spectral components at (b) a higher frequency and (c) a low frequency, shown in map view. On the thinner flanks, thin-bed tuning occurs at the higher-frequency components (in green). In the thalweg or center of the channel, thin-bed tuning occurs at the lower frequencies (in red). After Laughlin et al. (2002). AAPG©2002. Reprinted by permission of the AAPG whose permission is required for further use.

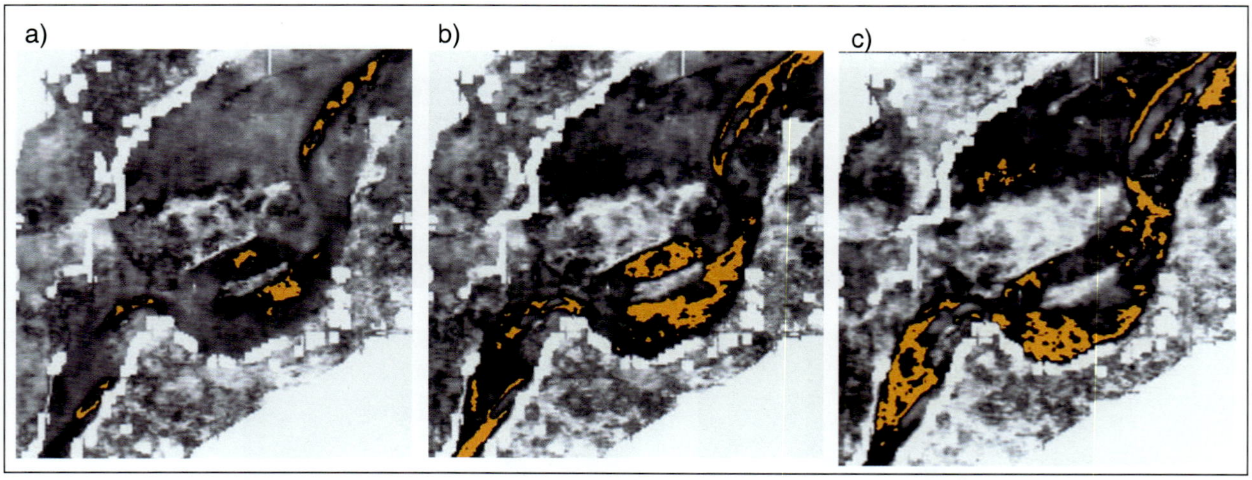

Figure 2.11. Three spectral components, increasing in frequency from (a) to (c), with the highest spectral amplitude plotted in yellow. Each spectral component responds uniquely to variations in reservoir thickness. The thickest part of the channel is seen in (a), thinner parts in (b), and the thinnest parts, including the flanks of the channel, in (c). Data are courtesy of Apache Corp. After Bahorich et al. (2002). Reprinted by permission of Hart's E&P.

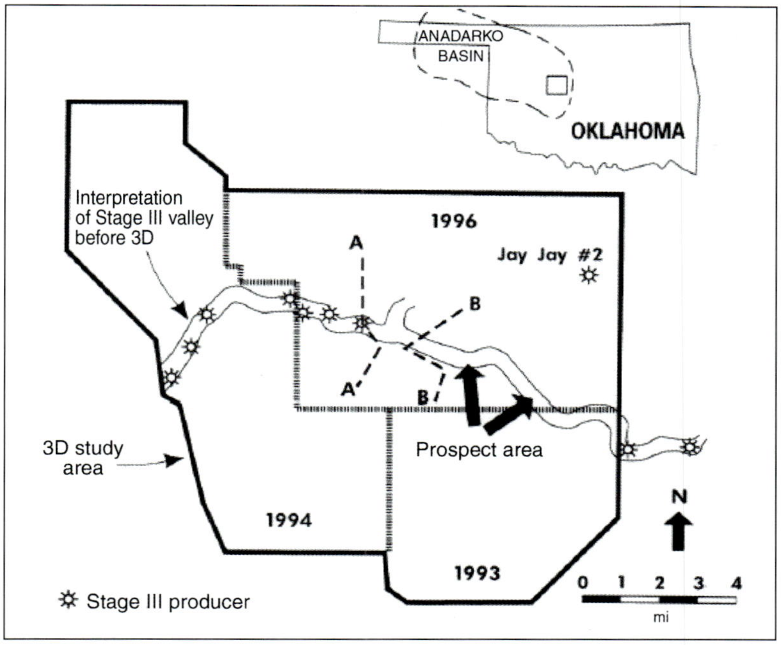

Figure 2.12. A location map showing 3D survey outlines, lines of cross section, wells producing from upper Red Fork Stage III valley fill, and an interpreted outline of the Stage III valley before acquisition of 3D seismic data. After Peyton et al. (1998).

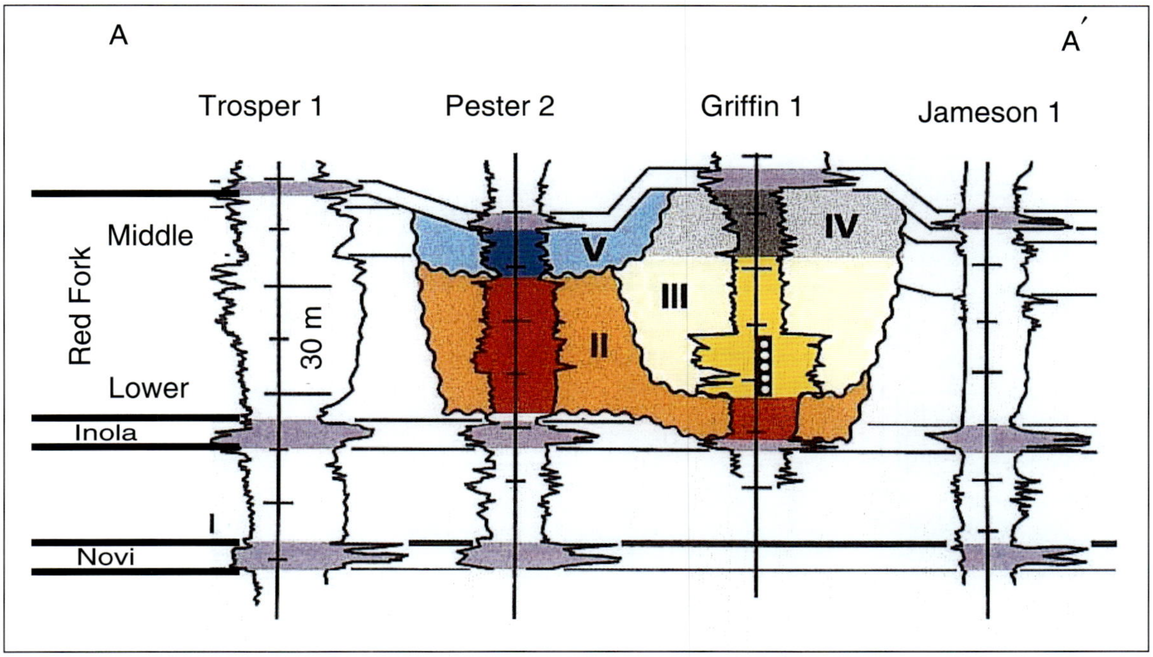

Figure 2.13. Stratigraphic well-log cross section AA′ from Figure 2.12., showing Stages II, III, IV, and V valley fill. The cross-section datum is the top of the Novi Limestone. Upper Red Fork valley stages and regionally correlative limestones are shaded. The Griffin 1 well (perforations shown) has produced 1.03 bcf of gas and 16,400 bbl of oil from Red Fork Stage III sandstones. After Peyton et al. (1998).

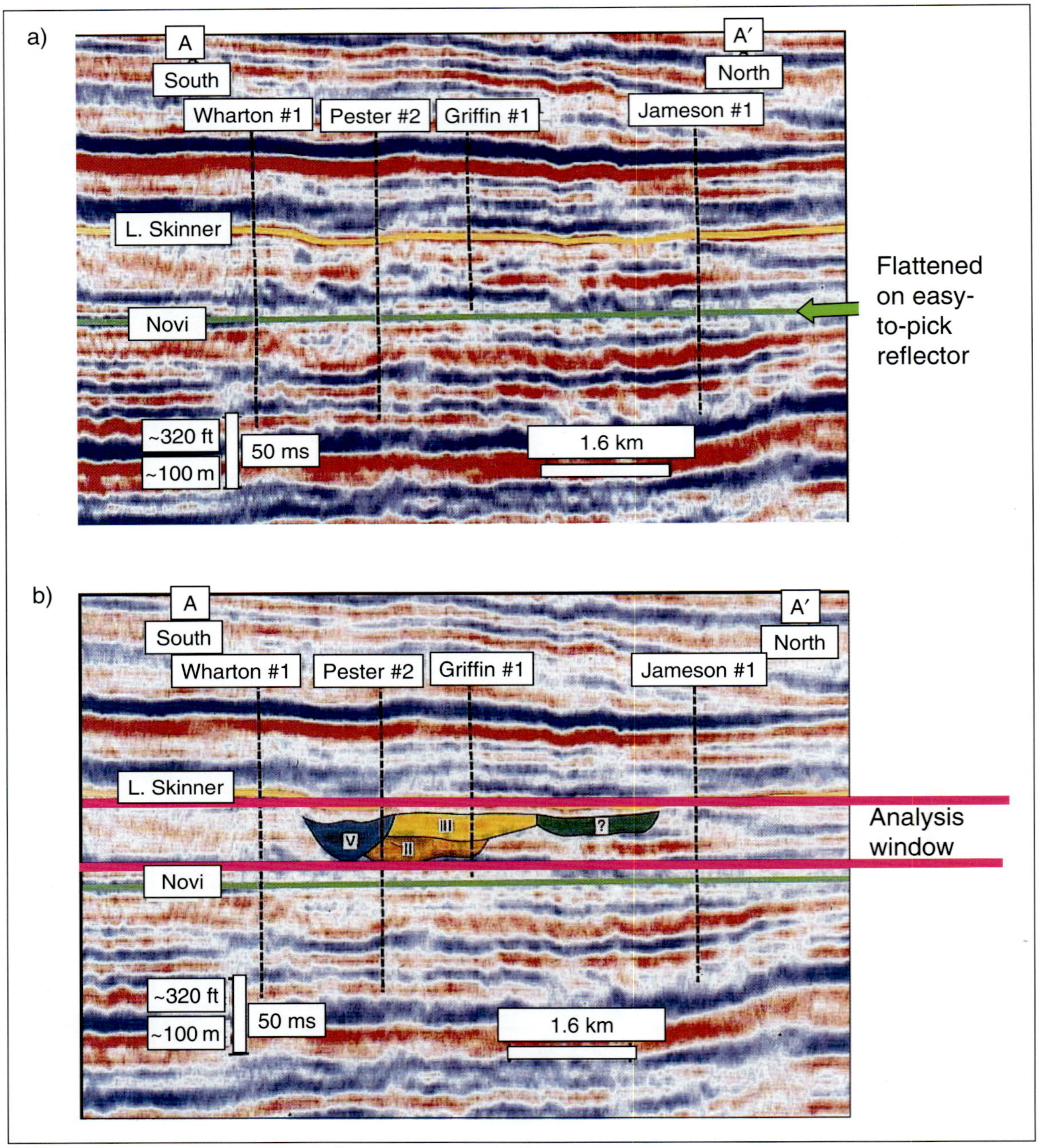

Figure 2.14. Seismic cross section AA′, flattened on the Novi horizon (a) without and (b) with channel interpretation. After Peyton et al. (1998).

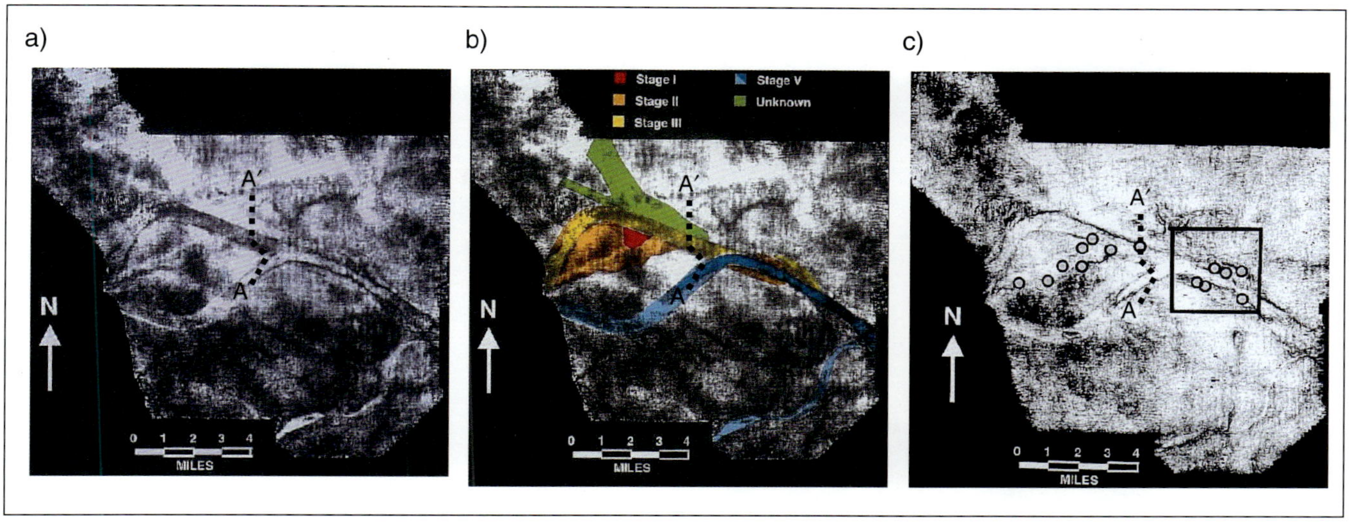

Figure 2.15. A 36-Hz amplitude slice from Red Fork strata spectral decomposition (a) before and (b) after interpretation. (c) A coherence slice 36 ms below the lower Skinner horizon, showing wells with faults in the Red Fork and Inola intervals. After Peyton et al. (1998).

Figure 2.16. High-resolution spectral decomposition, showing details of a complex offshore West Africa reservoir image. After Bahorich et al. (2002). Reprinted by permission of Hart's E&P.

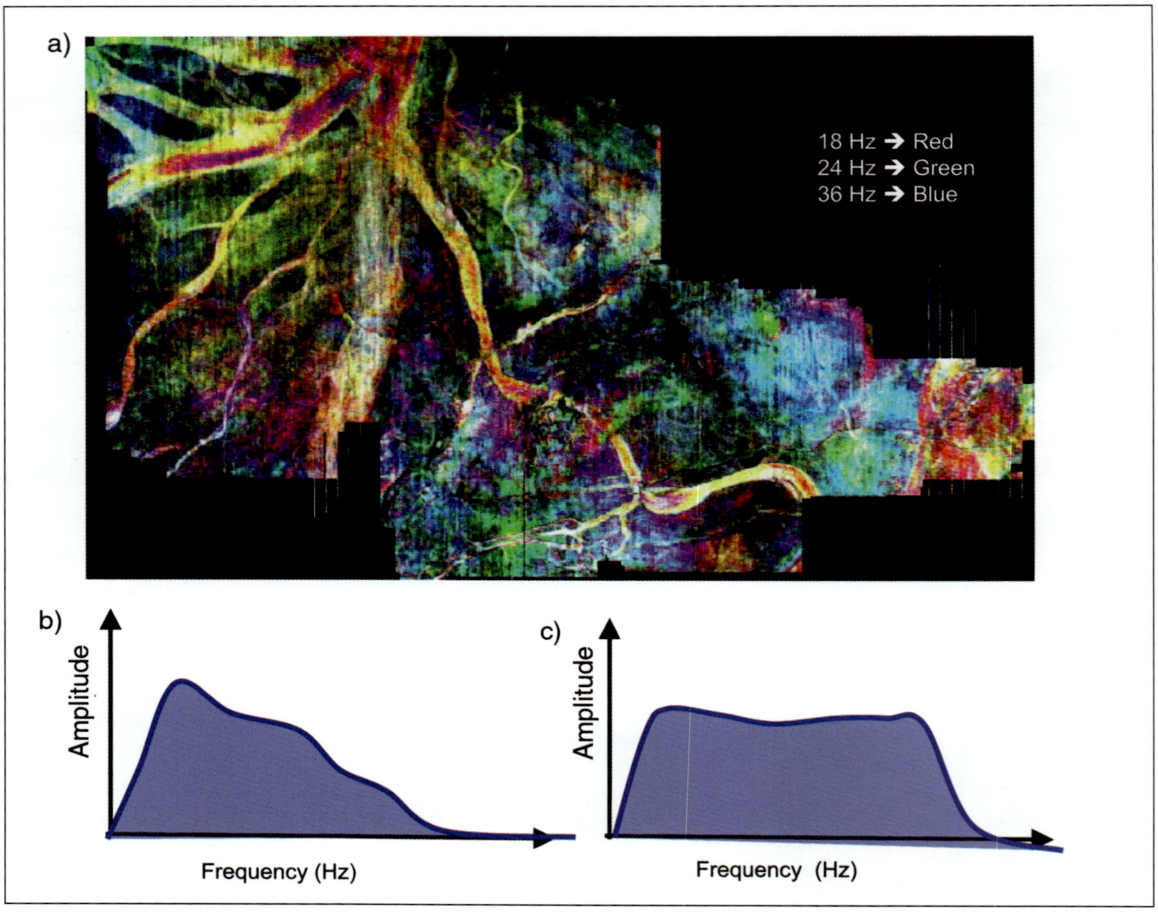

Figure 2.17. Spectral balance of frequency components in spectral decomposition. (a) A composite image of three spectral components: 18 Hz is mapped to red, 24 Hz to green, and 36 Hz to blue, to show the effect of tuning. Features that are tuned in at 18 Hz appear in red and indicate thicker channels. Moderately thick channels show equal amounts of 18-Hz and 24-Hz components and therefore map as yellow. Moderately thin zones show equal amounts of 24-Hz and 36-Hz components and map as cyan. The most thinly tuned zones map as blue. Zones with high amplitude and equal spectral components map as white. Generation of such an image requires that the spectral components be balanced. Typically, higher frequencies are attenuated faster as we go deeper, thereby resulting in a spectrum that is biased toward the low end, such as that shown in (b). We rectify this situation by using a process similar to that in spectral balancing of the seismic data discussed in Figure 2.6. However, instead of computing (b) as the spectrum of a single trace, we compute (b) as the average spectrum along a time slice. If we define the average spectrum as $<a(f)>$ and its maximum as a_{max}, we proceed as in Figure 2.7 and estimate a noise level as a fraction, ε, of the peak spectral amplitude. Then we rescale each spectral component by $1./[<a(f) + \varepsilon a_{max}]$, thereby obtaining the flattened spectrum shown in (c).

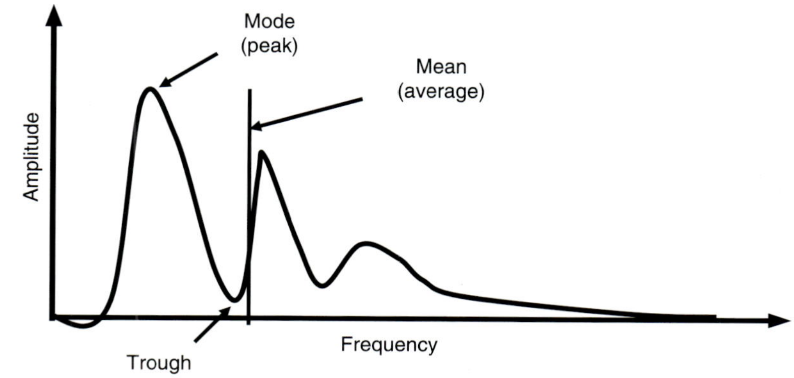

Figure 2.18. Schematic diagram showing a typical spectrum resulting from spectral decomposition. In general, the mode of the spectrum changes strongly and the mean frequency changes only moderately with changes in thin-bed tuning.

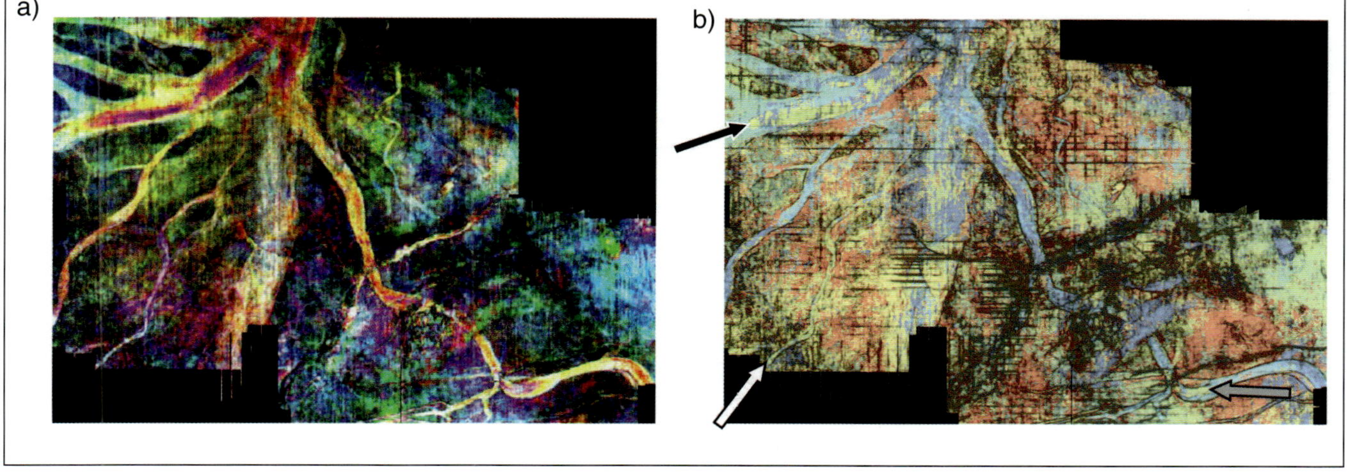

Figure 2.19. Alternative means of plotting multiple spectral information corresponding to the images shown in Figure 2.9. (a) A blended image using the RGB (red-green-blue) color model. R corresponds to 16 Hz, G to 26 Hz, and B to 36 Hz. After Wessels et al. (1996). (b) A blended image of the mode of the spectrum (plotted against hue) and coherence (plotted as a gray scale, with low coherence being black). The low-frequency (blue) areas correspond to a thicker channel, whereas the high-frequency (yellow) areas correspond to a thinner channel (white arrow), a point bar (gray arrow), and a longitudinal bar (black arrow). After Marfurt and Kirlin (2001).

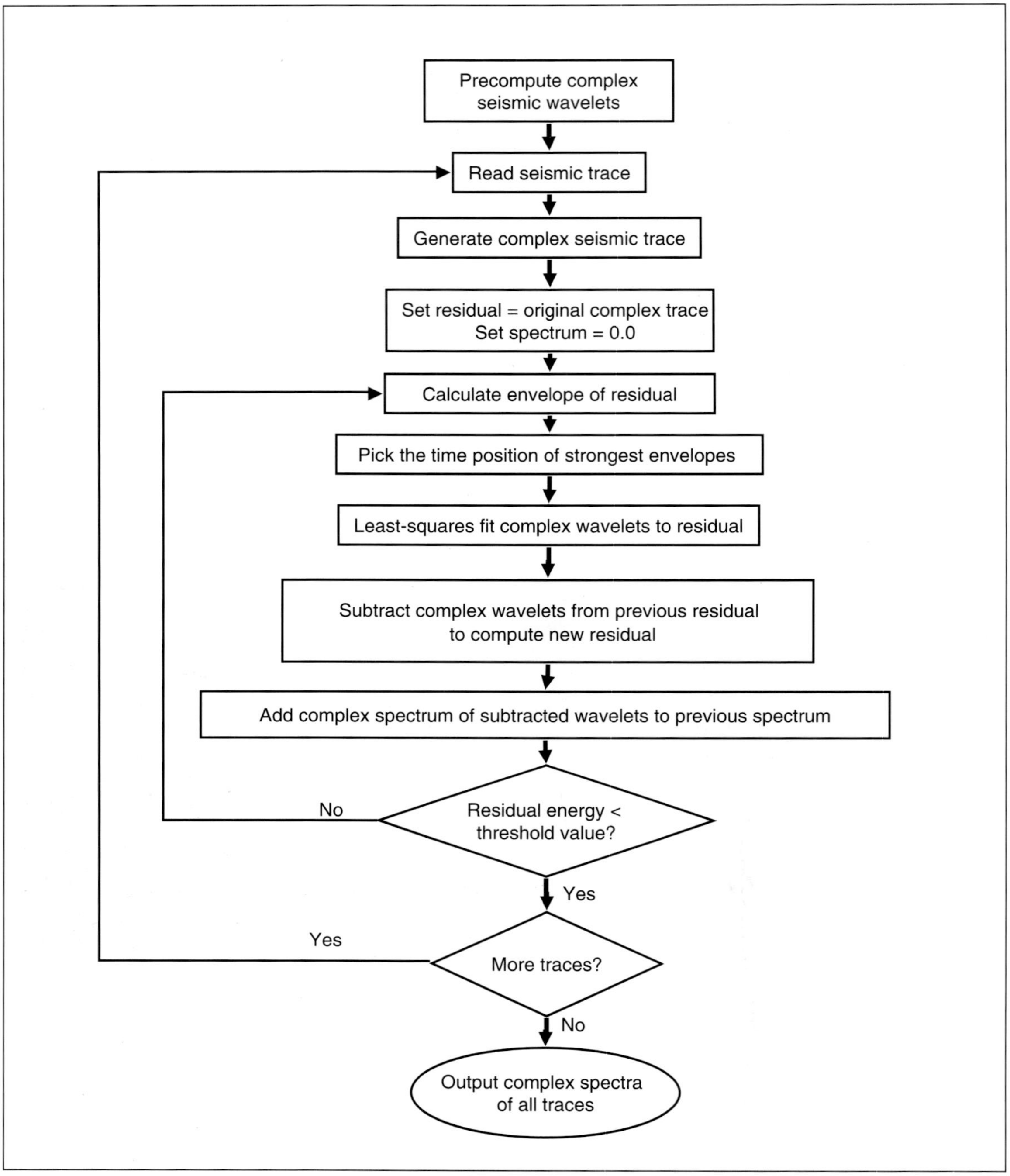

Figure 2.20. A flowchart for a wavelet decomposition algorithm using a matched-pursuit technique. After Liu and Marfurt (2005).

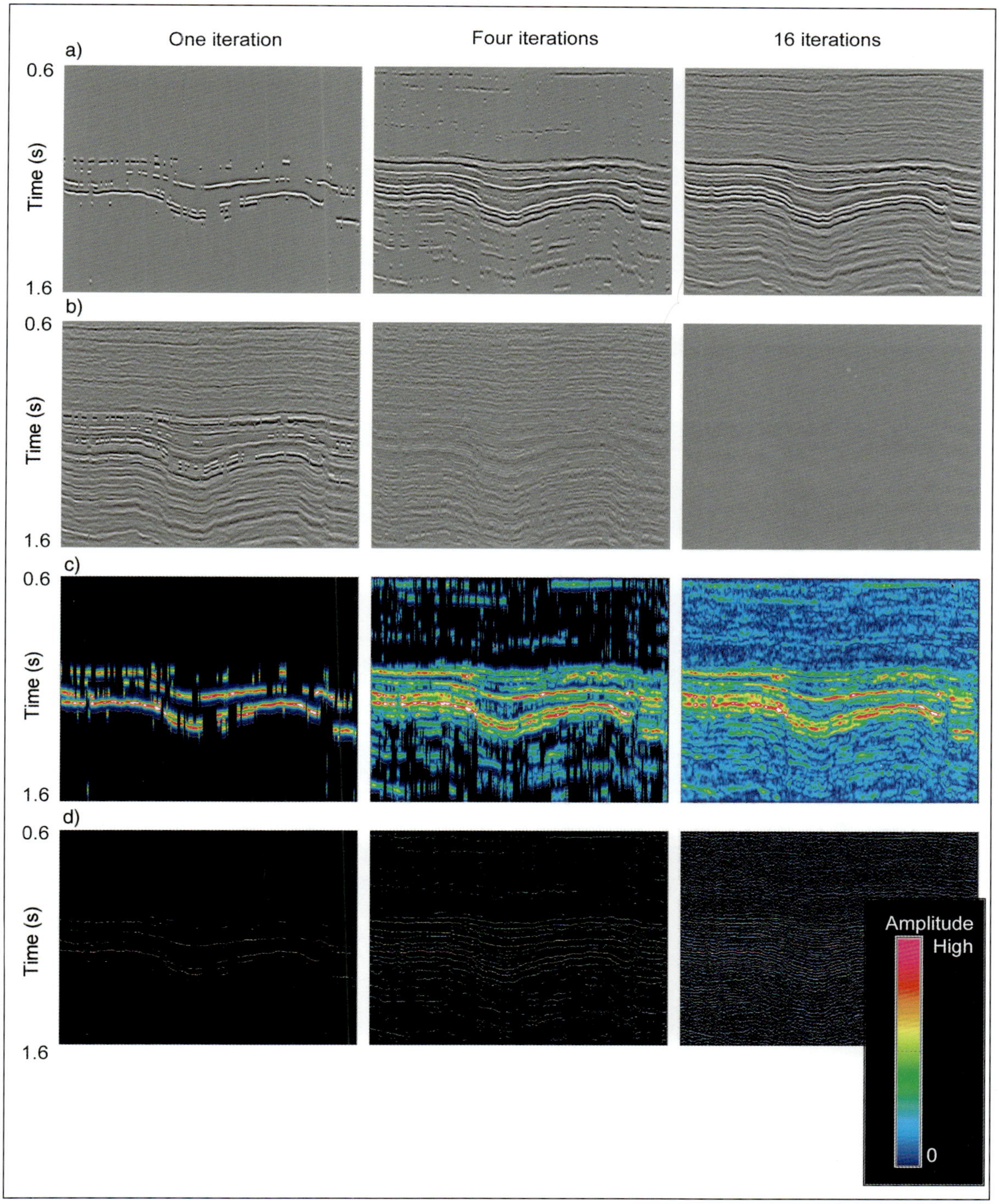

Figure 2.21. An example showing matched-pursuit wavelet decomposition of a seismic line from a survey acquired over the Central Basin Platform, west Texas, U.S.A. Columns represent algorithm results after the first, fourth, and 16th iterations of (a) modeled data, (b) residual (unmodeled) data, (c) the 40-Hz component of the modeled data, and (d) the wavelet location and envelope of the modeled data. The frequency and phase of the modeled wavelets are not displayed. After Liu and Marfurt (2005). Data are courtesy of Burlington Resources.

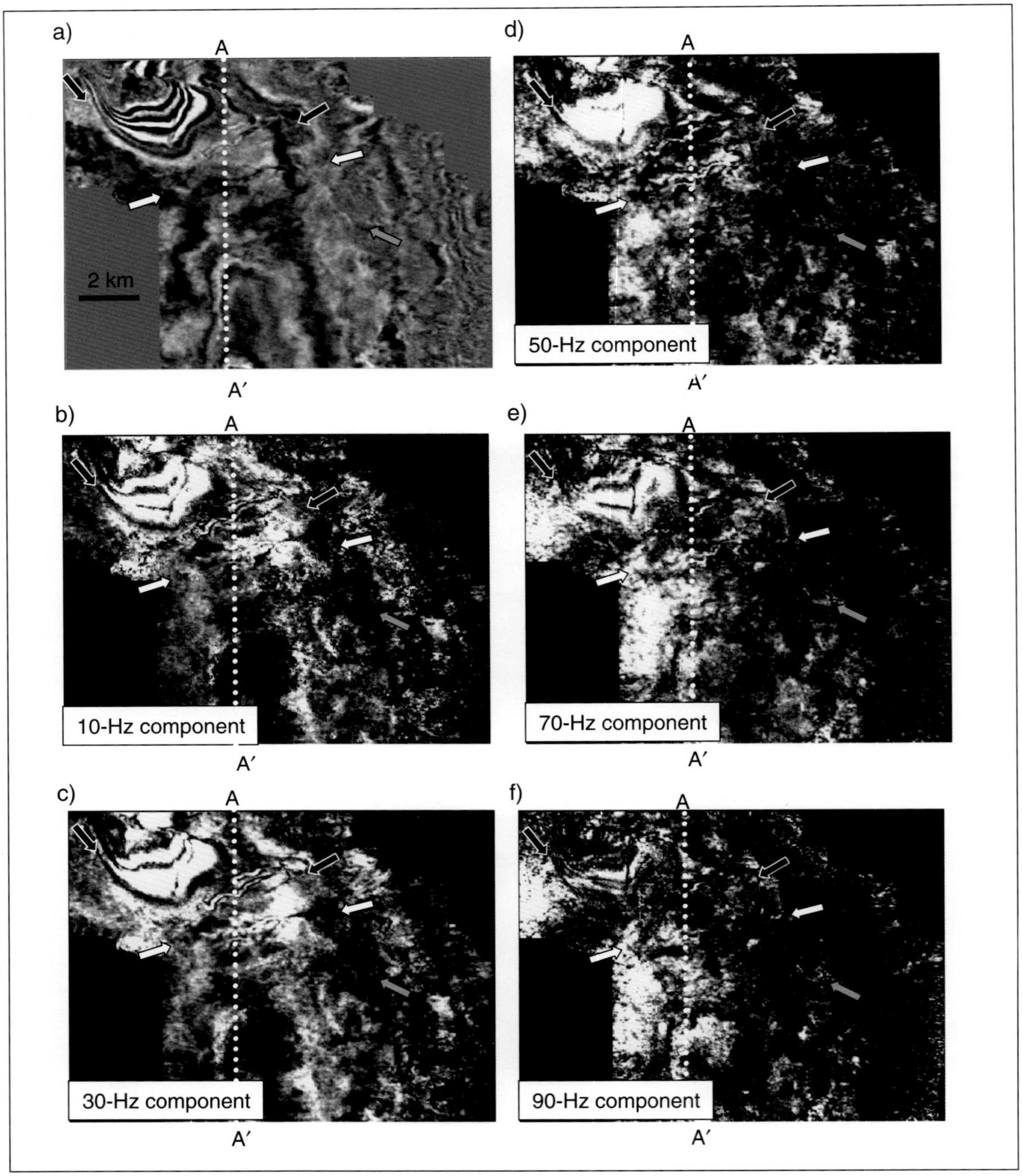

Figure 2.22. (a) A time slice, at $t = 1.060$ s, through a survey acquired in the Central Basin Platform, Texas, U.S.A. Arrows indicate channels that drained the platform during the Pennsylvanian Atoka unconformity. (b)-(f), Spectral components at (b) 10 Hz, (c) 30 Hz, (d) 50 Hz, (e) 70 Hz, and (f) 90 Hz. Note how the channels tune in and out with frequency as we pass through their tuning thickness. Channels that have their strongest response at 10 Hz are generally thicker than those that have their strongest response at 90 Hz. To display the tuning phenomenon better, we will plot the peak frequency and the amplitude of the peak frequency in Figure 2.23. Seismic data are courtesy of Burlington Resources.

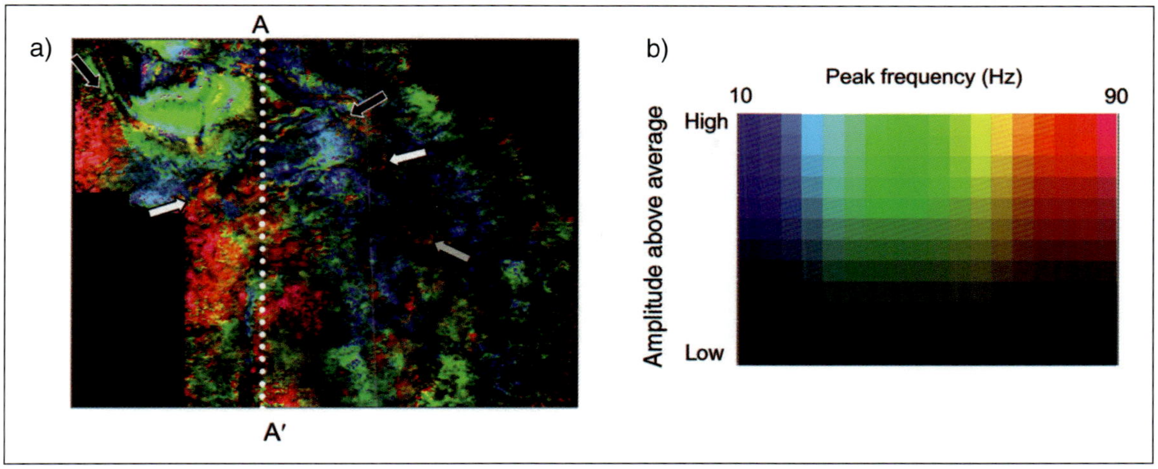

Figure 2.23. (a) A time slice, at $t = 1.060$ s, through a composite image of peak frequency and amplitude of the peak frequency above the average spectral amplitude, corresponding to the seismic data and spectral components shown in Figure 2.22. (b) The 2D color bar used in the display was discussed previously in Unit 1. The channel delineated by the black arrows in (a) is thin (yellow) in the west and becomes thicker (green) to the east, although a few intervals appear very thin (red) and may be associated with channel infill. The gray channel is quite diffuse at this time level. A map of these channels at this and neighboring time levels is shown on coherence in Figure 4.43.

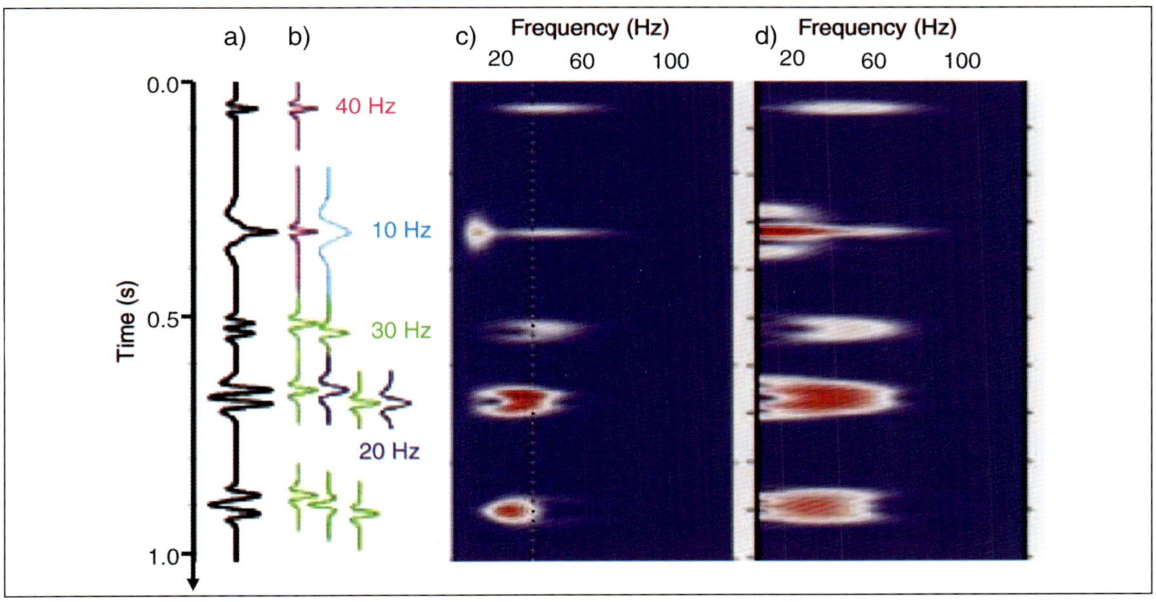

Figure 2.24. A synthetic waveform with (a) transient arrivals (black) and (b) constituent wavelets (color-coded by center frequency). Comparison of (c) the instantaneous spectral analysis (ISA, or wavelet transform) with (d) the short-time-window discrete Fourier transform analysis (SWDFT). Although the SWDFT has excellent vertical resolution, the frequency spectrum has been smoothed by convolution with the spectrum of the window, and false events are associated with side lobes of transient arrivals. After Castagna et al. (2003).

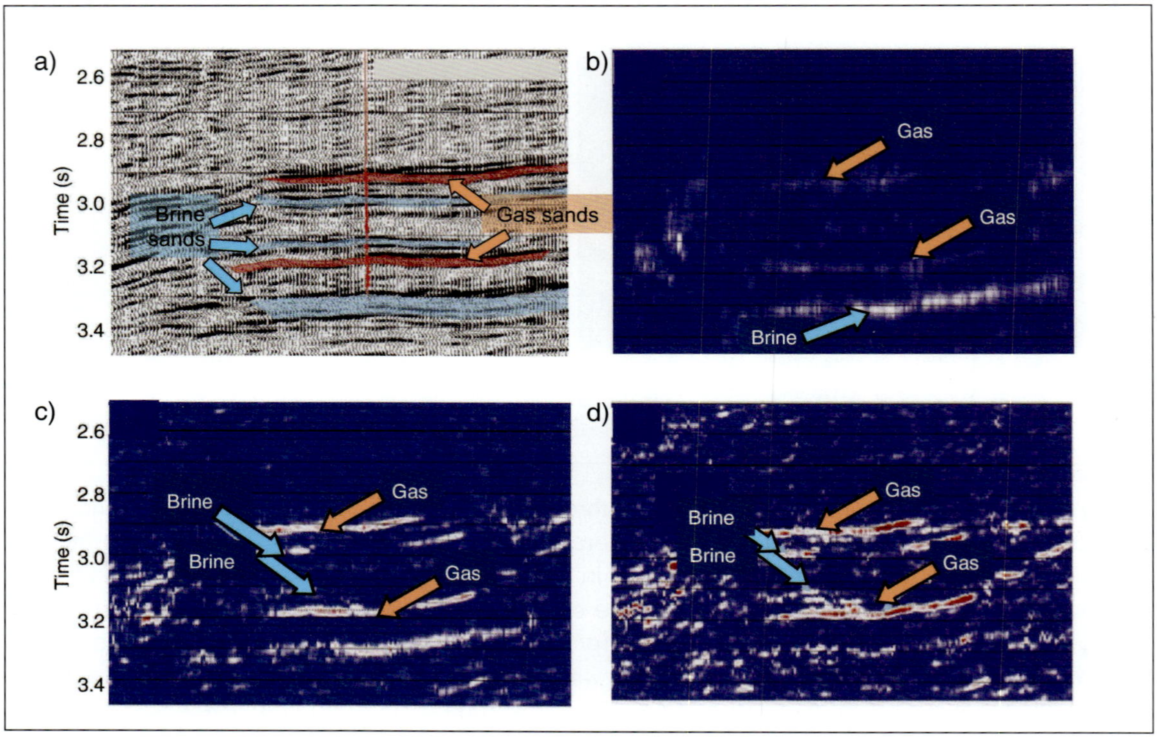

Figure 2.25. (a) A broadband seismic section from the North West Shelf of Australia. Gas sands are pink and brine sands are blue. (b)-(d). Spectral components at (b) 10 Hz, (c) 20 Hz, and (d) 30 Hz. The low-frequency shadow indicated by the cyan arrow in (b) beneath the lower gas sand is the strongest event at 10 Hz, but it is weaker than the overlying gas sands at 20 Hz and it fades away at 30 Hz. After Castagna et al. (2003).

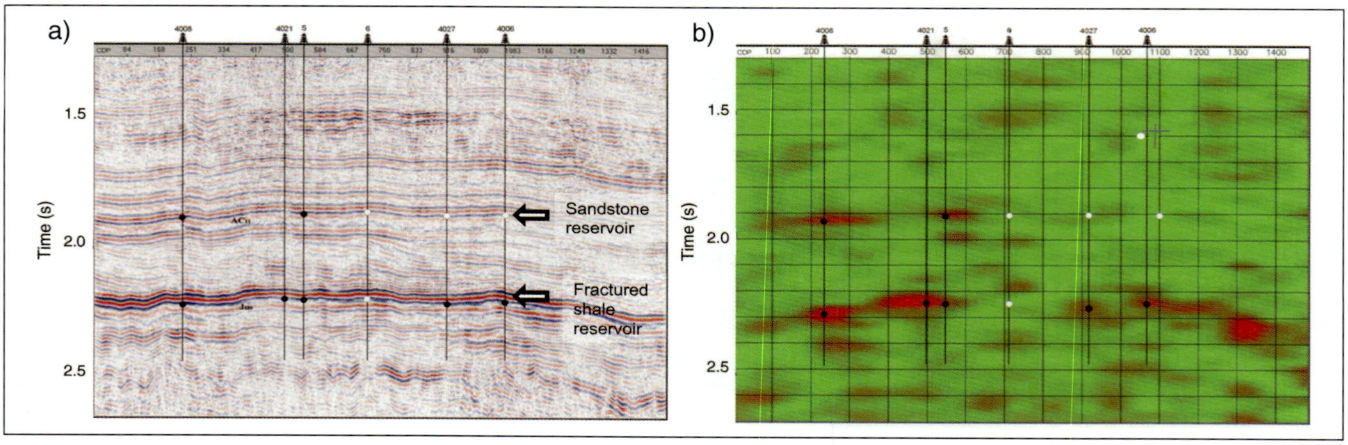

Figure 2.26. (a) A 2D seismic line from western Siberia and (b) a corresponding attribute proportional to fluid mobility, *m*, given by $m \sim \left(\frac{\partial a}{\partial \omega}\right)^2 \omega$, where *a* is the spectral magnitude and ω is the temporal frequency. Black circles indicate high oil saturation. White circles indicate no oil saturation. Some of these wells will be used to predict hydrocarbons, as shown in the next figure. After Goloshubin et al. (2002).

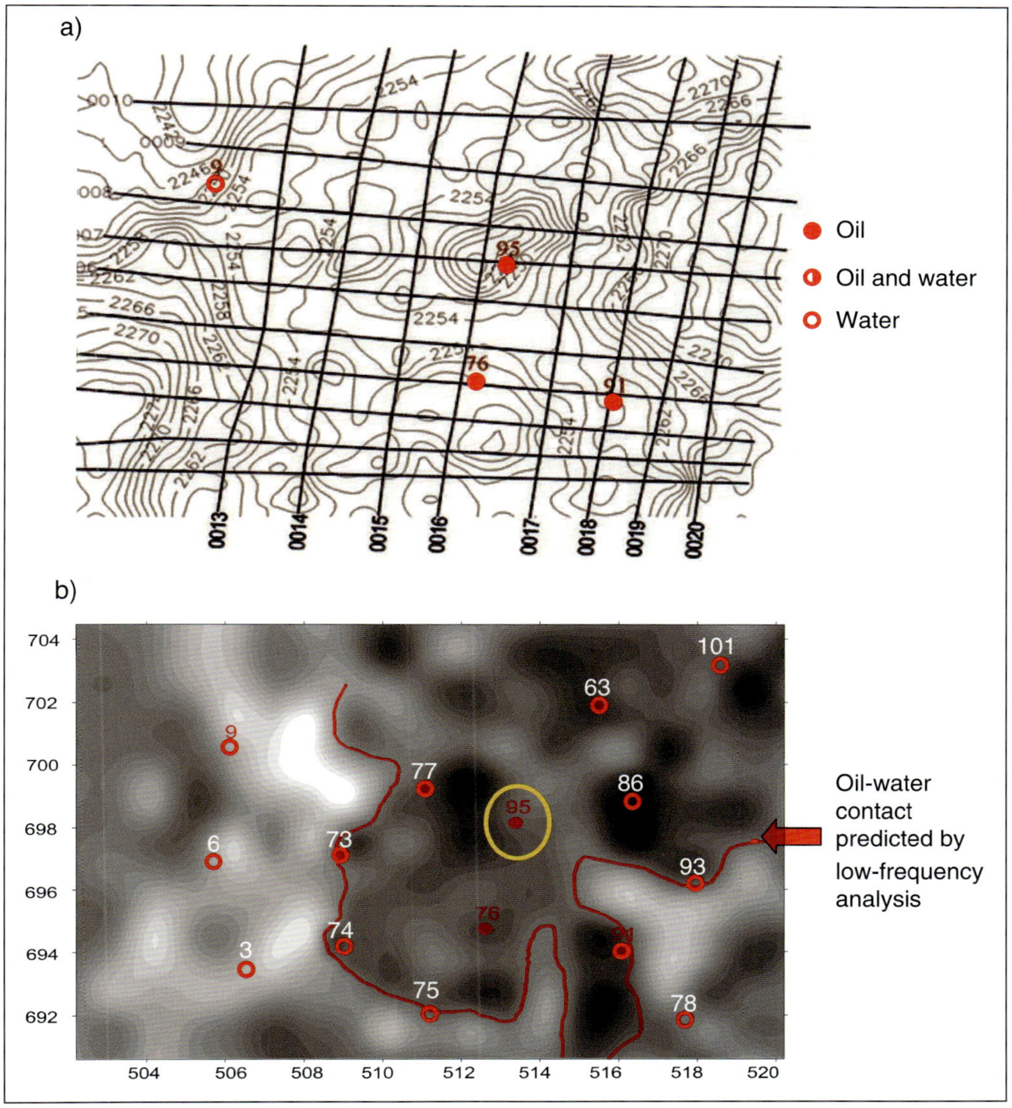

Figure 2.27. A time-structure map corresponding to the 2D survey shown in part (a) of the previous figure. (a) A blind test based on four calibration wells (three oil wells and one water well). (b) A map of the fluid mobility attribute corresponding to the vertical slice shown in Figure 2.26b. The orange-circled well is used in the calibration shown in the next figure. After Goloshubin et al. (2002).

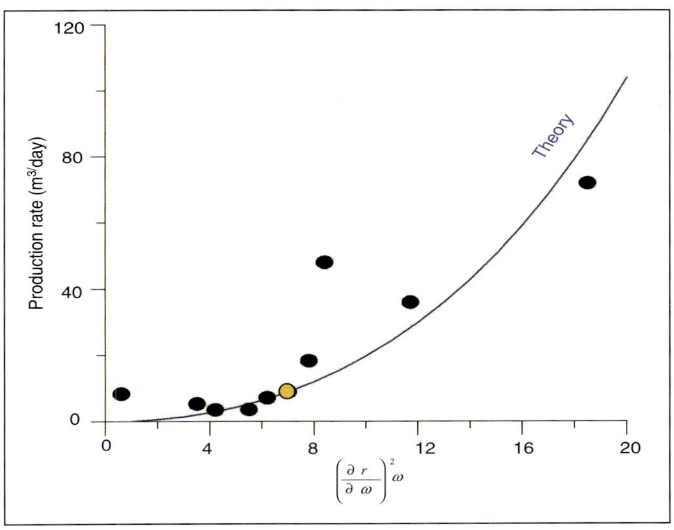

Figure 2.28. Production versus the mobility attribute fluid mobility, m, given by $m \sim \left(\frac{\partial a}{\partial \omega}\right)^2 \omega$, where a is the spectral magnitude and ω is the temporal frequency, for the wells shown in Figure 2.27b. The predicted hydrocarbon production (the blue curve) as a function of the mobility factor is calibrated to fit the production data seen at the orange-circled well, which corresponds to the orange data point in Figure 2.27b. Note how closely the production at the other wells follows this theoretical curve. After Goloshubin et al. (2002).

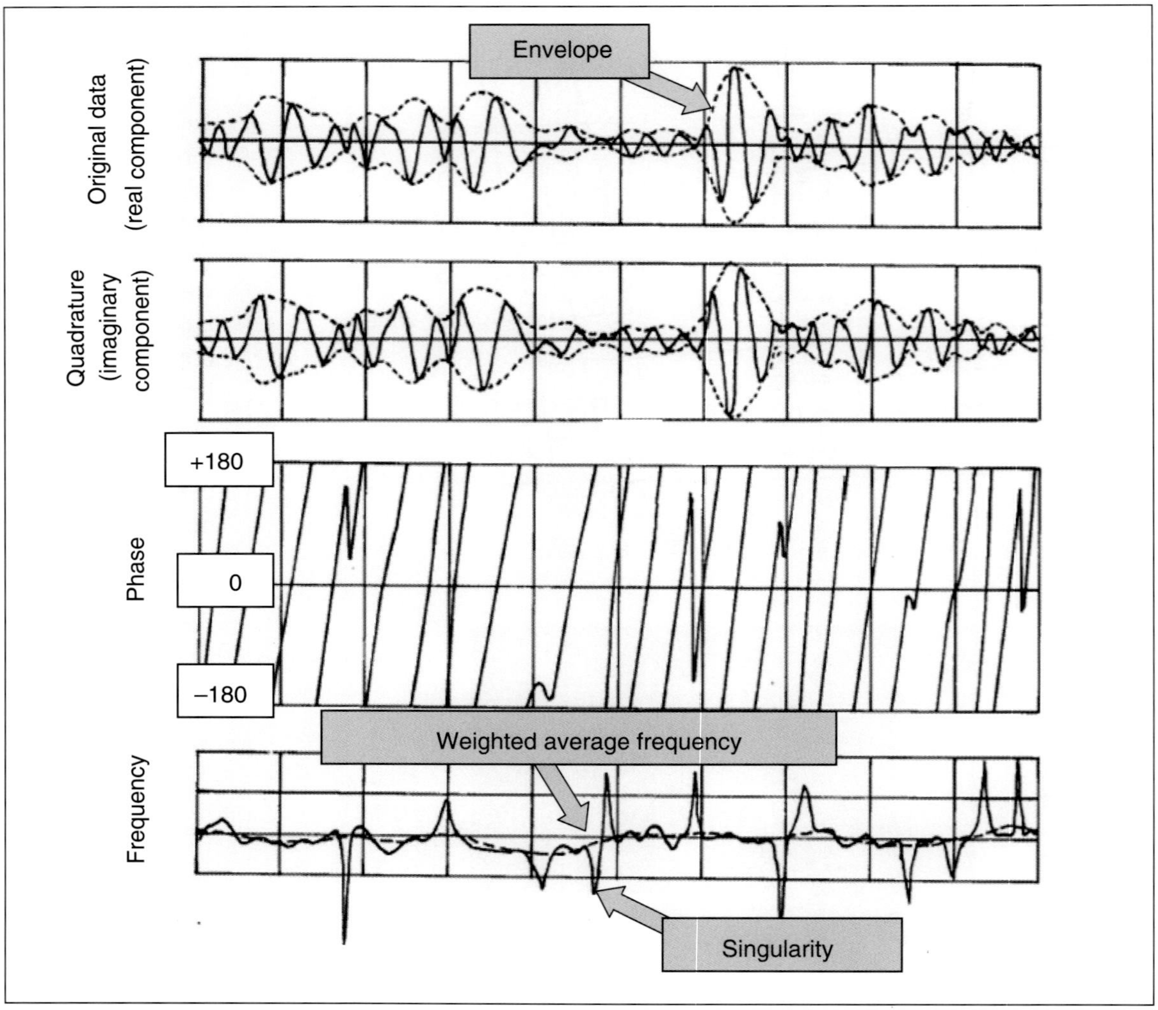

Figure 2.29. Complex trace attributes, including the real trace, its quadrature, its envelope, and its instantaneous phase and instantaneous frequency. Waveform interference results in singularities in the instantaneous frequency, such as the negative excursion indicated by the lowermost arrow. Such singularities can be suppressed by calculating a weighted-average frequency obtained by weighting the instantaneous frequency by the envelope and smoothing over a short time window. After Taner et al. (1979).

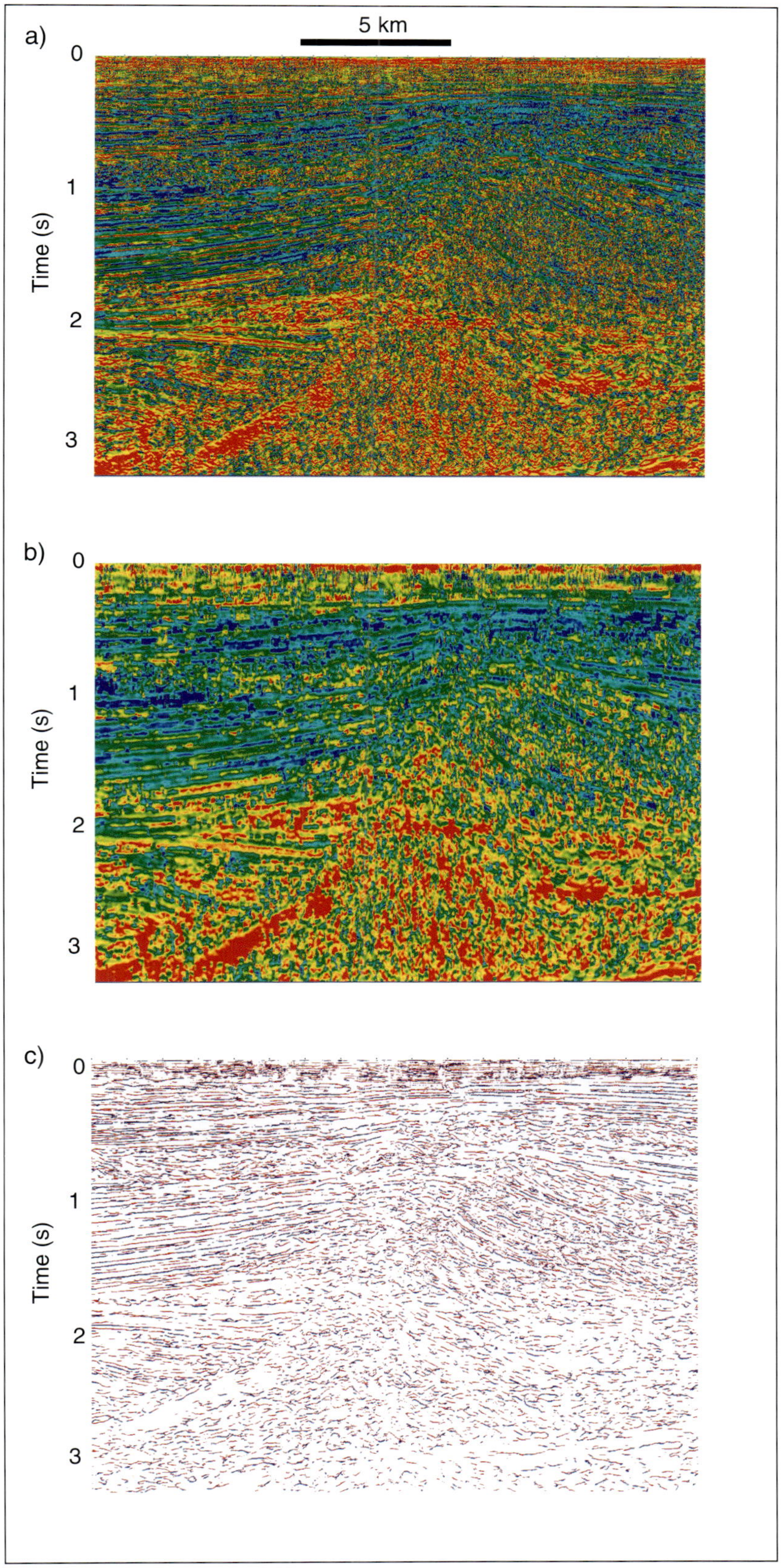

Figure 2.30. (a) An instantaneous frequency, (b) a weighted-average frequency, and (c) the difference between (a) and (b), giving a "thin-bed indicator" attribute. The image in (c) is plotted using a color bar that highlights the extreme values. This thin-bed indicator is based on the same physical phenomena as the SPICE algorithm displayed in Figure 2.35. After Taner (2000). Reprinted by permission of Rock Solid Imaging.

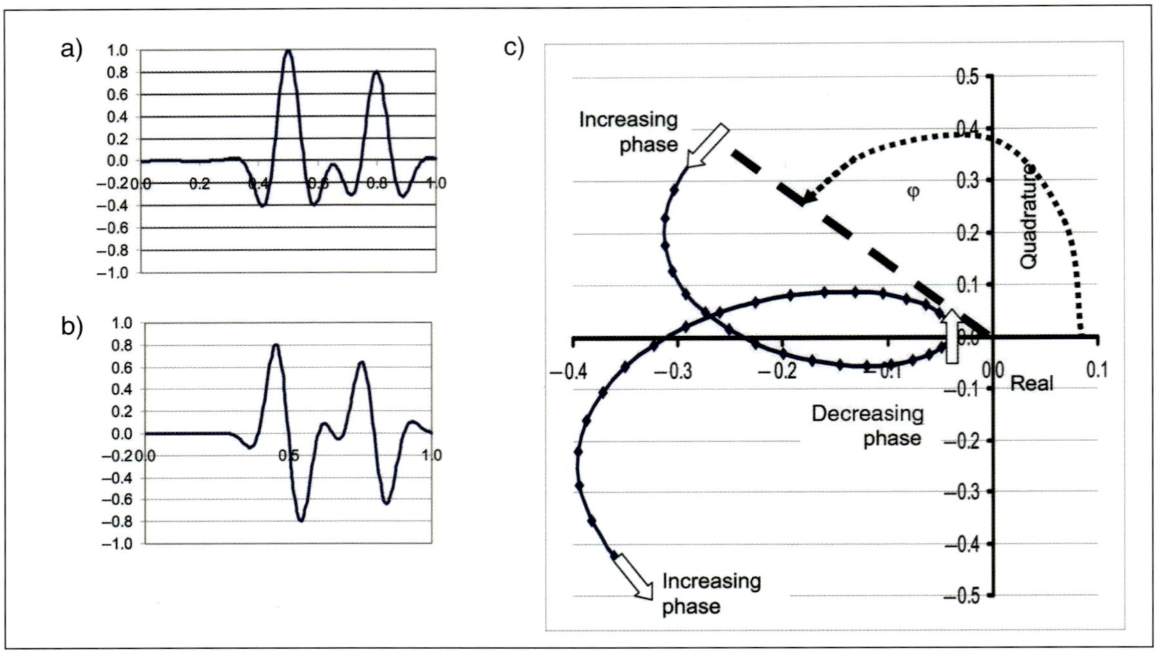

Figure 2.31. An illustration of singularities encountered when seismic wavelets interfere. (a) Two positive-polarity seismic wavelets, the first with a peak amplitude of 1.0, the second with a peak amplitude of 0.9. (b) The quadrature (or Hilbert transform) of (a). Peaks and troughs in (a) become zero crossings in (b). (c) The complex amplitude, with quadrature on the vertical axis and the original data on the horizontal axis. The instantaneous phase is the angle defined by these two components. In this example, it generally rotates counterclockwise. However, when the two waveforms interfere, we note that there is an area of decreasing phase. This decrease in phase with time creates negative values of instantaneous frequency.

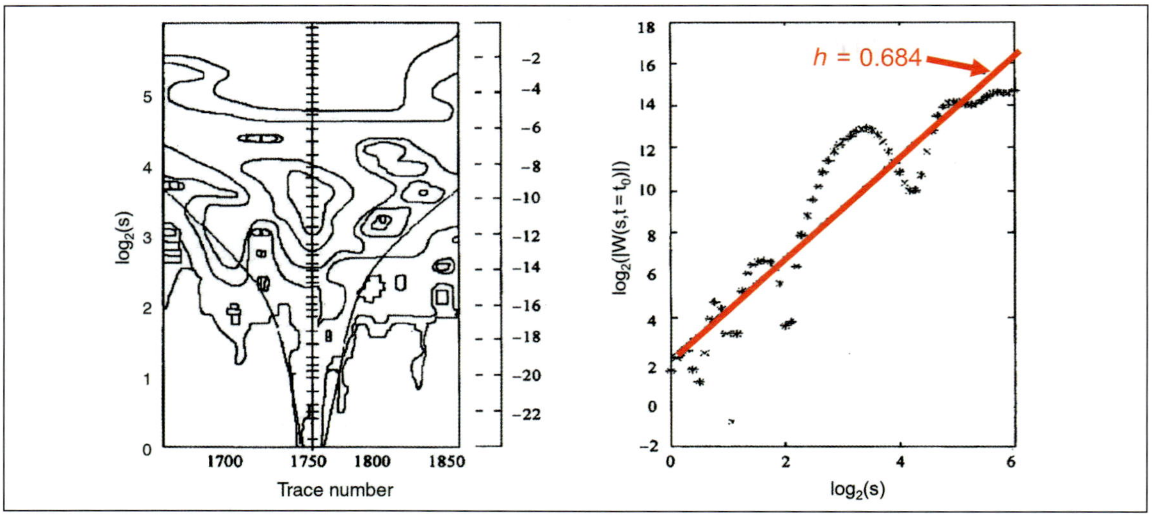

Figure 2.32. (a) A scalogram contour plot of wavelet spectral coefficients at a given time and for a suite of seismic traces. (b) Fitting of the power of the Holder exponent to wavelet coefficients. Each wavelet has its own spectrum (as shown in Figure 2.5) and does not correspond to a single carrier frequency, f_c. After Li and Liner (2004).

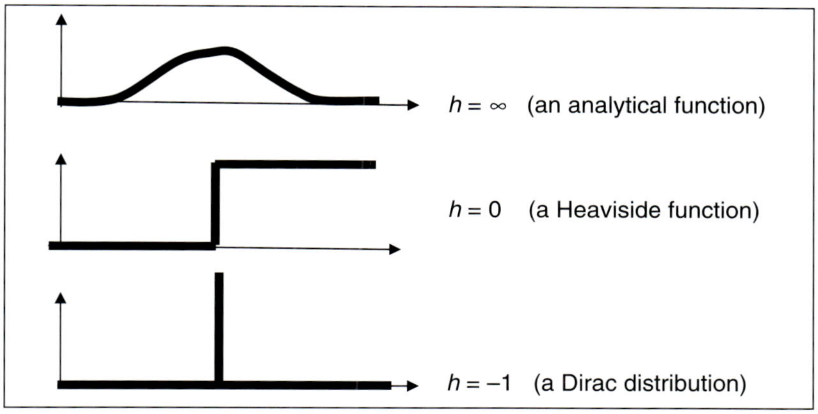

Figure 2.33. The Holder exponent h measures the strength of a singularity. The higher the value of h, the more regular the function (distribution) is.

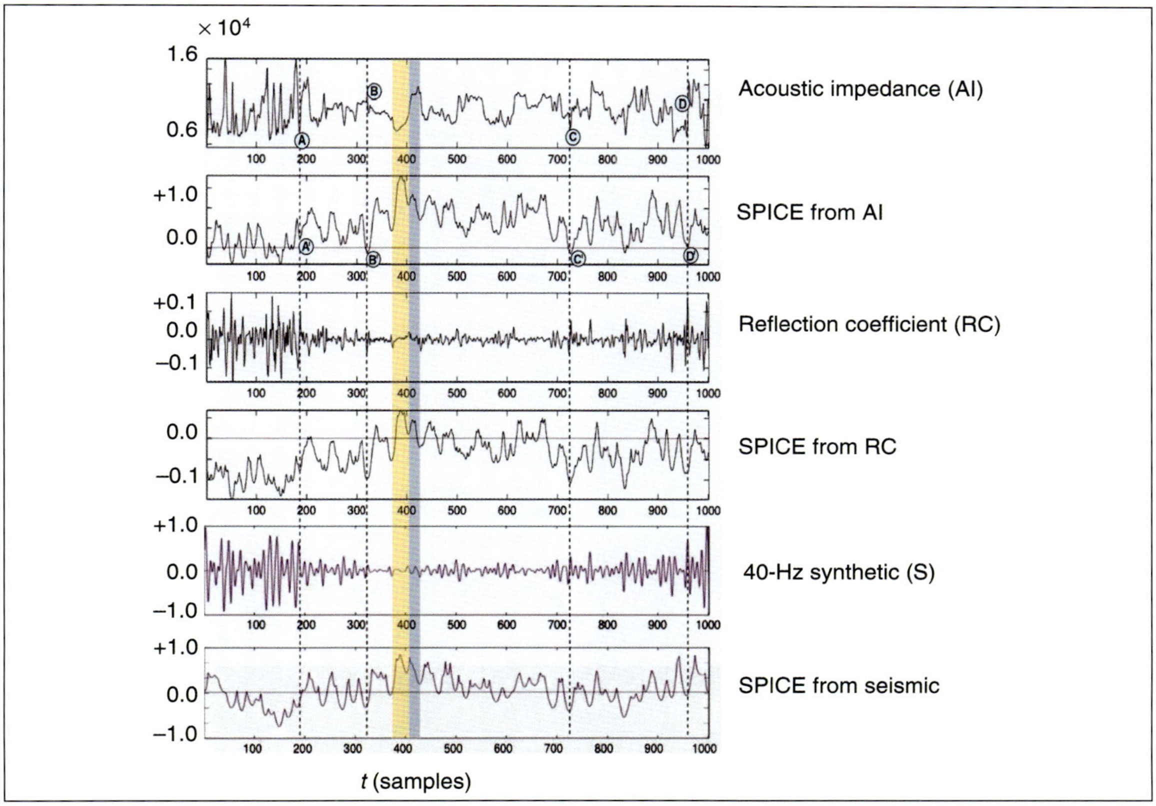

Figure 2.34. Singularities measured by the Holder exponent in the underlying acoustic impedance retain the same form if we convolve the data with smooth analytical functions. In this case, we begin with a discontinuity in the acoustic impedance, indicated between the yellow and blue lithologies. Because the derivative operator in the frequency (and scale) domain, $-i\omega$, is continuous, we do not change the singular nature of our acoustic-impedance profile. Likewise, because the source wavelet used in generating a synthetic trace also is continuous, the singularity on the synthetic trace is nearly identical to that of the acoustic impedance. Because the synthetic trace was designed to match the real seismic trace, we can show that the singularity on the seismic data is very close to the singularity seen on the acoustic-impedance log. Thus, the SPICE algorithm represents changes in the acoustic-impedance log. After Smythe et al. (2004).

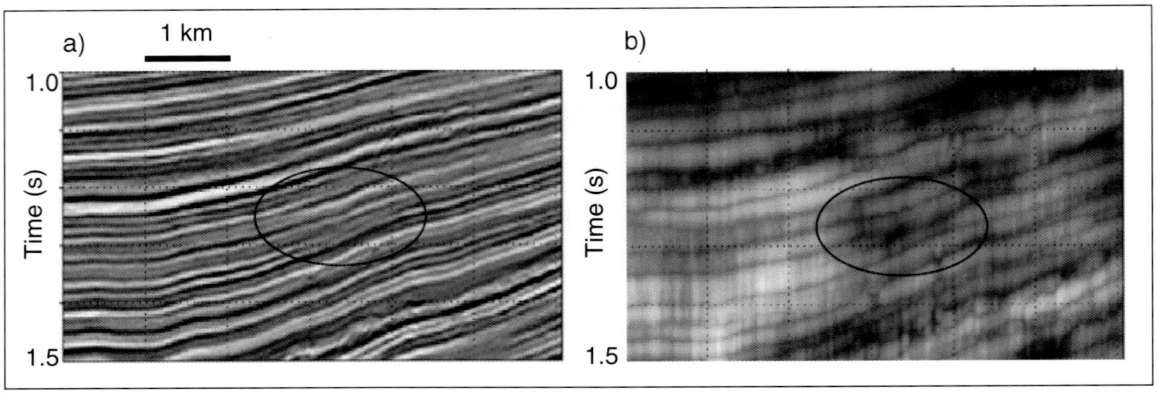

Figure 2.35. Vertical slices through (a) a seismic data volume and (b) a SPICE volume. The SPICE algorithm enhances singularities between waveforms, such as the subtle pinch-out circled. After Liner et al. (2004).

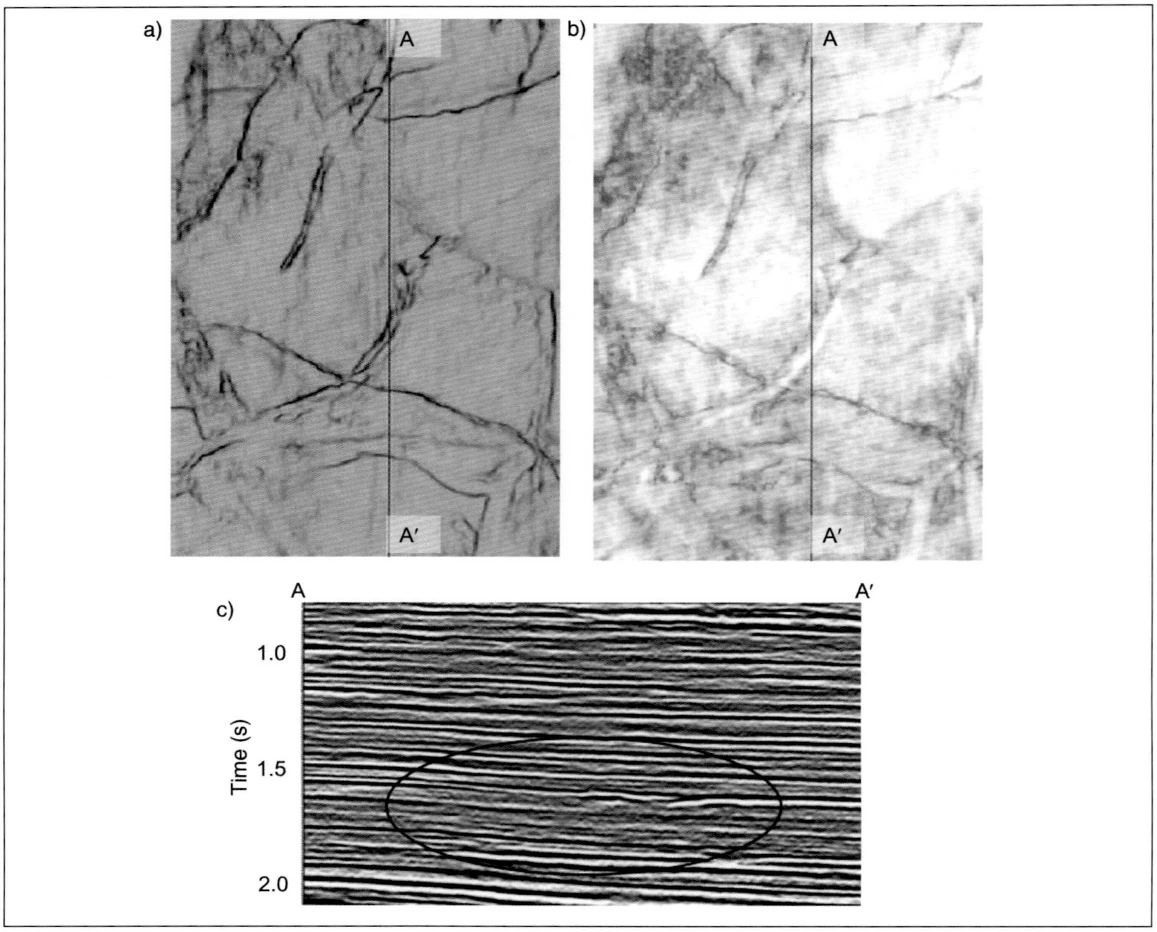

Figure 2.36. Miocene horizon slices from a West Cameron, Gulf of Mexico, U.S.A., survey, through (a) an eigenstructure coherence volume and (b) a SPICE volume. (c) A vertical section showing a known Miocene channel. After Liner et al. (2004).

Unit 3 Geometric Attributes

Overview

Geometric attributes define the morphology and lateral variability of a seismic reflector. Dip and azimuth maps of interpreted seismic reflectors are used routinely to map subtle faults in conventional seismic interpretations. Horizon-curvature maps are used to predict fractures, and amplitude-derivative maps can be used to map edges of hydrocarbon accumulations.

We begin by reviewing recently developed algorithms that allow us to calculate 3D volumes of reflector dip and azimuth without explicitly picking a given horizon. Besides being an important interpretation tool, dip and azimuth define a local reflector surface on which we can detect discontinuities, or conversely, along which we can filter the data to extract its coherent component.

Next, we discuss a second family of geometric attributes — coherence and related measures of similarity between adjacent seismic waveforms. Geologically, highly coherent seismic waveforms indicate laterally continuous lithologies or highly charged hydrocarbon reservoirs. In contrast, abrupt changes in waveform resulting in low coherence can indicate faults, fractures, salt or shale diapirism, karst, channel edges, and mass-transport complexes.

Reflector curvature and shape comprise the third family of geometric attributes. In 3D analysis, we define curvature in terms of two orthogonal circles tangent to a reflector surface. Therefore, an estimate of curvature is independent of bulk rotations and translations of the reflector. We define the circle with the minimum radius or tightest curvature as the maximum curvature. We define the second circle (which is perpendicular to the first circle and has an equal or larger radius) as the minimum curvature. Together, these two radii allow us to define the shape of the local reflector.

Finally, we introduce the fourth family of geometric attributes, all of which measure lateral changes in amplitude along a reflector. In this family, we include "texture" measurements of lateral changes in amplitude built on gray-level co-occurrence matrix (GLCM) textures.

We then illustrate the sensitivity of each of these geometric attributes to geology through a suite of examples.

Summary

Geometric attributes provide a quantitative measure of faults and fractures, and they enhance stratigraphic information that is tedious to extract using conventional seismic interpretation methodologies. Because geometric attributes are applied to uninterpreted seismic data volumes, they facilitate rapid interpretation of large 3D data volumes. Geometric attributes are relatively insensitive to seismic wavelets, so they provide robust images even when applied to merged surveys that have been subjected to different acquisition and processing.

Much of seismic stratigraphy is based on the morphology of seismic textures. Key to mapping these textures (and to structure-oriented filtering, discussed in Unit 5) are quantitative estimates of dip and azimuth throughout the seismic volume. Volumetric estimates of dip and azimuth are computed from a window of neighboring samples above and below each analysis point, avoiding discontinuities. Thus, such estimates generally are statistically more robust than those made from a picked surface of peaks or troughs.

Coherence computations should be performed along the local dip and azimuth of the best reflector at the analysis point. Failure to do so introduces into the image low-coherence artifacts that mask geologic discontinuities of interest. Crosscorrelation and eigenstructure estimates measure only changes in reflector waveform, whereas semblance, variance, and gradient structure tensor (GST) estimates of coherence are sensitive to waveform and lateral changes of reflector amplitude. By using the analytic trace rather than just the input trace, we can improve the fidelity of coherence images at zero crossings for small vertical analysis windows. Coherence algorithms are local — faults will not show up on coherence volumes if the faults have drag, are poorly migrated, separate two similar reflectors, or otherwise do not appear locally to be discontinuous.

Volumetric estimates of curvature and rotation are mathematically independent of coherence and amplitude gradient measures. Nevertheless, certain features, such as well-focused normal faults, show up equally well on all geometric attribute volumes. In the absence of differential compaction, curvature attributes generally provide poor images of channels. In contrast, curvature provides excellent images of subtle flexures, folds, and collapse features not commonly seen on coherence volumes. Reflector rotation can be a good indicator of data quality and of scissor movement along a fault.

Energy-weighted coherent-amplitude gradients are sensitive to lateral changes in amplitude for a fixed waveform. Such gradients are particularly effective in detecting thin channels that are not seen by eigenstructure coherence algorithms, and they provide insight into reservoir heterogeneity. Because energy-weighted coherent-amplitude gradients work only on the coherent component of data, they highlight channels that have stronger reflectivity, and they deemphasize incoherent fault zones that may confuse our stratigraphic interpretation.

Long-wavelength estimates of amplitude variability can enhance lateral changes in reservoir thickness and porosity that are more gradational and that have scales larger than the typical five- or nine-trace coherence algorithm. Long-wavelength second-derivative estimates of amplitude variability, calculated by applying the most-positive-curvature and most-negative-curvature algorithms to amplitude gradients, also can enhance such subtle changes in reflectivity.

Texture-attribute studies commonly are not associated with seismic attribute studies. We consider texture analysis, in its most general form, to be a superset of the geometric attributes that are the major focus this short course. Texture analysis uses second-order statistics to extract attributes that help describe the textural properties of classes. Textures based on GLCMs work well if the granularity of textures being examined is of the order of voxel size, which commonly is the case with seismic data.

Geometric attributes such as volumetric dip and azimuth or seismic-amplitude textures form the building blocks of computer-assisted pattern recognition. In that procedure, an interpreter identifies several geologic features that have a particular seismic configuration, trains the computer to mimic this identification by using a combination of attributes, verifies the identification on a feature not used in the training, and then asks the software to find all such features in the seismic volume.

Whenever possible, we recommend using algorithms that are as mathematically decoupled from each other as possible. Such algorithms should give the interpreter orthogonal (independent) views of the data.

Figure 3.1. Mathematical, geologic, and seismic nomenclature used in defining reflector dip. By convention, **n** = unit vector normal to the reflector, **a** = unit vector dip along the reflector, θ = dip magnitude, ϕ = dip azimuth, ψ = strike, θ_x = the apparent dip in the xz plane, and θ_y = the apparent dip in the yz plane.

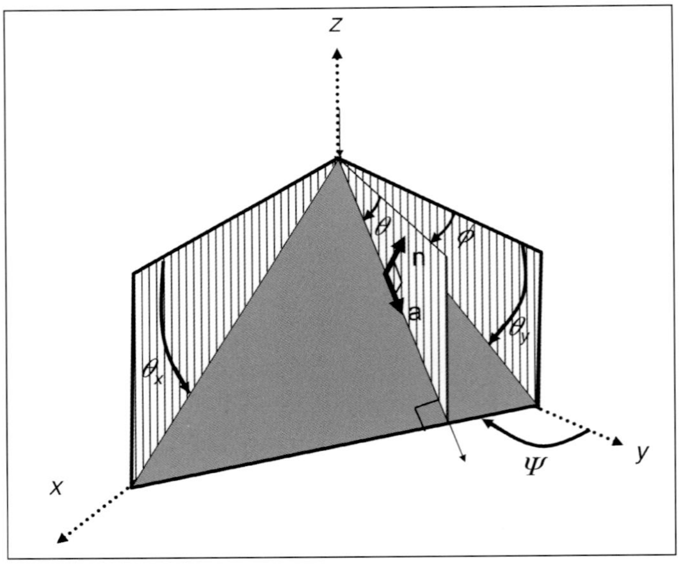

Hilbert transform

Instantaneous phase $$\phi = \tan^{-1}(d^H / d)$$

Instantaneous frequency $$\omega = 2\pi f = 2\pi \frac{\partial \phi}{\partial t} = 2\pi \frac{\frac{\partial d}{\partial t} d + \frac{\partial d^H}{\partial t} d^H}{d^2 + \left(d^H\right)^2}$$

Instantaneous inline wavenumber $$k_x = \frac{\partial \phi}{\partial x} = \frac{\frac{\partial d}{\partial x} d + \frac{\partial d^H}{\partial x} d^H}{d^2 + \left(d^H\right)^2}$$

Instantaneous crossline wavenumber $$k_y = \frac{\partial \phi}{\partial y} = \frac{\frac{\partial d}{\partial y} d + \frac{\partial d^H}{\partial y} d^H}{d^2 + \left(d^H\right)^2}$$

Instantaneous apparent dips $$p = \frac{k_x}{\omega}; q = \frac{k_y}{\omega}$$

Figure 3.2. Computation of apparent dip, using the analytic trace. Numerically, the instantaneous phase, ϕ, is the angle between the seismic data and its Hilbert transform. We begin by computing the instantaneous frequency, ω, which is the time derivative of the instantaneous phase, ϕ. Next, we compute the instantaneous wavenumbers, k_x and k_y, which are the x- and y-derivatives of the instantaneous phase, ϕ. Finally, we compute the apparent dips, p and q, measured in milliseconds per meter, by forming the ratio of k_x and k_y with ω.

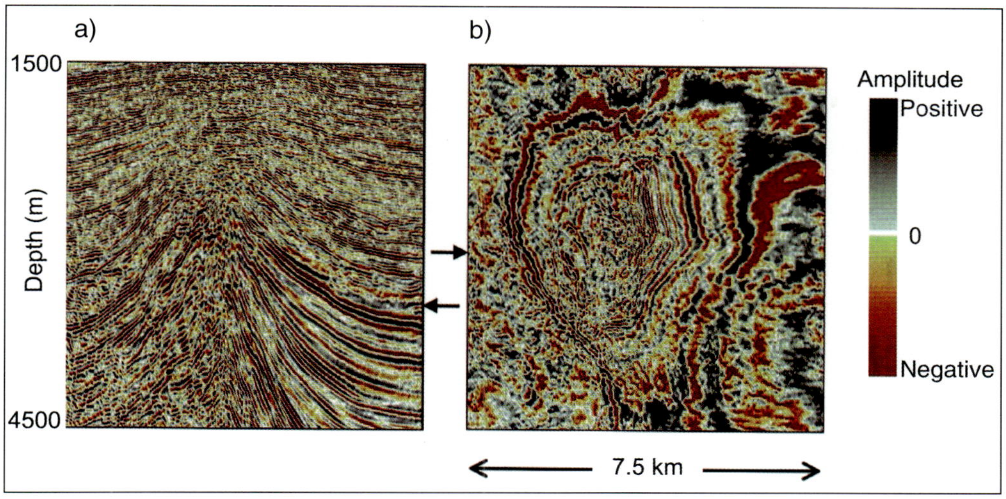

Figure 3.3. A vertical slice and a horizontal depth slice through a seismic data volume gathered over a salt dome. Note the characteristic "cut-onion" appearance on the depth slice. After Barnes (2000a).

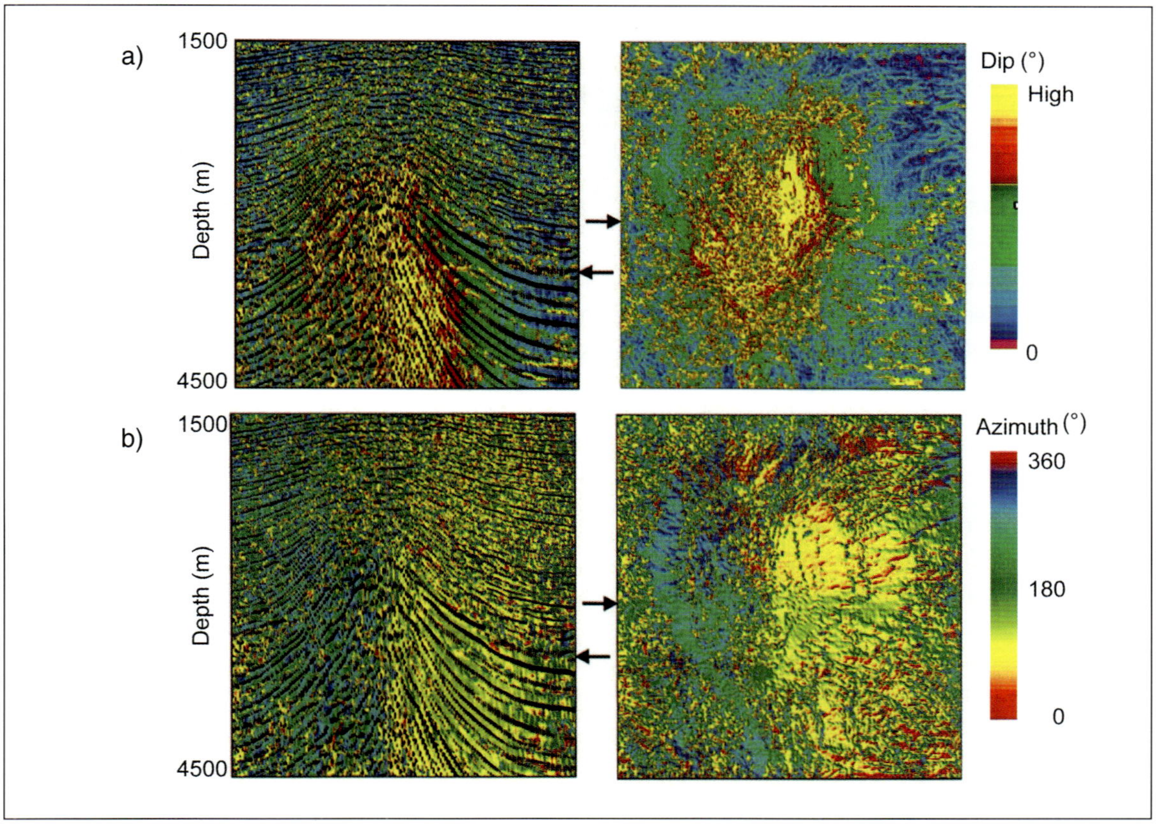

Figure 3.4. Vertical and horizontal slices corresponding to those shown in Figure 3.3 through (a) instantaneous dip magnitude, given by the equation in Figure 3.2, and (b) instantaneous dip azimuth, also given by the equation in Figure 3.2. The speckled appearance of adjacent conflicting dip magnitude and dip azimuth is the result of singularities in the complex-trace analysis estimation of k_x, k_y, and ω. After Barnes (2000a).

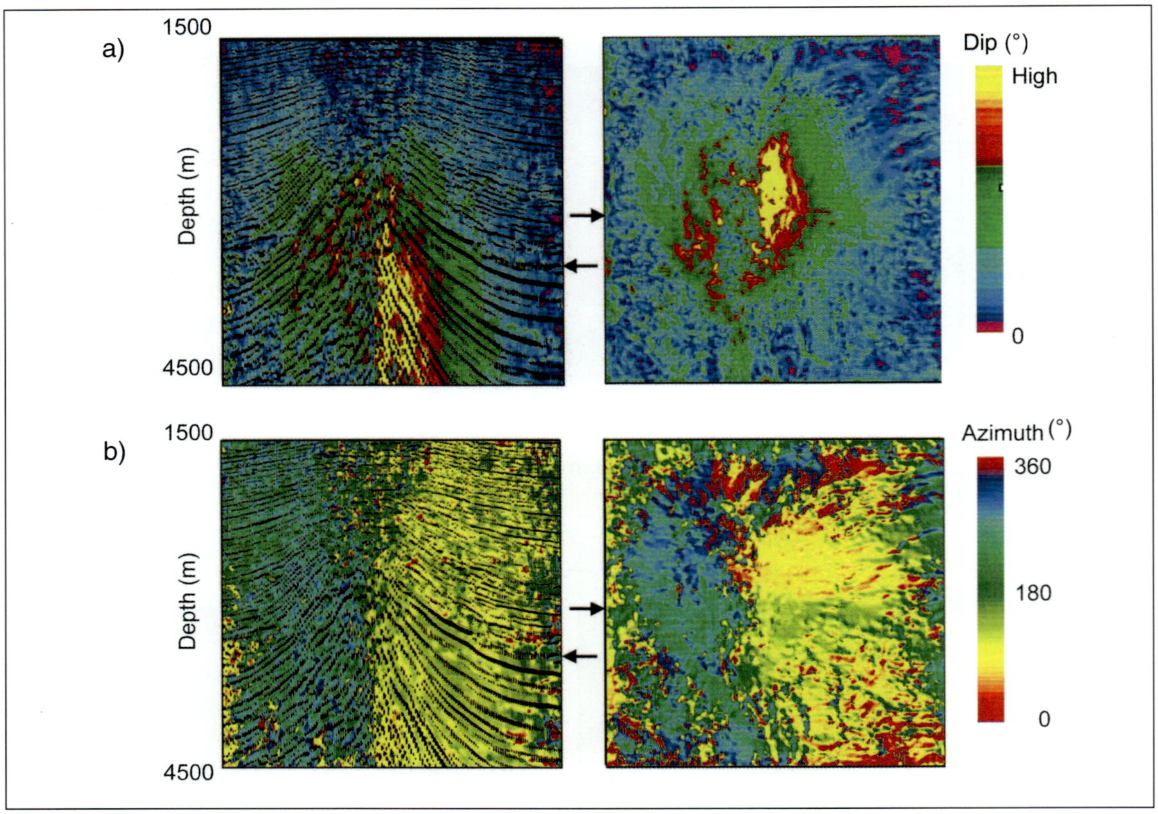

Figure 3.5. Vertical and horizontal slices corresponding to those shown in Figure 3.3, here through (a) a smoothed instantaneous dip magnitude and (b) a smoothed instantaneous dip azimuth. Smoothing was performed using an envelope-weighted running mean filter (composed of five inlines, five crosslines, and seven depth samples) on each of the constituent k_x, k_y, and ω components. Note that the singularities seen in Figure 3.3 are diminished. After Barnes (2000a).

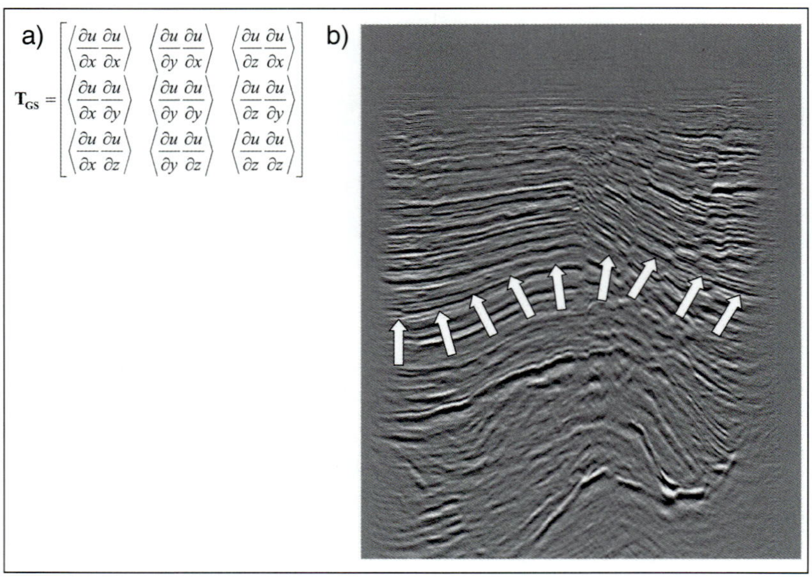

Figure 3.6. Construction of the gradient structure tensor, GST, from the seismic data, *u*. First we compute a series of directional derivatives, *du/dx, du/dy*, and *du/dz*. Next we crosscorrelate these derivatives with themselves and each other within a small volumetric analysis window, thereby generating the elements of the matrix shown in (a). The 3×1 principal-component eigenvector of this matrix will be aligned parallel to the direction of maximum change in the wavefield, which generally is a good approximation to the normal to the reflector dip, such as that indicated by the white arrows in (b).

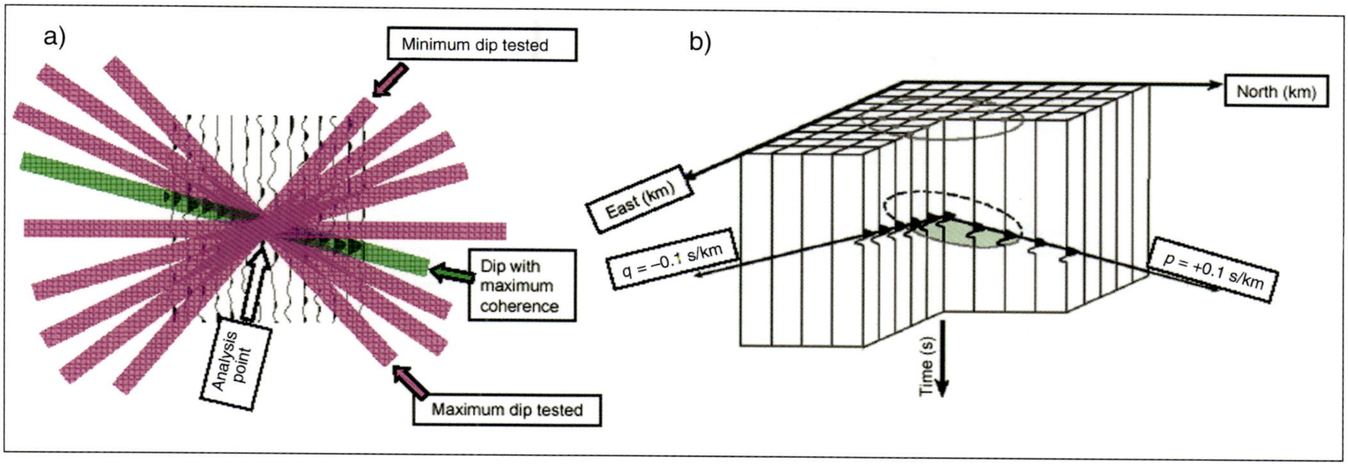

Figure 3.7. (a) A schematic diagram showing a 2D search-based estimate of coherence. First the algorithm estimates coherence using semblance, variance, principal component, or some other statistical measure along a discrete number of candidate dips (shown in magenta and green). In this example, the maximum coherence is calculated along the dip (shown in dark green). Next the algorithm passes an interpolation curve through the coherence measures estimated by the peak value and two or more neighboring dips (shown here in light green). The peak value of this curve gives an estimate of coherence, whereas the dip value of this peak gives an estimate of instantaneous dip. (b) A schematic diagram showing a 3D search-based estimate of coherence, in which *p* indicates the inline and *q* the crossline components of vector time dip. The technique is analogous to that shown in (a). After Marfurt et al. (1998).

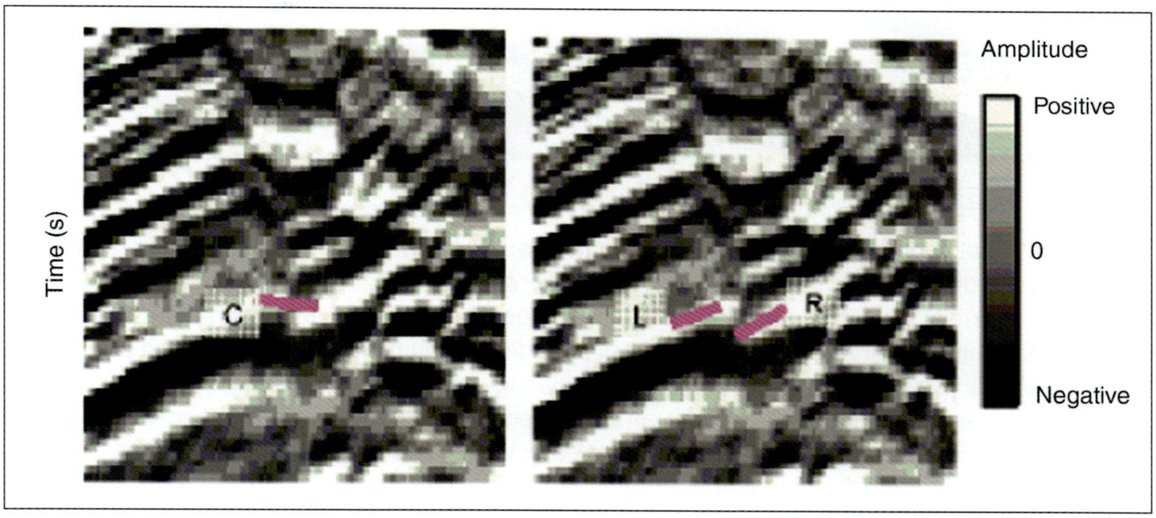

Figure 3.8. A schematic diagram showing the apparent dip (represented by the magenta lines), estimated using the discrete search algorithm shown in Figure 3.7b, in (a) a window, C, centered about the analysis point, and in (b) windows L and R, shifted to the right and left of the analysis point. Clearly, estimates of apparent dip obtained in either of these two shifted windows are superior to the "smoothed" apparent dip given by the centered window.

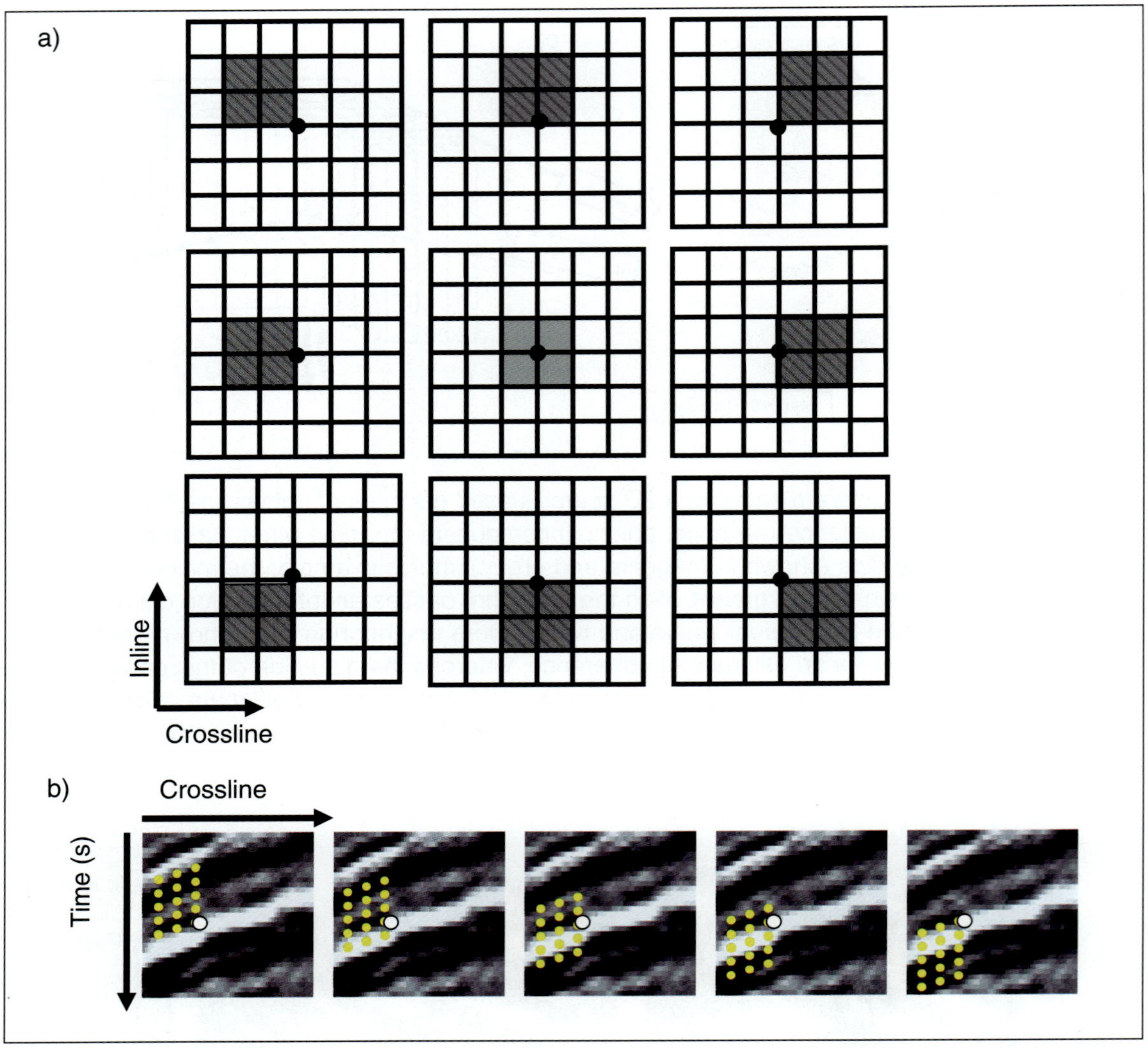

Figure 3.9. (a) A plan view of a nine-trace, nine-lateral-window search for dip and azimuth. Solid circles indicate the analysis point, and shaded rectangles indicate alternative analysis windows. (b) A vertical view of the first (northwest) window shown in (a), illustrating the search over five vertical windows containing the analysis point indicated by the white dot (only the three crossline traces are displayed). We begin by calculating and then interpolating for the coherence and dip and azimuth within each window, as described in Figure 3.7b. This process is repeated for all 9 × 5 = 45 overlapping analysis windows, each containing the desired analysis point. The dip and azimuth at the analysis point are defined to be the dip and azimuth of the window that encompasses the analysis point that has the maximum coherence. Use of such temporally and laterally shifted analysis windows helps preserve angular unconformities and other features of geologic interest. This dip and azimuth and the traces comprising the most coherent window are used for the covariance matrix and principal-component data projection calculations given by Figure 3.35 and in Unit 5.

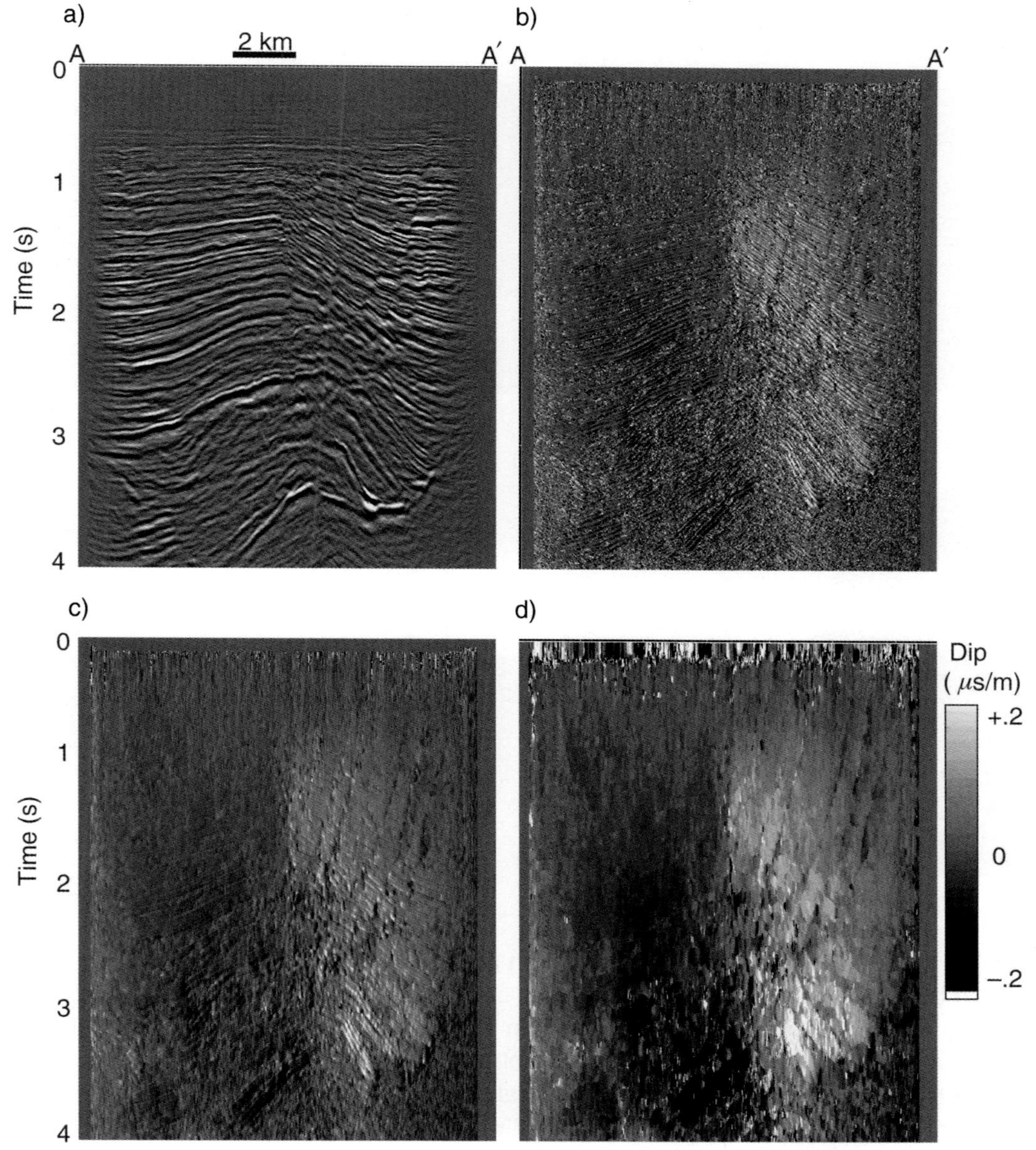

Figure 3.10. Vertical slices along line AA′ from a survey over Vinton Dome, Louisiana, U.S.A., through (a) the original seismic volume, and (b)-(d) through the east-west components of apparent dip, *p*, calculated (b) using the instantaneous dip estimation given by equations shown in Figure 3.2; (c) using smoothed values of k_x and ω over a five-trace by five-trace by five-sample window, and (d) using the multiwindow dip scan described in Figure 3.9. Note the inconsistent estimate of interleaved positive and negative dips in (b), probably resulting from singularities in the frequency and wavenumber calculations (Liner et al., 2004). This inconsistency is averted by smoothing in (c), but at the cost of decreased lateral resolution. The multiwindow dip scan method is less sensitive to aliasing and honors the abrupt changes in dip seen along the faults. Data are courtesy of OPEX.

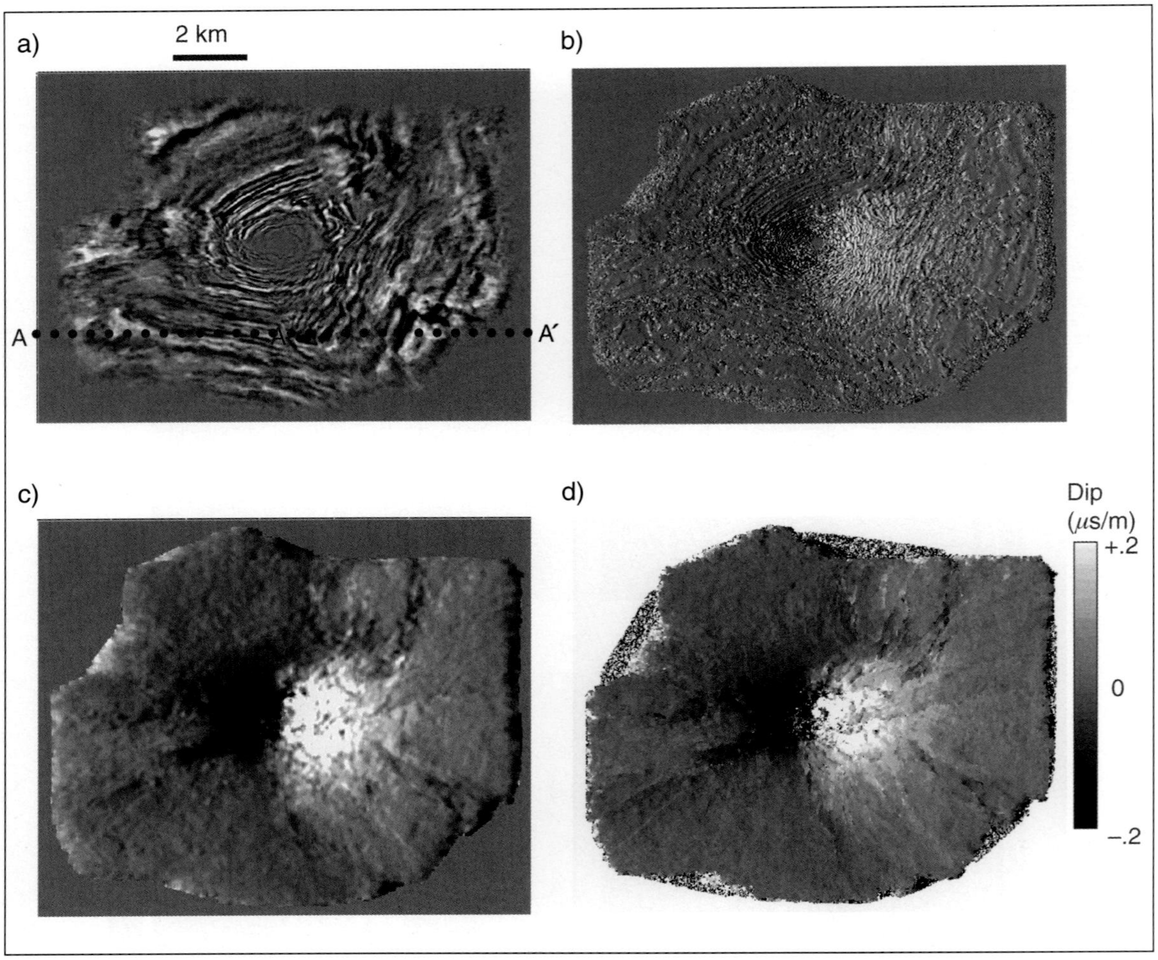

Figure 3.11. Time slices at t = 1.000 s, through (a) the original Vinton Dome seismic volume, and (b)-(d) through the inline (east) components of apparent dip calculated (b) using the instantaneous dip estimation given by equations shown in Figure 3.2, (c) using smoothed values of k_x and ω over a five-trace by five-trace by five-sample window, and (d) using the multiwindow dip scan. Note the aliasing at steep dips in (b). This aliasing is averted by smoothing in (c), but at the cost of decreased lateral resolution. The multiwindow dip scan method is less sensitive to aliasing and honors the abrupt changes in dip seen along the faults. Line AA′ indicated on this image is shown in Figure 3.10.

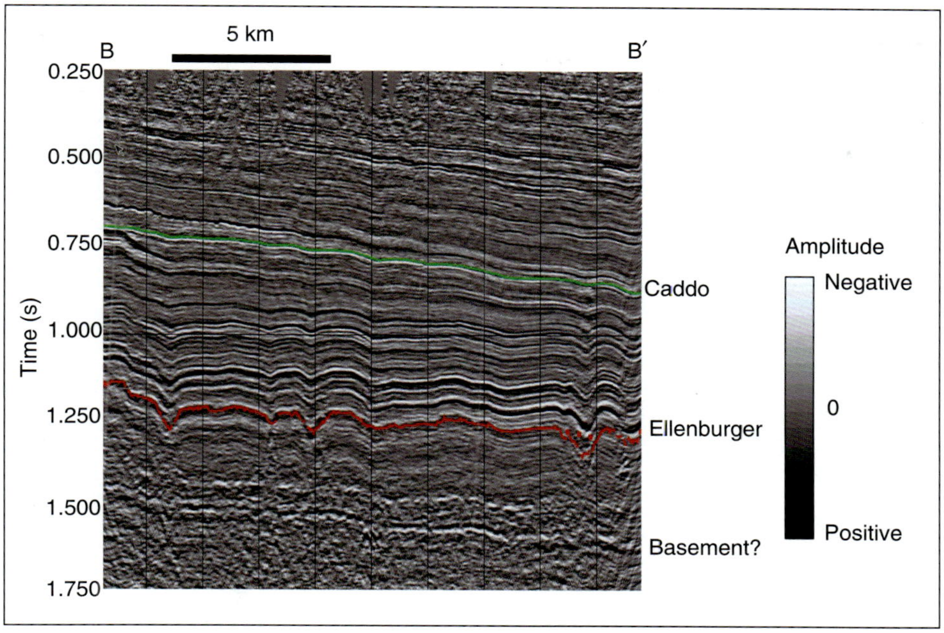

Figure 3.12. Line BB′, through a survey in Fort Worth Basin, Texas, U.S.A. The green Pennsylvanian Caddo Limestone horizon is strong, coherent, and easy to pick. In contrast, the red "karsted" Cambrian-Ordovician Ellenburger Limestone horizon is quite difficult to pick. Seismic data are courtesy of Devon Energy.

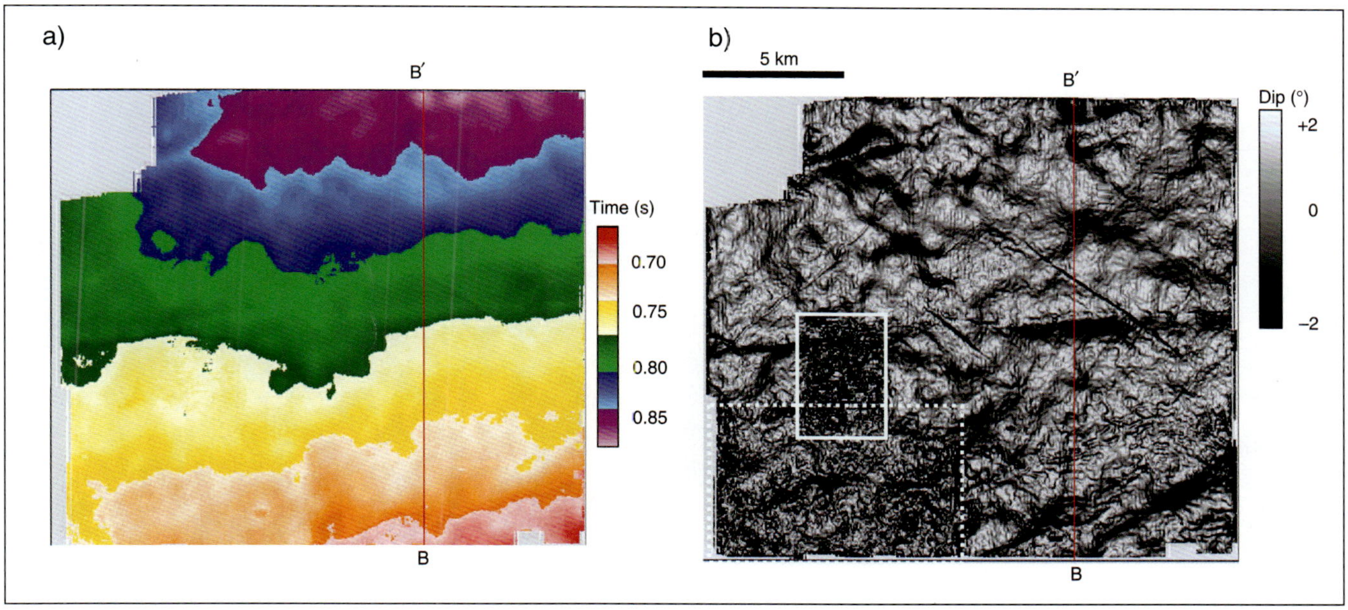

Figure 3.13. (a) A time-structure map corresponding to the Caddo horizon shown in Figure 3.12. (b) A conventional dip-magnitude map formed from the inline and crossline components of dip shown in (a). Note the poor quality of the dip-magnitude estimates in the region of the 1995 survey (solid white rectangle) and the 1997 survey (dotted white rectangle). The data outside these two zones were acquired in 1999. The same person reprocessed all surveys using the same software.

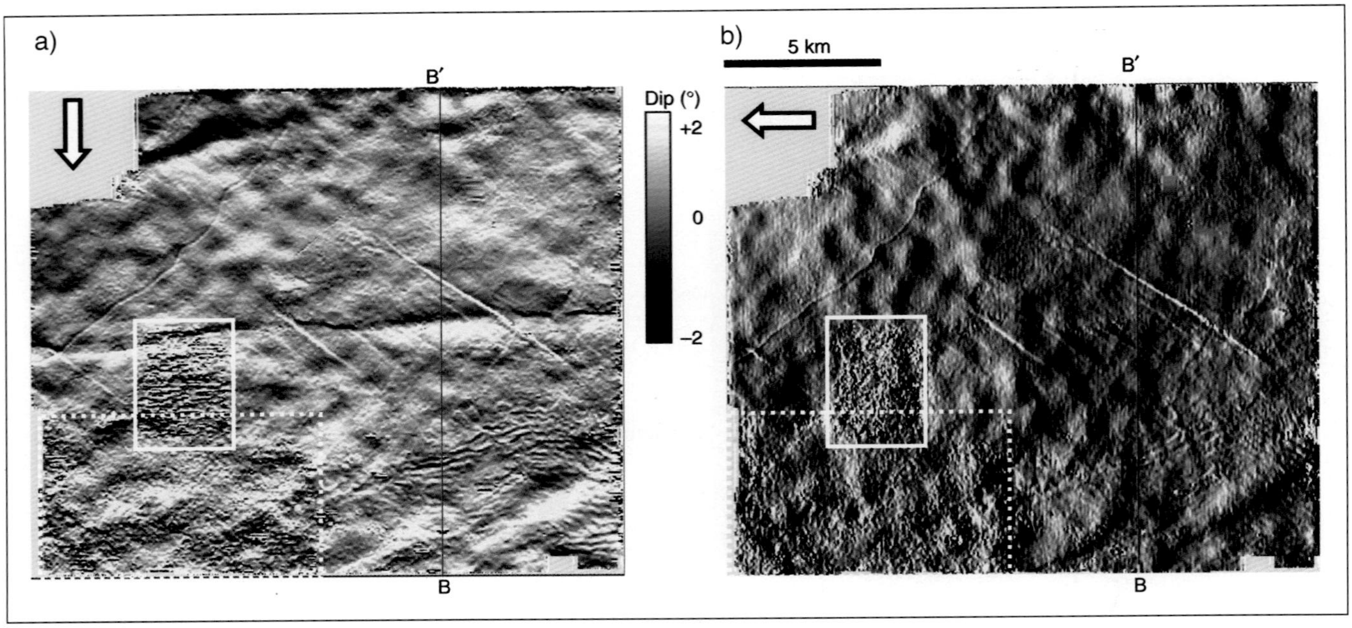

Figure 3.14. (a) A north-south component of dip, *q*, and (b) an east-west component of dip, *p*, computed from the Caddo horizon time-structure map displayed in Figure 3.13a.

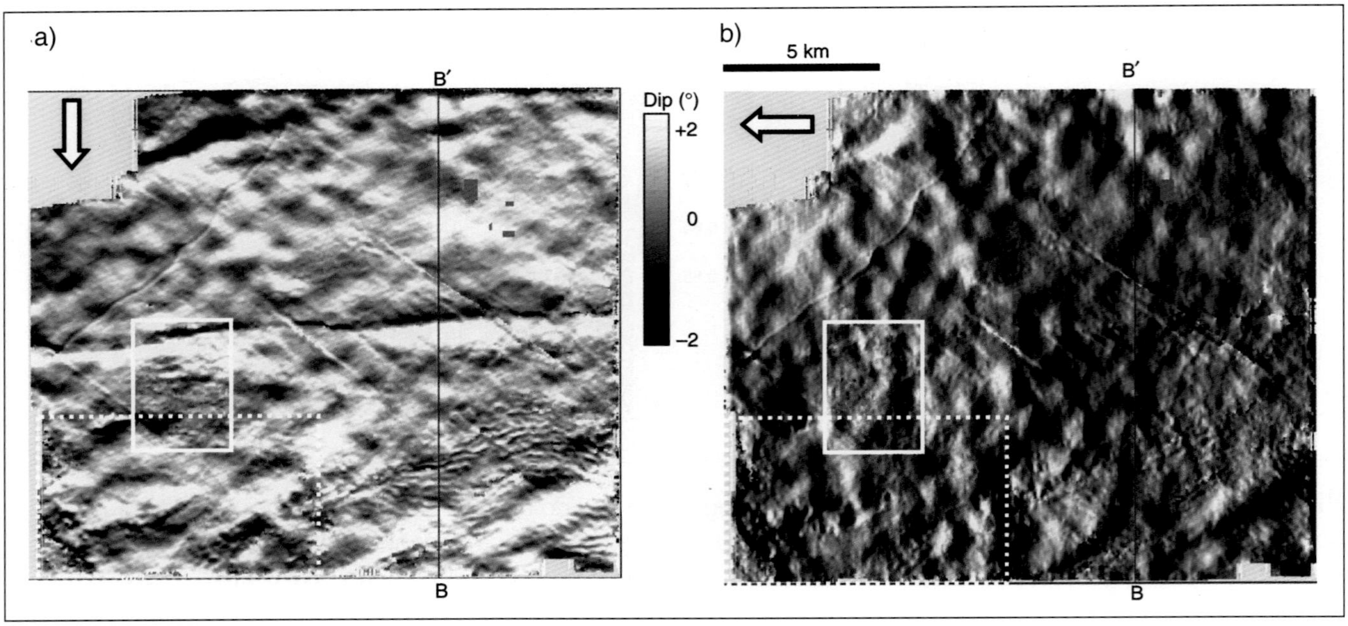

Figure 3.15. Horizon slices along the Caddo horizon, corresponding to Figure 3.14, of (a) the north-south component of dip, *q*, and (b) the east-west component of dip, *p*, using the multiwindow volumetric dip scan shown in Figure 3.9. Note the considerable improvement in the dip estimate seen in the area of the older surveys.

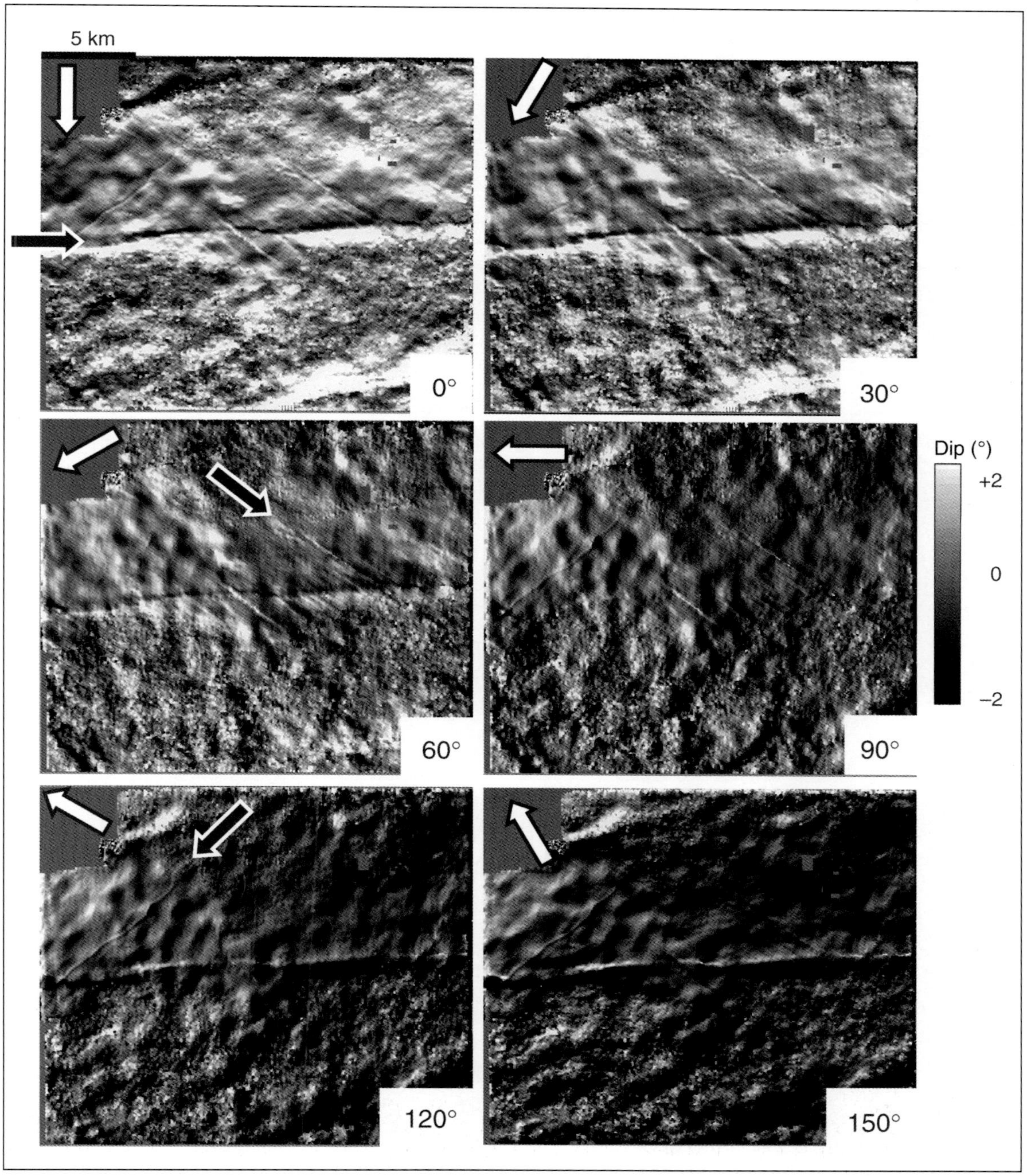

Figure 3.16. Apparent dip at 0°, 30°, 60°, 90°, 120°, and 150° from north, along a time slice at t = 0.800 s, corresponding approximately to the Caddo horizon shown in Figure 3.12. The color bar and orientations of the white arrows were chosen to be consistent with commercial horizon-derivative polarity and to interpret the black areas as shadows. Black arrows indicate faults and collapse features that are illuminated best by derivatives perpendicular to strike.

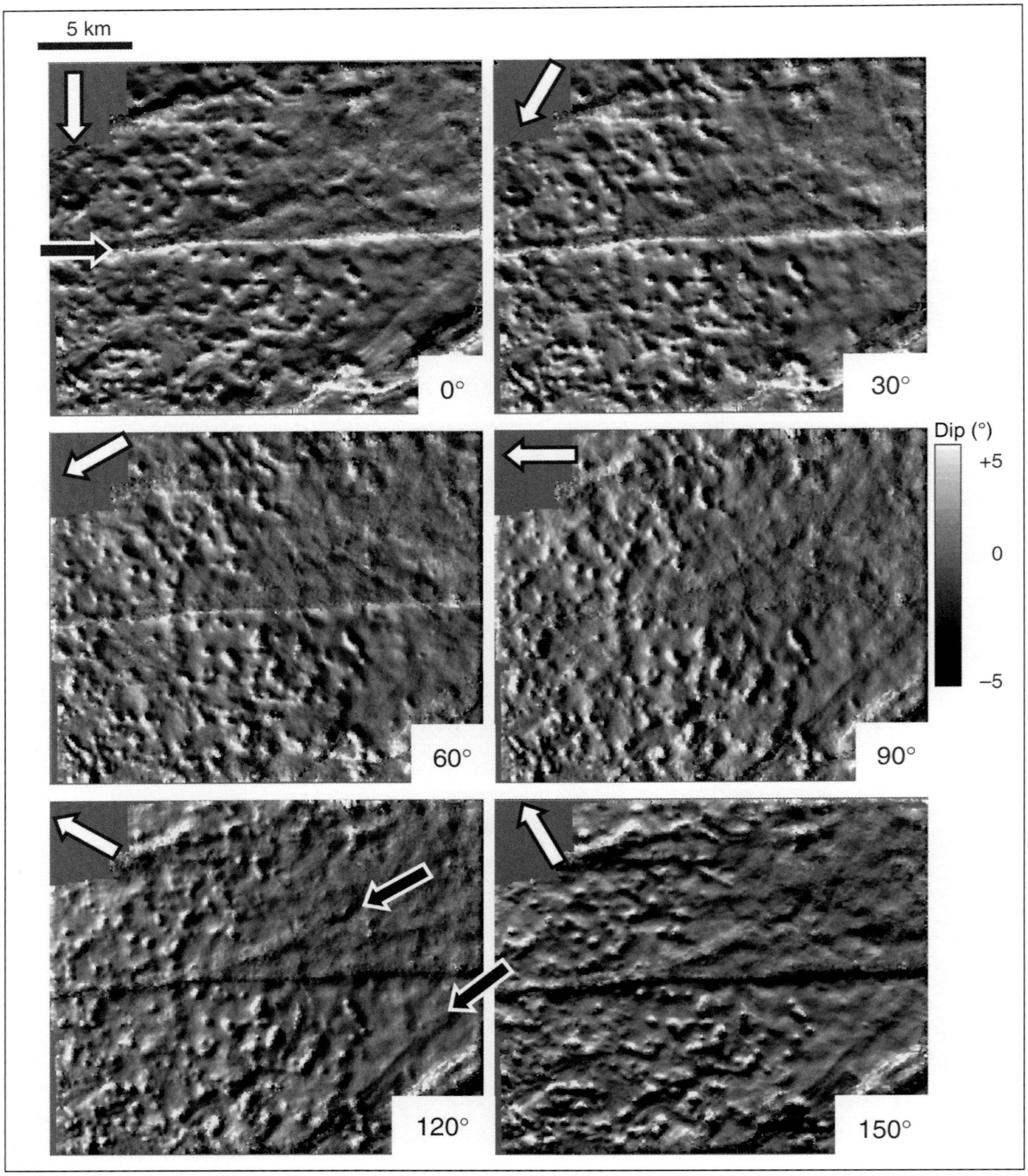

Figure 3.17. Apparent dip at 0°, 30°, 60°, 90°, 120°, and 150° from north, along a time slice at $t = 1.100$ s, corresponding approximately to the Ellenburger level shown in Figure 3.12. By interpreting the dark areas as shadows, we see that the circular anomalies are collapse features. Black arrows indicate faults and collapse features that are illuminated best by derivatives perpendicular to strike.

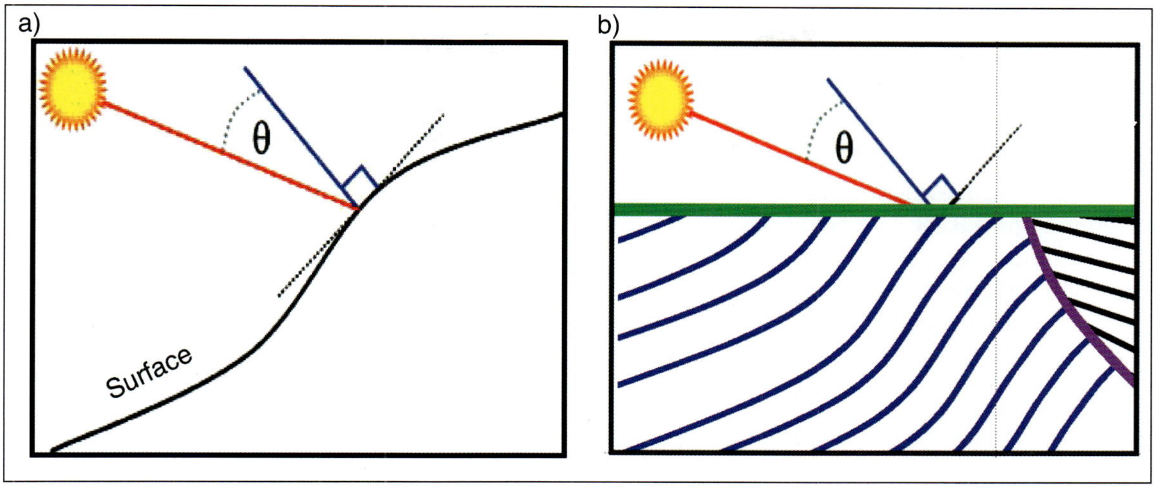

Figure 3.18. Diagrams illustrating the generation of a shaded-relief map (a) of an interpreted surface and (b) along a time slice cut through a dip and azimuth cube. The intensity of the image is proportional to the angle between the solar illumination and the normal to the dip of the interpreted surface in (a) or to the volumetric dip estimated at the sample along the time slice in (b). After Barnes (2002). AAPG©2002. Reprinted by permission of the AAPG whose permission is required for further use.

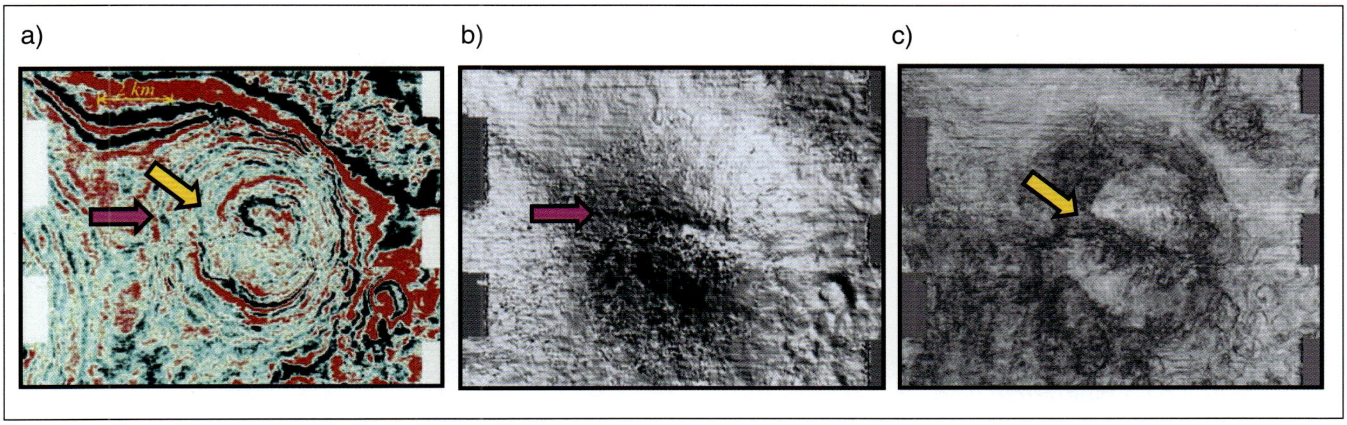

Figure 3.19. Times slices through (a) a seismic data volume, (b) a shaded-relief image of dip and azimuth, and (c) a continuity volume. The shaded-relief dip and azimuth slice is defined in Figure 3.18b, and continuity is a measure of waveform similarity or coherence, which we will discuss later. Although they are connected through geology, it is important to realize that dip and azimuth, coherence, and seismic amplitude measurements are mathematically independent and therefore illuminate different features. The salt dome appears to be folded along the axis defined by the magenta arrow and faulted along the axis defined by the yellow arrow. After Barnes (2002). AAPG©2002. Reprinted by permission of the AAPG whose permission is required for further use.

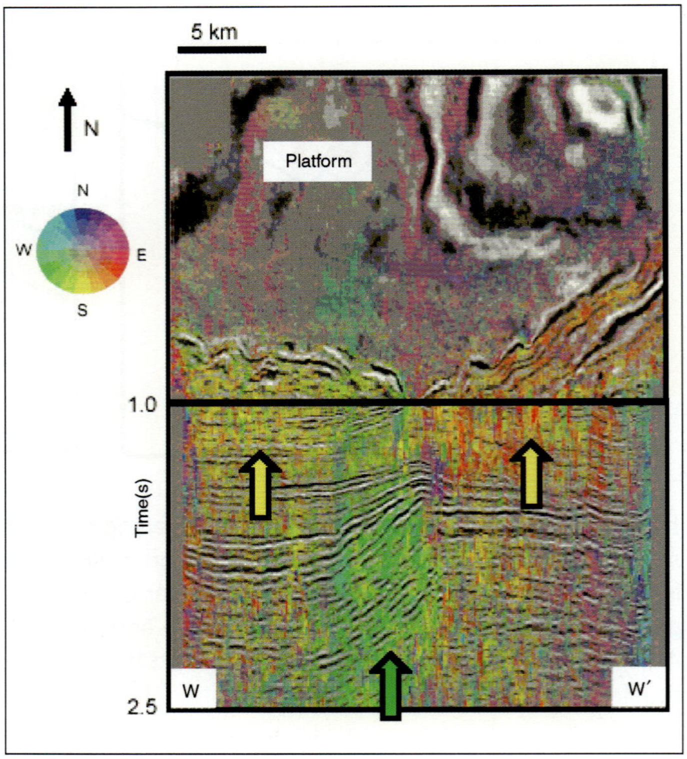

Figure 3.20. A composite image of dip and azimuth and seismic data for a survey from the Delaware Basin, New Mexico, U.S.A. The time slice, at t = 1.000 s, cuts a carbonate platform. Location of WW′ is given in Figure 3.23. The color display technique is described in Figure 1.26. Because the platform is flat, it appears in colors that have a low saturation. Yellow arrows indicate the Brushy Canyon clastic slump deposits in front of the platform, which were deposited later. On the western half of the image, the slumps dip to the south (yellow), and on the eastern half, they dip to the southeast (orange). The green arrow indicates a deeper structure dipping to the west-southwest. Data are courtesy of Marathon Oil Co.

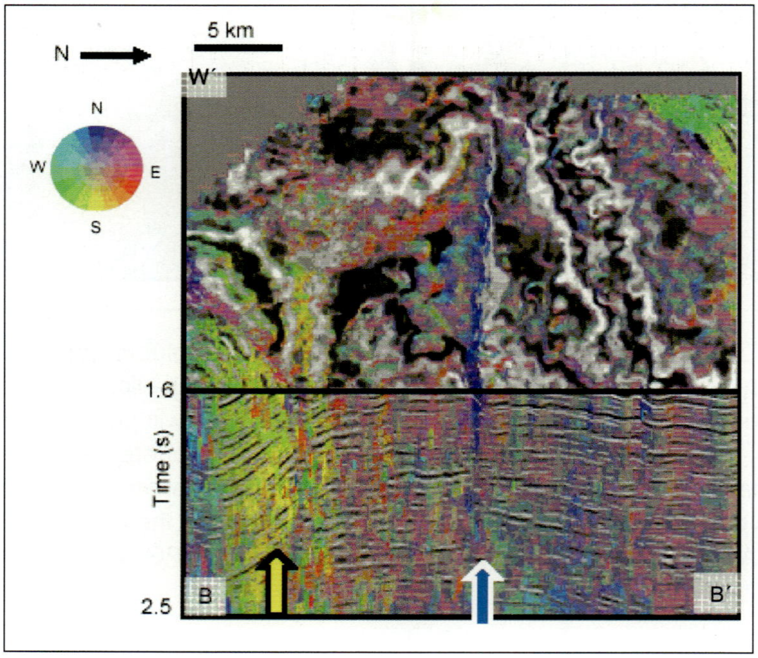

Figure 3.21. A composite image of dip and azimuth and seismic data from a survey from the Delaware Basin, New Mexico, U.S.A. The time slice, at t = 1.600 s, displays a discrete flexure to the north, shown in blue and indicated by a blue arrow. The location of line BB′ is shown in Figure 3.23. The Brushy Canyon clastic slumps dip to the south and southwest, appear as yellow through green, and are indicated by a yellow arrow. Data are courtesy of Marathon Oil Co.

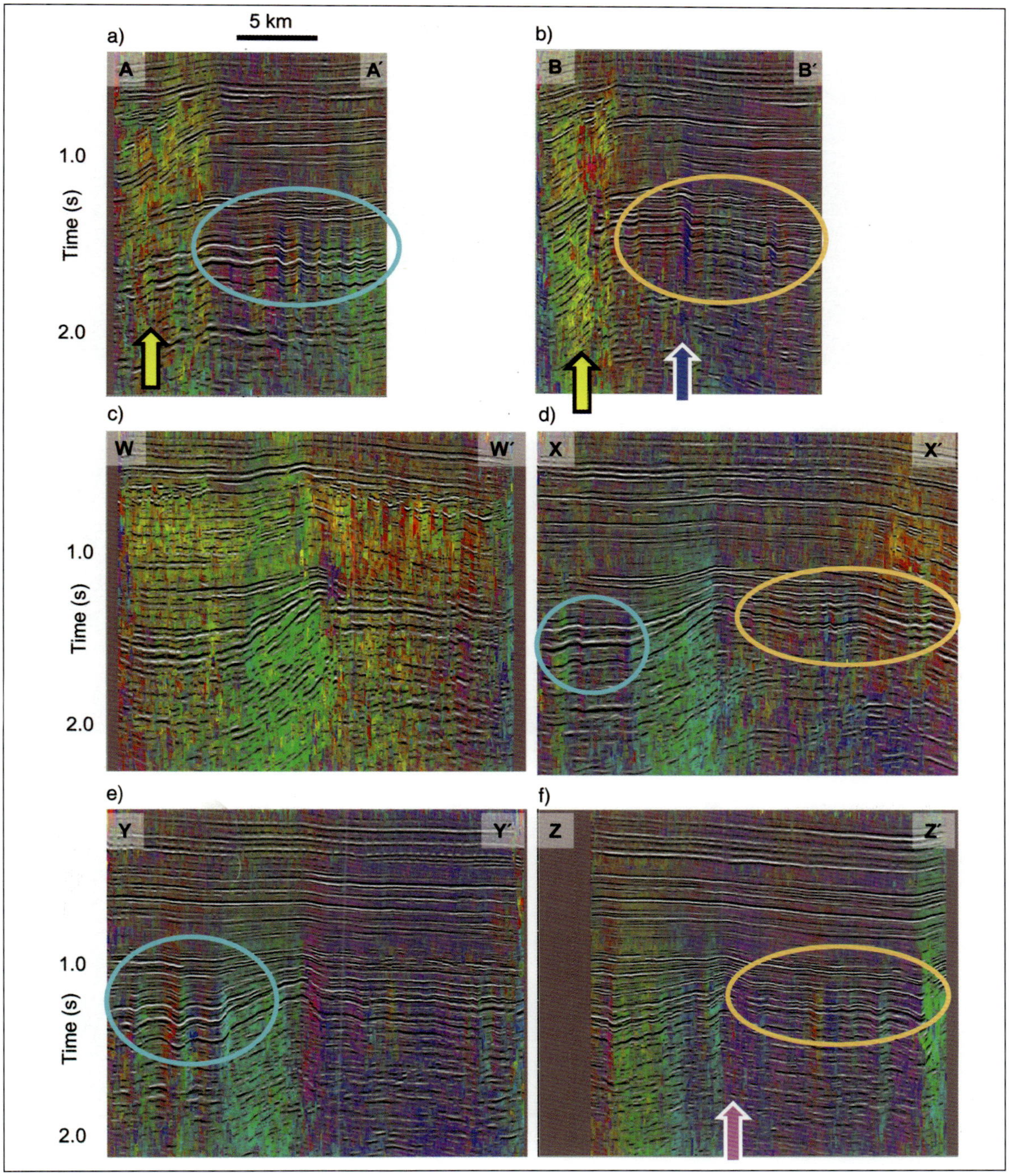

Figure 3.22. Vertical slices through a composite image of dip and azimuth and seismic data, corresponding to the survey shown in Figures 3.20, 3.21, and 3.23. Blue and magenta arrows correspond respectively to the north- and northeast-dipping flexures seen in Figure 3.23. The highly deformed zones bordered by cyan and orange ellipses also are indicated on Figure 3.23.

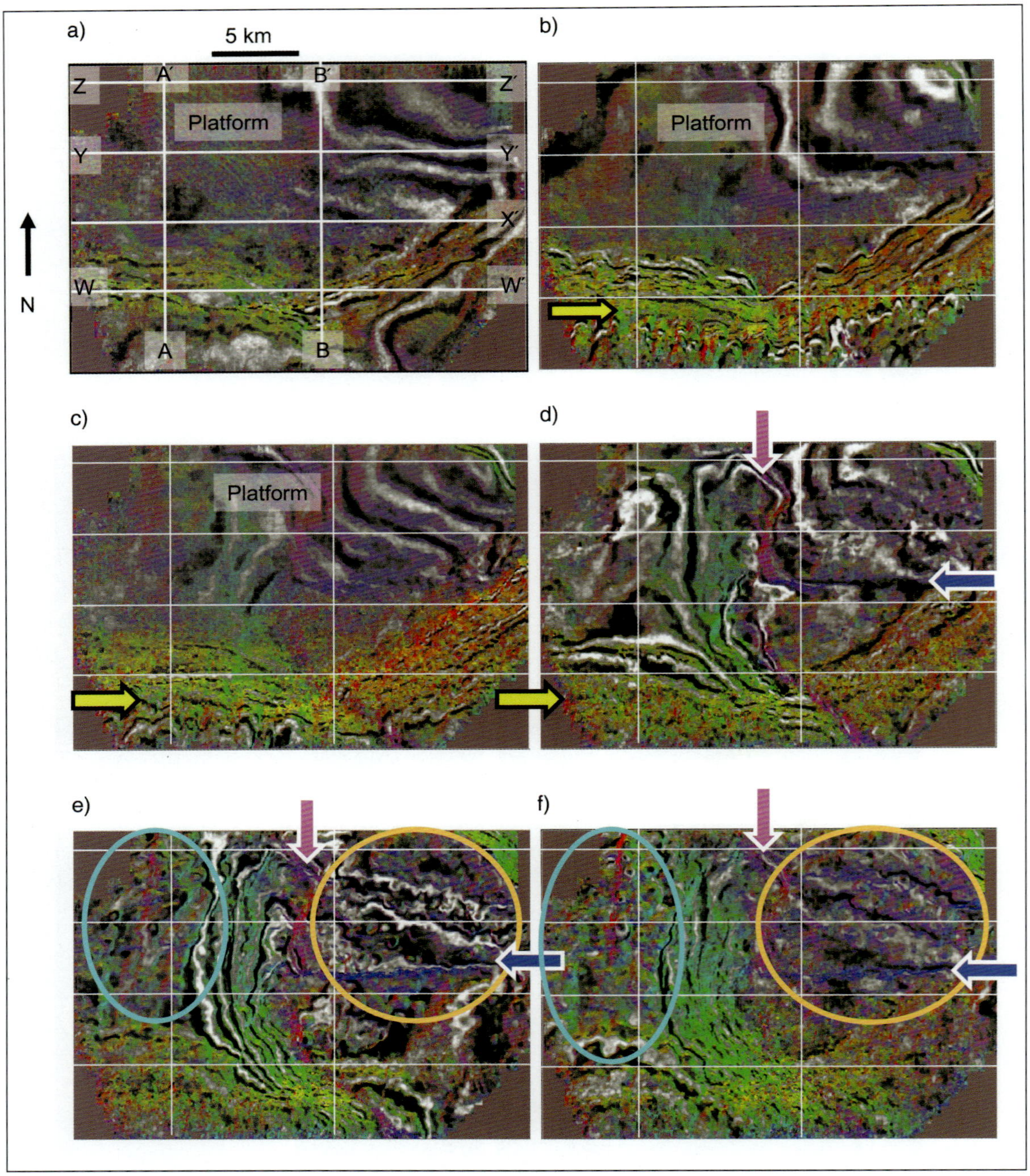

Figure 3.23. Time slices at t = (a) 0.800, (b) 1.000, (c) 1.200, (d) 1.400, (e) 1.600, and (f) 1.800 s, through the same survey shown in Figures 3.20 through 3.22. Yellow arrows indicate the Brushy Canyon clastic slumps, blue arrows point to a discrete flexure toward the north, and the magenta arrow identifies a less pronounced flexure toward the northeast. The cyan ellipse indicates a highly deformed zone of juxtaposed dips. Vertical sections are displayed in Figure 3.22. Colors are the same as in Figure 3.20.

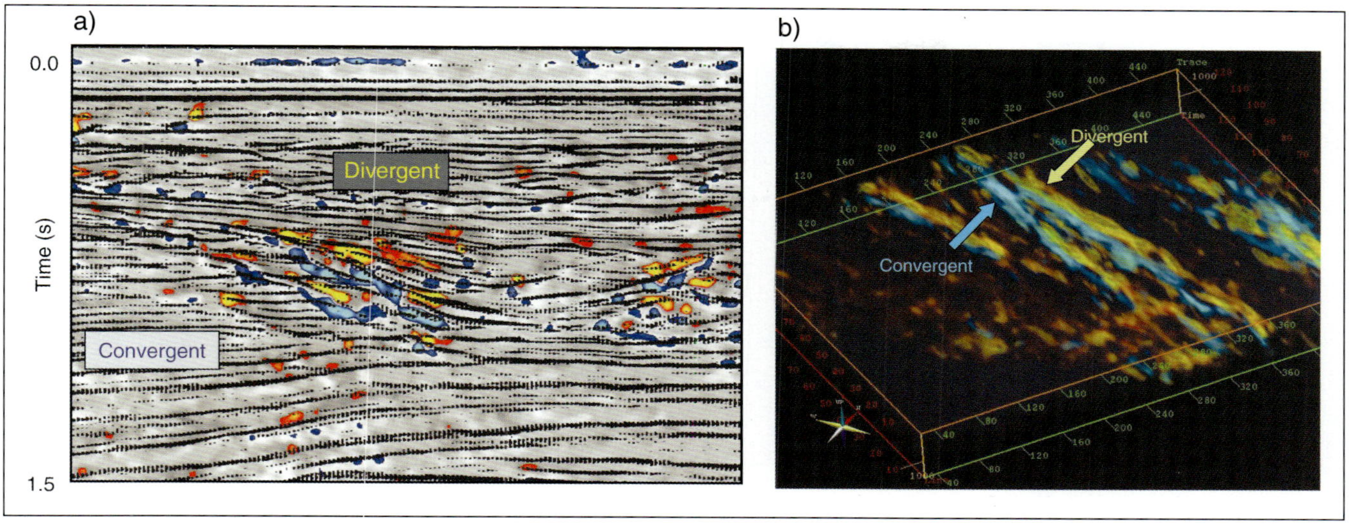

Figure 3.24. Calculation of reflector convergence and divergence is the first step toward 3D computer-aided seismic stratigraphy. In these images, positive convergence is shown in blue, and negative convergence (divergence) is shown in yellow. Like dip and azimuth, reflector convergence is a vector. Here we display the component of convergence/divergence in the inline direction. After Barnes (2000b).

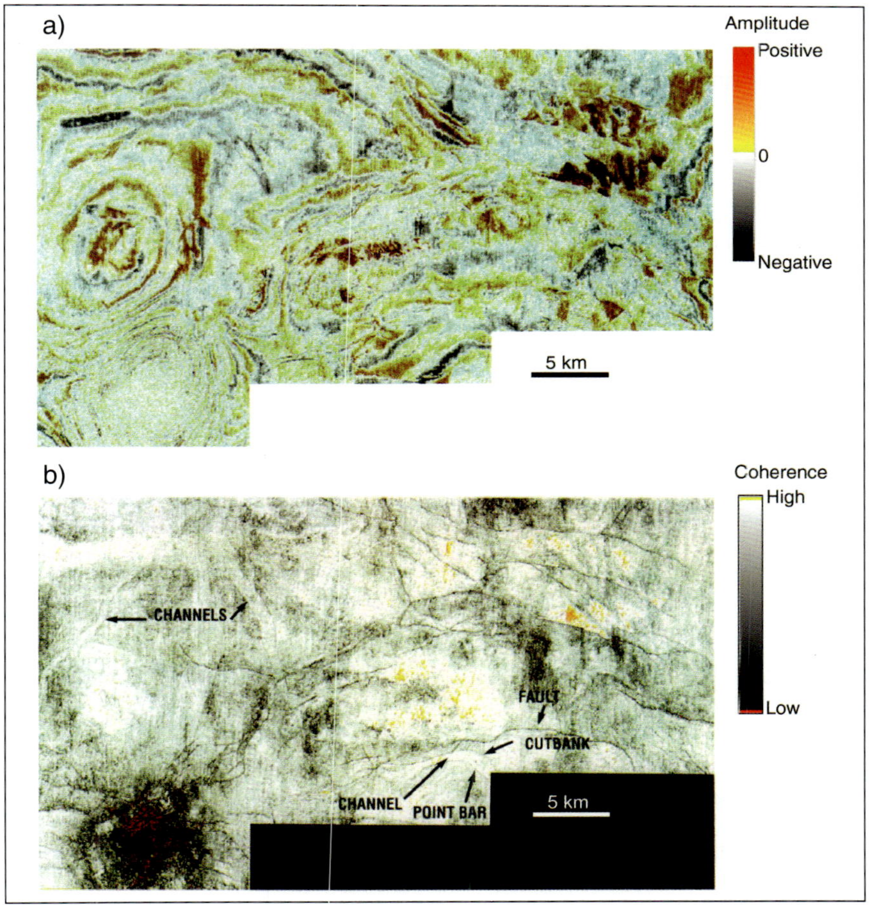

Figure 3.25. One of the first applications of coherence, showing a time slice at $t = 1.200$ s, through (a) the seismic data volume and (b) the crosscorrelation-based coherence volume. The color scale is red (for lowest coherence), black, gray, white, and yellow (for highest coherence). After Bahorich and Farmer (1995).

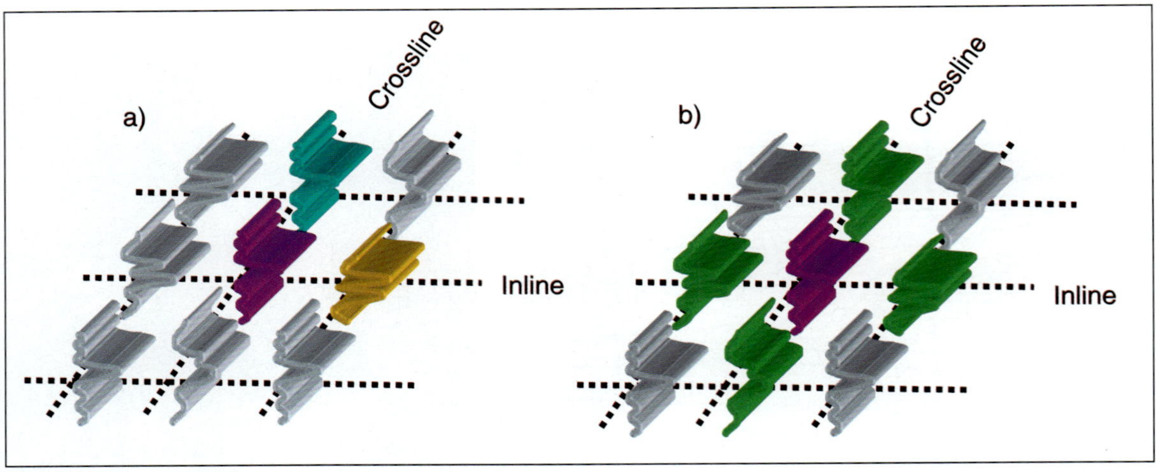

Figure 3.26. Spatial (or multitrace) analysis windows commonly used in coherence calculations for (a) the crosscorrelation algorithm and (b) the semblance and eigenstructure algorithms. In the crosscorrelation algorithm, we first crosscorrelate the target trace (in magenta) with the inline trace (in orange) over a suite of temporal lags. Then we repeat that process between the target trace and the crossline trace (in cyan). The coherence estimate is obtained by computing the geometric mean of the two estimates. In the semblance and eigenstructure algorithms, we first estimate dip and azimuth, as described in Figure 3.7, and then calculate either the semblance or a covariance matrix between the target trace (in cyan) and its nearest neighbors. Here we show four nearest neighbors in green and eight nearest neighbors in green and gray. These sets of nearest neighbors give rise to five- and nine-trace coherence algorithms, respectively.

Figure 3.27. A schematic diagram showing crosscorrelation between two traces. Trace number 1 is held fixed while a window of trace number 2 (here, 40 ms) is slid along a suite of time lags and is crosscorrelated. The lag that has the maximum signed crosscorrelation is a crude measure of inline (or crossline) dip. The crosscorrelation value that corresponds to this peak then is used to generate a 3D estimate of coherence.

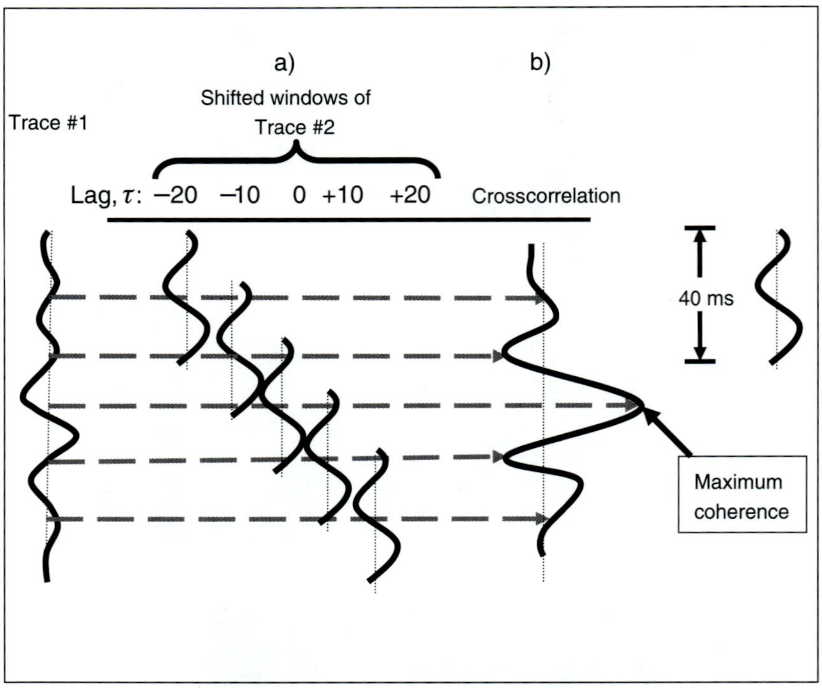

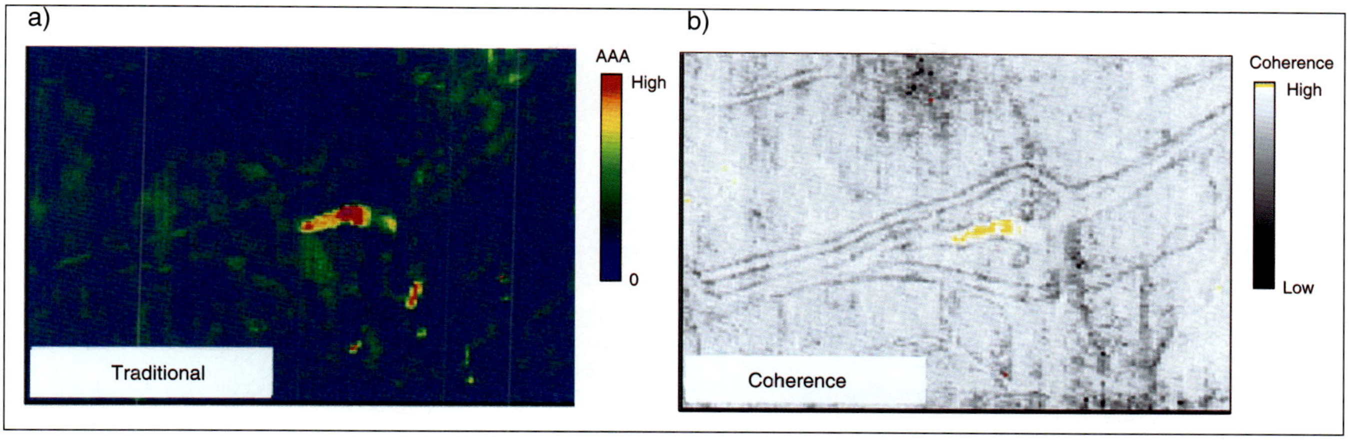

Figure 3.28. (a) A horizontal time slice of average absolute amplitude (AAA) calculated over a 40-ms vertical analysis window. (b) The corresponding three-trace 40-ms crosscorrelation coherence slice through the same volume. The bright spot seen in (a) corresponds to the highest coherence value displayed in yellow in (b), lying within the channel. Because bright spots generally have a high signal-to-noise ratio, it is common for them to show up with very high coherence. After Bahorich et al. (1995).

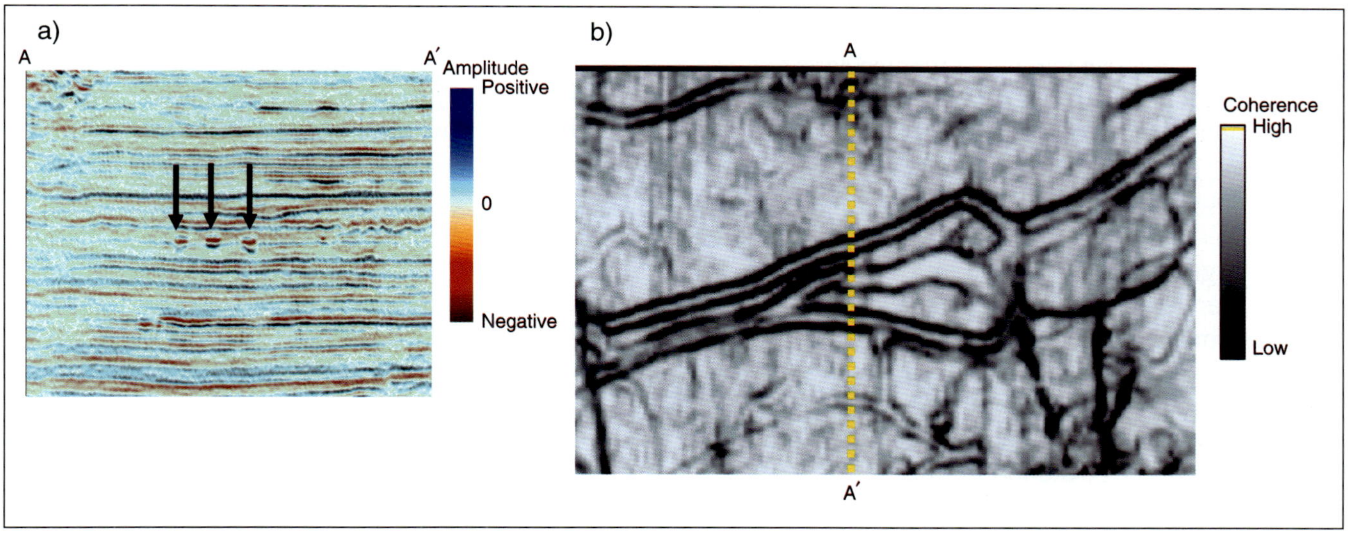

Figure 3.29. (a) Channels (indicated by arrows) seen on a conventional vertical seismic slice through the same volume as that in the previous figure. (b) Channels in map view generated by an 11-trace 40-ms semblance-based coherence algorithm. Note the higher signal-to-noise ratio of the image here, compared with the three-trace crosscorrelation coherence shown in Figure 3.28b. The map view helps us differentiate between separate channels (perhaps having been formed at different geologic times) and a single meandering channel. After Haskell et al. (1995).

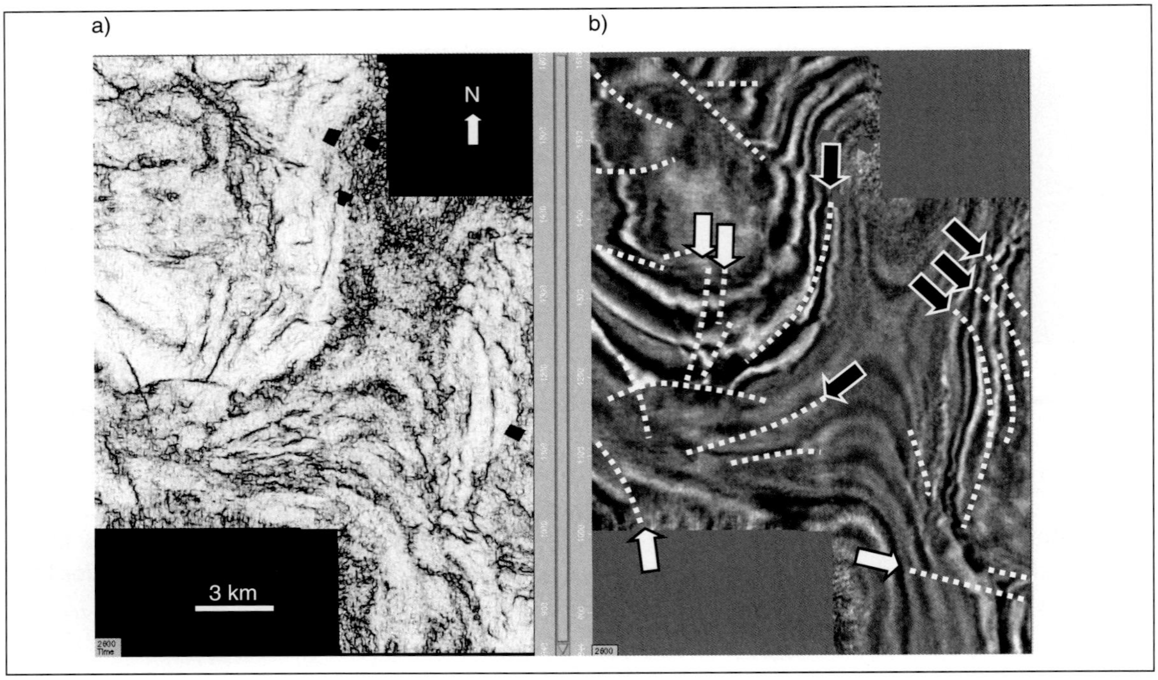

Figure 3.30. Time slices at $t = 2.600$ s, through (a) a coherence volume and (b) the corresponding seismic volume, for a survey acquired in northwestern Louisiana, U.S.A. White arrows indicate faults that cut perpendicularly to strike and are easy to see on the seismic time slice. Black arrows indicate faults that cut nearly parallel to strike and are difficult to differentiate from normal waveform changes seen on the time slice. Small black rectangles in (a) and corresponding gray rectangles in (b) indicate acreage that was not permitted for exploration. The data are courtesy of Seitel.

Figure 3.31. A schematic diagram showing the steps used in semblance estimation of coherence. (a) First we calculate the energy of the five input traces within an analysis window, (b) then we calculate the average trace, and (c) finally, we replace each trace by the average trace and calculate the energy of the five average traces. The semblance is the ratio of the energy of (c) to the energy of (a). If each windowed trace in (a) has the identical waveform and amplitude, the semblance = 1.0; otherwise, it is less than 1.0.

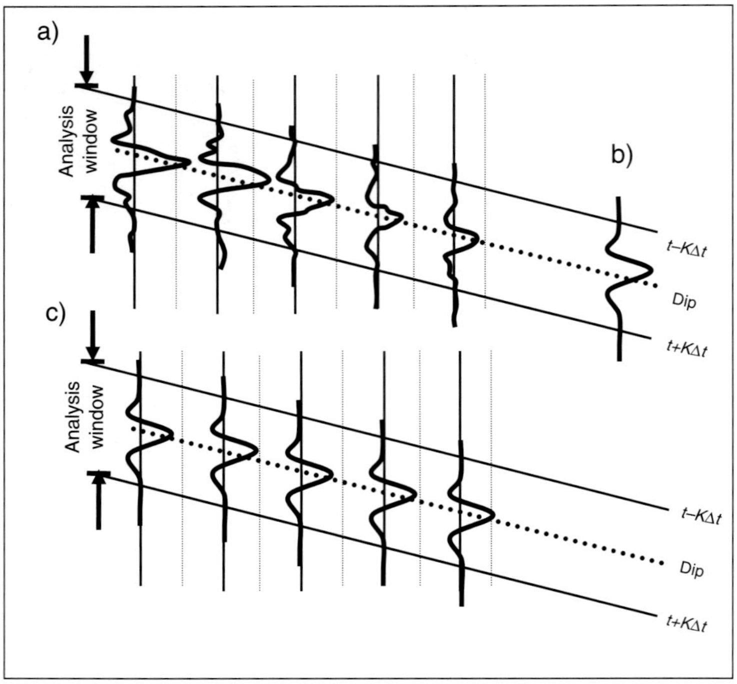

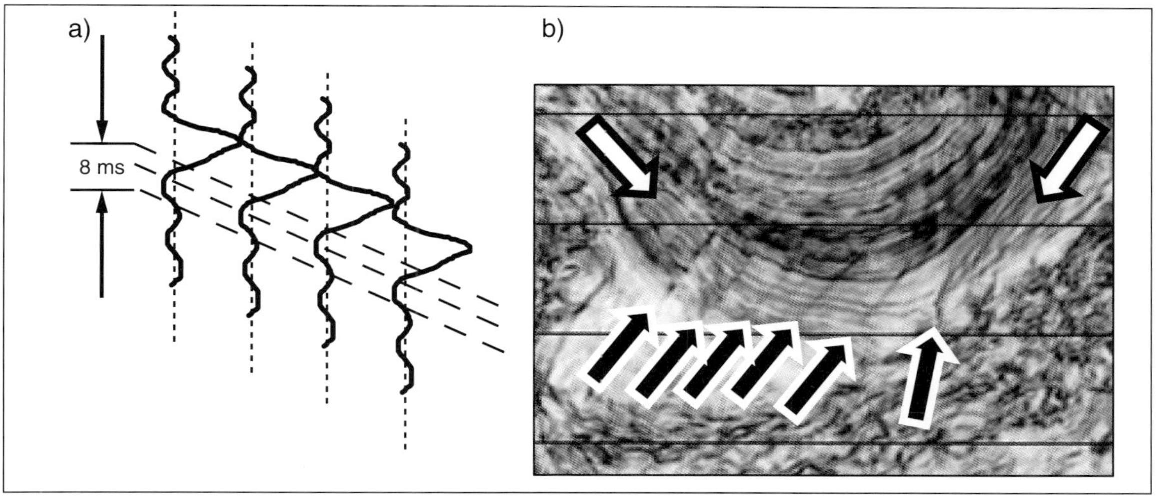

Figure 3.32. The effect of zero crossings in the signal, using the traditional semblance algorithm. (a) The signal-to-noise ratio is low when a short analysis window (here, 8 ms, corresponding to three samples) straddles a zero crossing. (b) The resulting coherence image, seen on a time slice at $t = 2.164$ s. White arrows indicate artifacts associated with low signal-to-noise about zero crossings. Black arrows indicate geologic faults of interest.

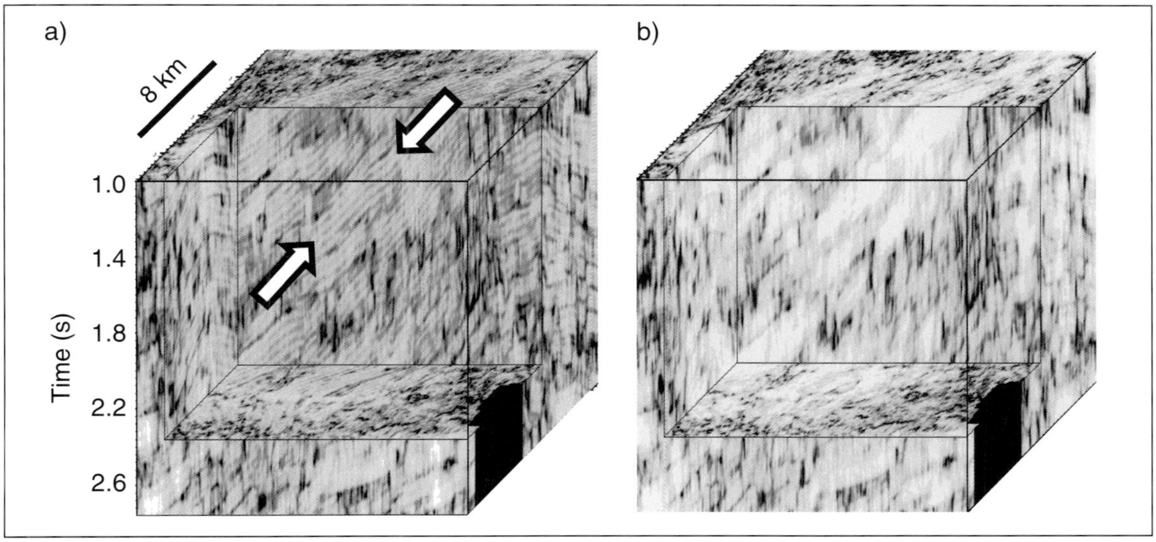

Figure 3.33. Semblance estimates of coherence generated using (a) the input data trace and (b) the analytic trace. The vertical window length is 20 ms. For a fixed level of noise, the signal-to-noise ratio can become low near reflector zero crossings, thereby resulting in low-coherence artifacts that follow the structure (arrows). Using the analytic trace avoids that problem. When the magnitude of the real input trace is low, the magnitude of the quadrature component is high. Likewise, when the magnitude of the quadrature component is low, the magnitude of the real input trace is high, thereby maintaining a good signal-to-noise ratio in the presence of strong reflectors. It still is common to see low-coherence trends following structure when we have low-reflectivity (and hence low signal-to-noise ratio) shale-on-shale events, when we have truly incoherent geology such as that encountered with erosional and angular unconformities, or when we encounter karst, mass-transport complexes, and turbidites.

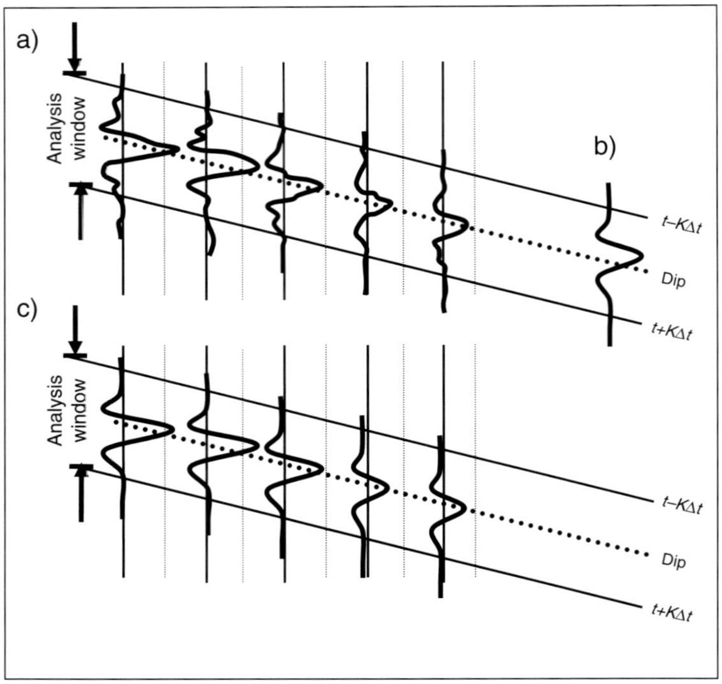

Figure 3.34. A schematic diagram showing the steps used in eigenstructure estimation of coherence. (a) First we calculate the energy of the input traces within an analysis window, (b) next we calculate the seismic waveform that best approximates the waveform of each input trace, and (c) finally, we replace each trace by a scaled version of (b) that best fits the input trace. The eigenstructure coherence is the ratio of the energy of (c) to the energy of (a). If each windowed trace in (a) has the same waveform (but perhaps a different amplitude), the coherence = 1.0; otherwise, it is less than 1.0.

Figure 3.35. (on page 73). The steps for generating the data-covariance matrix used in calculating coherence (discussed in Figure 3.34), coherent amplitude gradients (discussed in Figure 3.82), and structurally oriented principal-component filtering (discussed in Unit 5). In this illustration, $K = 1$, which results in three sample-data vectors. In step 1, we extract a suite of sample vectors from the data (Figure 3.12). If our analysis window includes nine traces, the data sample vector will be of length nine. There will be one sample vector, corresponding to a suite of time-interpolated data samples, for each value of k ranging between $\pm K$ falling within our vertical analysis window and extracted parallel to a plane defined by apparent dips p and q. Each sample vector forms a row in the data matrix. In step 2, we take each column of the data matrix (which is simply a shifted window of each seismic trace) and crosscorrelate each column to itself and to every other column, forming what we call a covariance matrix. The covariance matrix contains all possible unnormalized crosscorrelations and autocorrelations of all nine windowed traces with one another. These correlations then are ordered systematically to construct a covariance matrix. Specifically, crosscorrelating (without normalization) a window of trace 2 (column 2 of the data matrix) with the corresponding window of trace 8 (column 8 of the data matrix) give us the C_{28} element of the 9×9 covariance matrix (step 3).

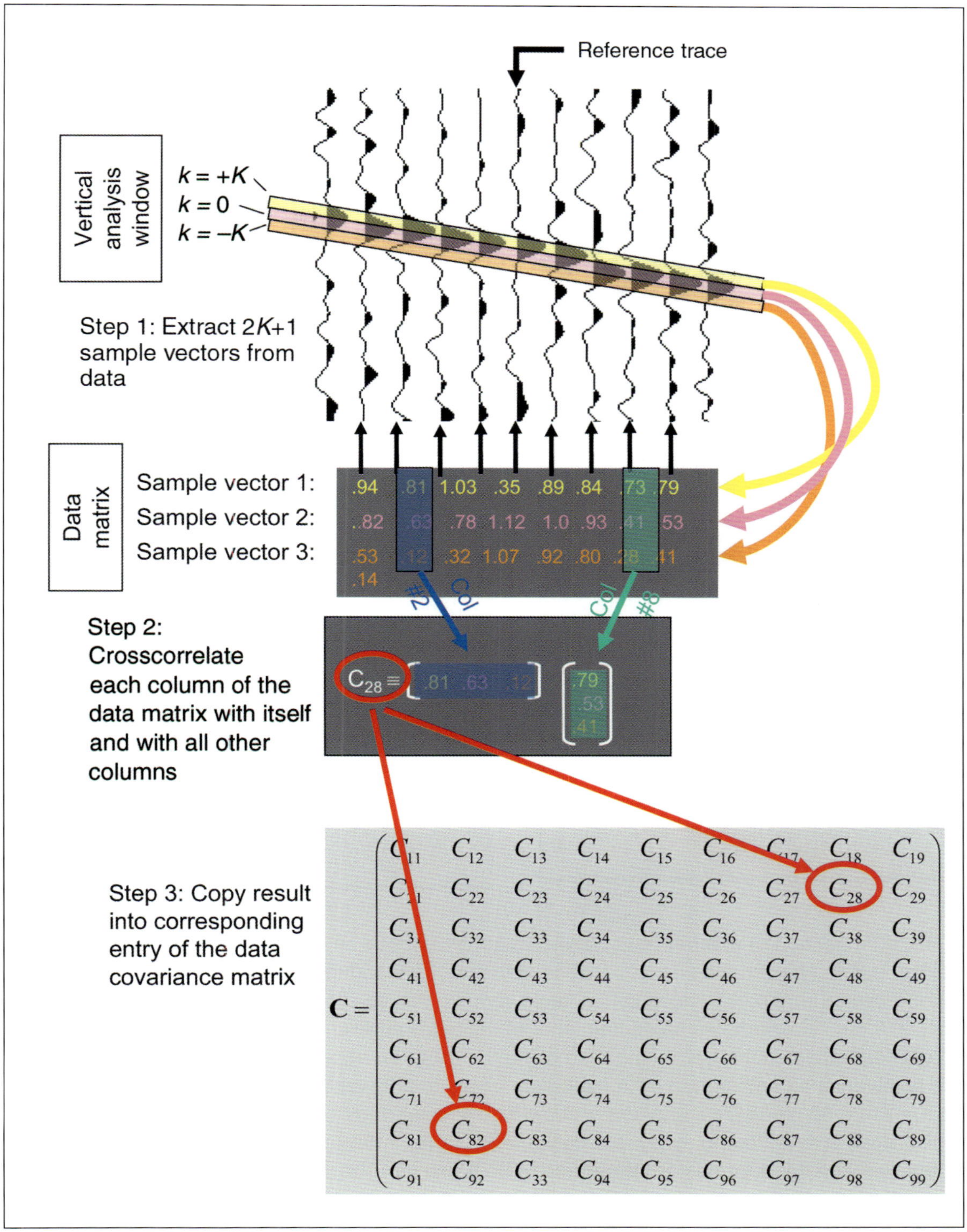
Reference trace
Vertical analysis window
$k = +K$
$k = 0$
$k = -K$
Step 1: Extract $2K+1$ sample vectors from data
Data matrix
Sample vector 1: .94 .81 1.03 .35 .89 .84 .73 .79
Sample vector 2: ..82 .63 .78 1.12 1.0 .93 .41 .53
Sample vector 3: .53 .12 .32 1.07 .92 .80 .28 .41 .14
Col #2
Col #8
Step 2: Crosscorrelate each column of the data matrix with itself and with all other columns
$C_{28} \equiv$ [.81 .63 .12] [.79 .53 .41]
Step 3: Copy result into corresponding entry of the data covariance matrix
$\mathbf{C} =$
C_{11} C_{12} C_{13} C_{14} C_{15} C_{16} C_{17} C_{18} C_{19}
C_{21} C_{22} C_{23} C_{24} C_{25} C_{26} C_{27} C_{28} C_{29}
C_{31} C_{32} C_{33} C_{34} C_{35} C_{36} C_{37} C_{38} C_{39}
C_{41} C_{42} C_{43} C_{44} C_{45} C_{46} C_{47} C_{48} C_{49}
C_{51} C_{52} C_{53} C_{54} C_{55} C_{56} C_{57} C_{58} C_{59}
C_{61} C_{62} C_{63} C_{64} C_{65} C_{66} C_{67} C_{68} C_{69}
C_{71} C_{72} C_{73} C_{74} C_{75} C_{76} C_{77} C_{78} C_{79}
C_{81} C_{82} C_{83} C_{84} C_{85} C_{86} C_{87} C_{88} C_{89}
C_{91} C_{92} C_{33} C_{94} C_{95} C_{96} C_{97} C_{98} C_{99}

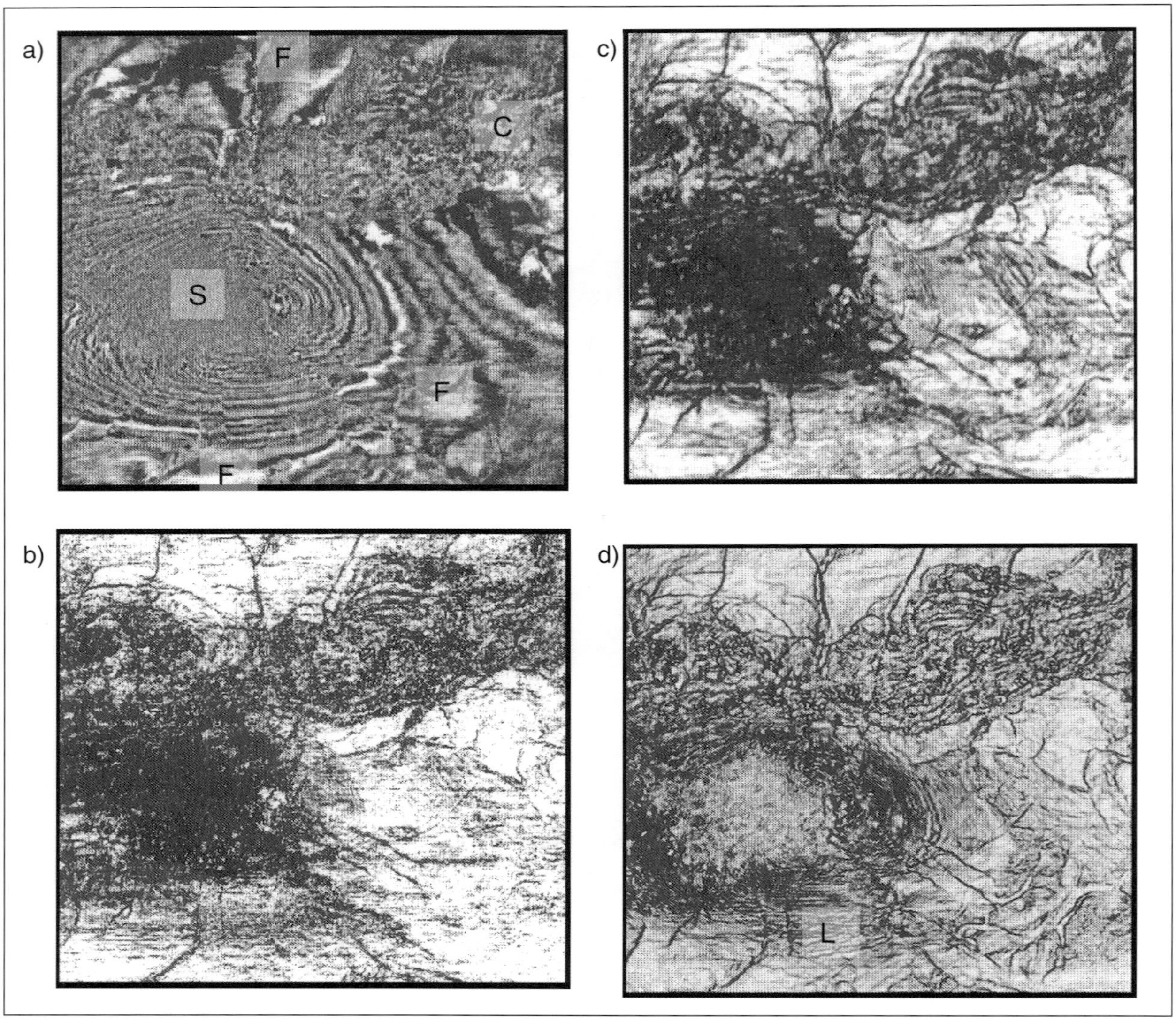

Figure 3.36. Comparison of alternative coherence algorithms used on data from South Marsh Island, Gulf of Mexico, U.S.A. The volume shown contains both structural and stratigraphic features associated with deposition over a terrain influenced by salt tectonism. (a) A time slice through the time-migrated seismic data at 1.8 s. "S" denotes a salt dome, and "F" indicates several radial faults. Corresponding slices through coherence cubes were generated using the (b) three-trace crosscorrelation algorithm, (c) a five-trace semblance algorithm, and (d) a five-trace eigenstructure algorithm. All coherence computations used the same 80-ms vertical analysis window. The circular rings seen in (a) correspond to sediments dipping against a salt dome, which are cut by radial faults indicated by "F." The disorganized feature indicated by "C" in the northeast is interpreted to be a canyon. The salt dome and faults appear to be incoherent (black) in (b) through (d). Note that there is considerably less speckle noise in the five-trace semblance algorithm than in the three-trace crosscorrelation algorithm. An even greater improvement in the signal-to-noise ratio and in lateral resolution accompanies the five-trace eigenstructure algorithm. The structural artifacts (leakage) about the salt dome indicated by "L" and the overall grayer level of the image in (d) are the result of a failure in this early work to search over structural dip. After Gersztenkorn and Marfurt (1996).

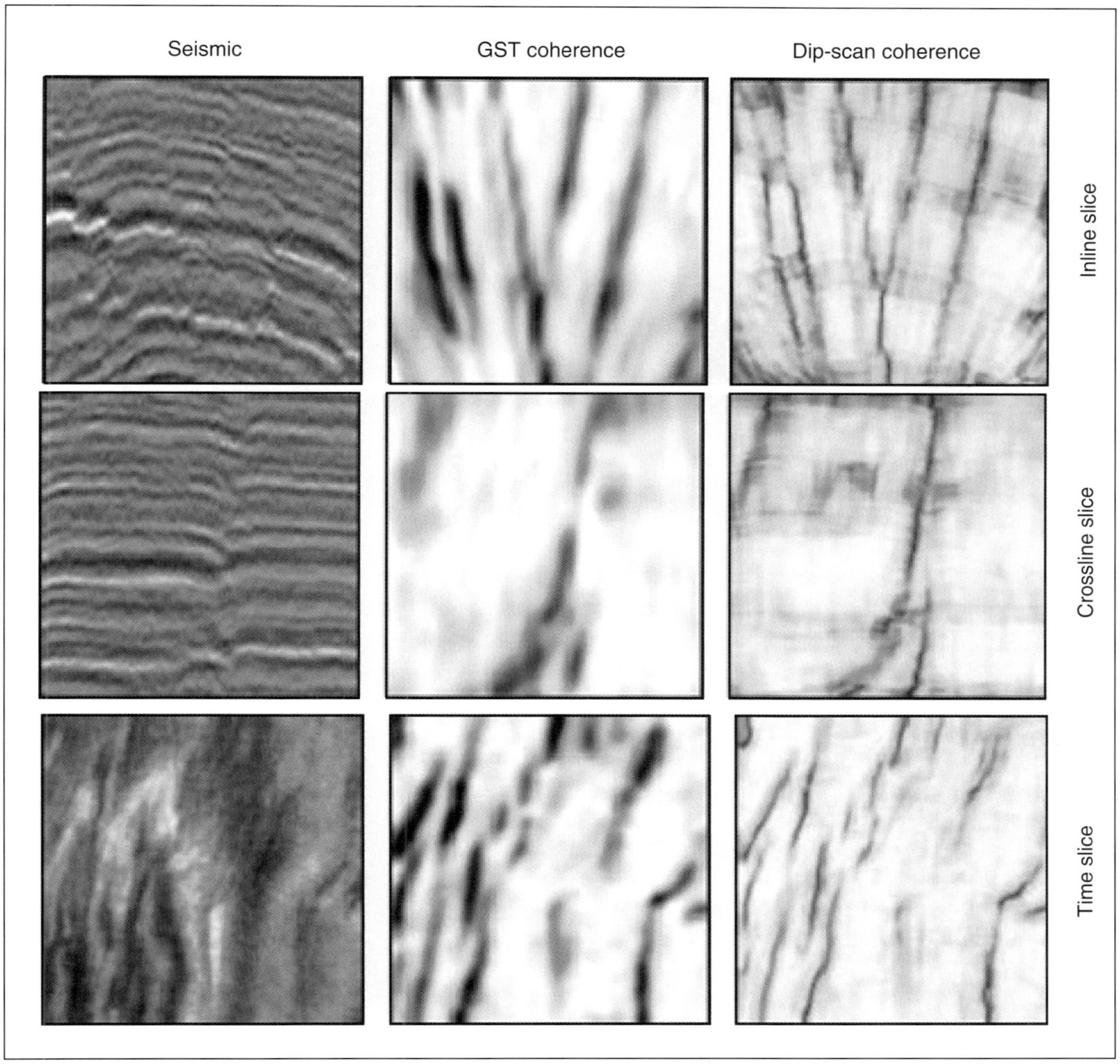

Figure 3.37. Orthogonal vertical and time slices through a seismic data cube, an eigenstructure coherence generated using the gradient structure tensor (GST), and an eigenstructure coherence generated using a discrete dip scan. In these images, we see greater lateral resolution in the eigenstructure dip-scan coherence. After Bakker (2003).

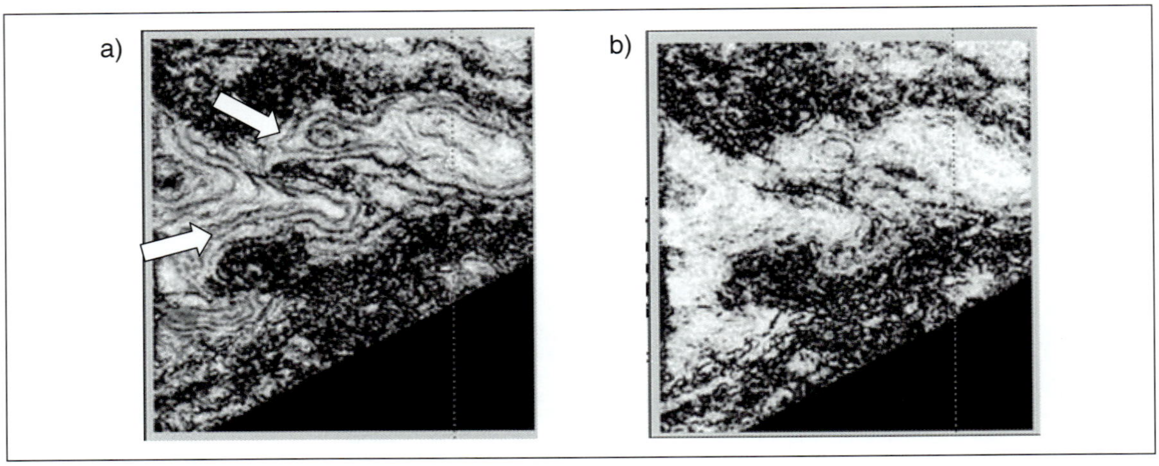

Figure 3.38. Comparison of coherence volumes generated (a) without and (b) with a dip search, using an identical eigenstructure estimate of coherence. The computational effort needed to generate the coherence volume in (a) is only 10% of that needed to generate the coherence volume in (b). However, failure to estimate waveform similarity along reflector dip results in artifacts sometimes called structural leakage, thereby giving rise to contourlike features indicated by arrows in (a). The low-coherence "contours" result from the structural dip being such that the lateral analysis window spans events from different geologic horizons, each of which has its own waveform. The contours in this image mask the fault that is seen more clearly in (b). After Chopra (2002).

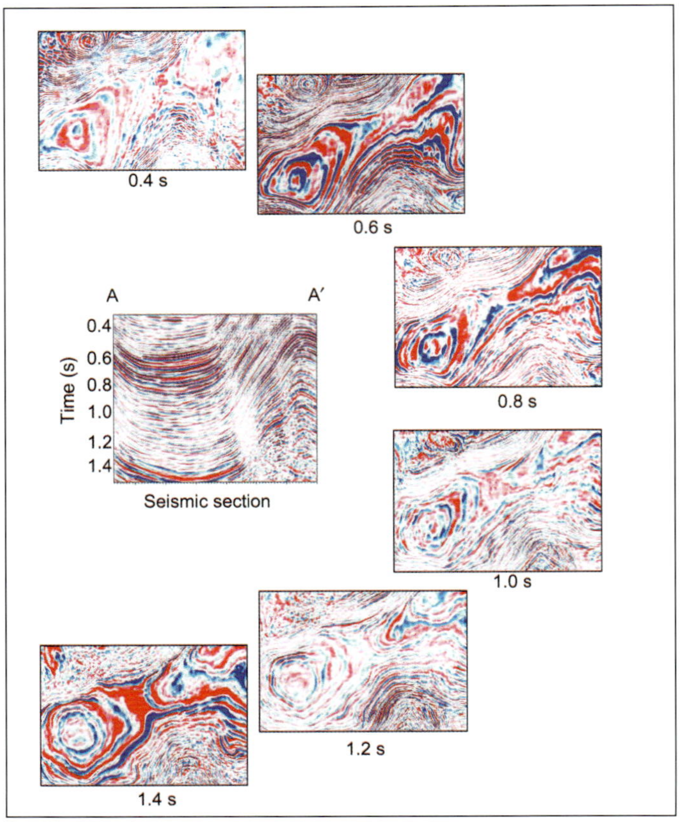

Figure 3.39. A vertical seismic slice AA′ and time slices at 0.2-s intervals through a time-migrated seismic volume from the western United States. We note a central syncline bounded by two anticlines.

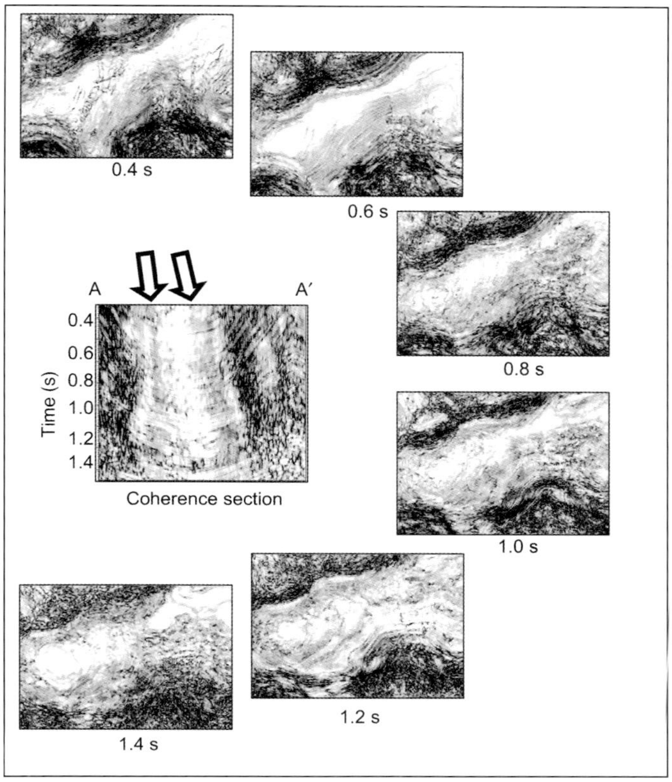

Figure 3.40. A vertical slice AA′ and time slices through the eigenstructure coherence volume corresponding to the slices through the seismic volume shown in the previous figure. Coherence was calculated assuming a flat dip. Arrows indicate structural-leakage artifacts that might be misinterpreted as low-coherence fracture zones.

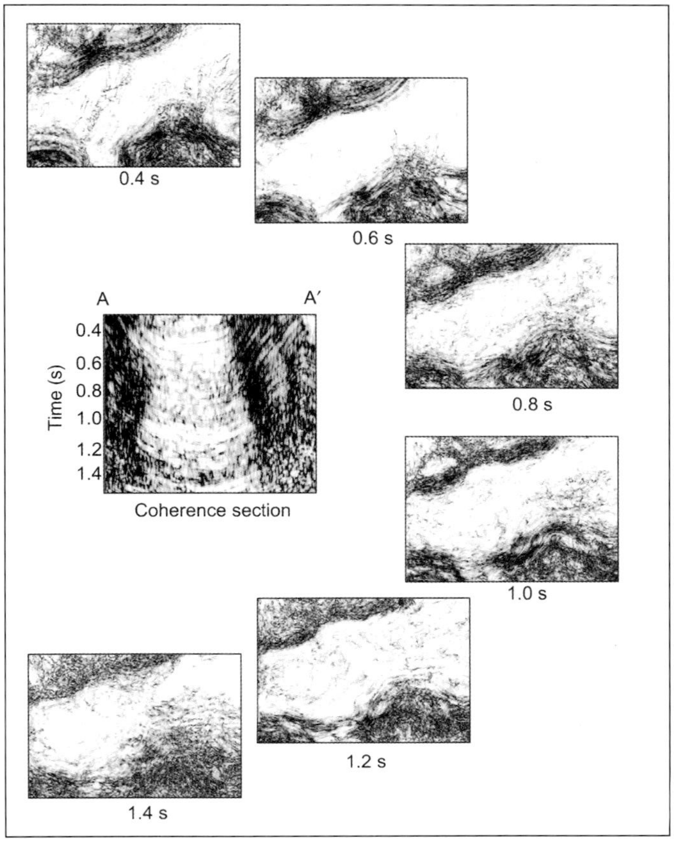

Figure 3.41. A vertical slice AA′ and time slices through the eigenstructure coherence volume corresponding to those in the previous two figures. Coherence was calculated along reflector dip. Note that the structural leakage has disappeared, although we do see bands of low-coherence shales following structure.

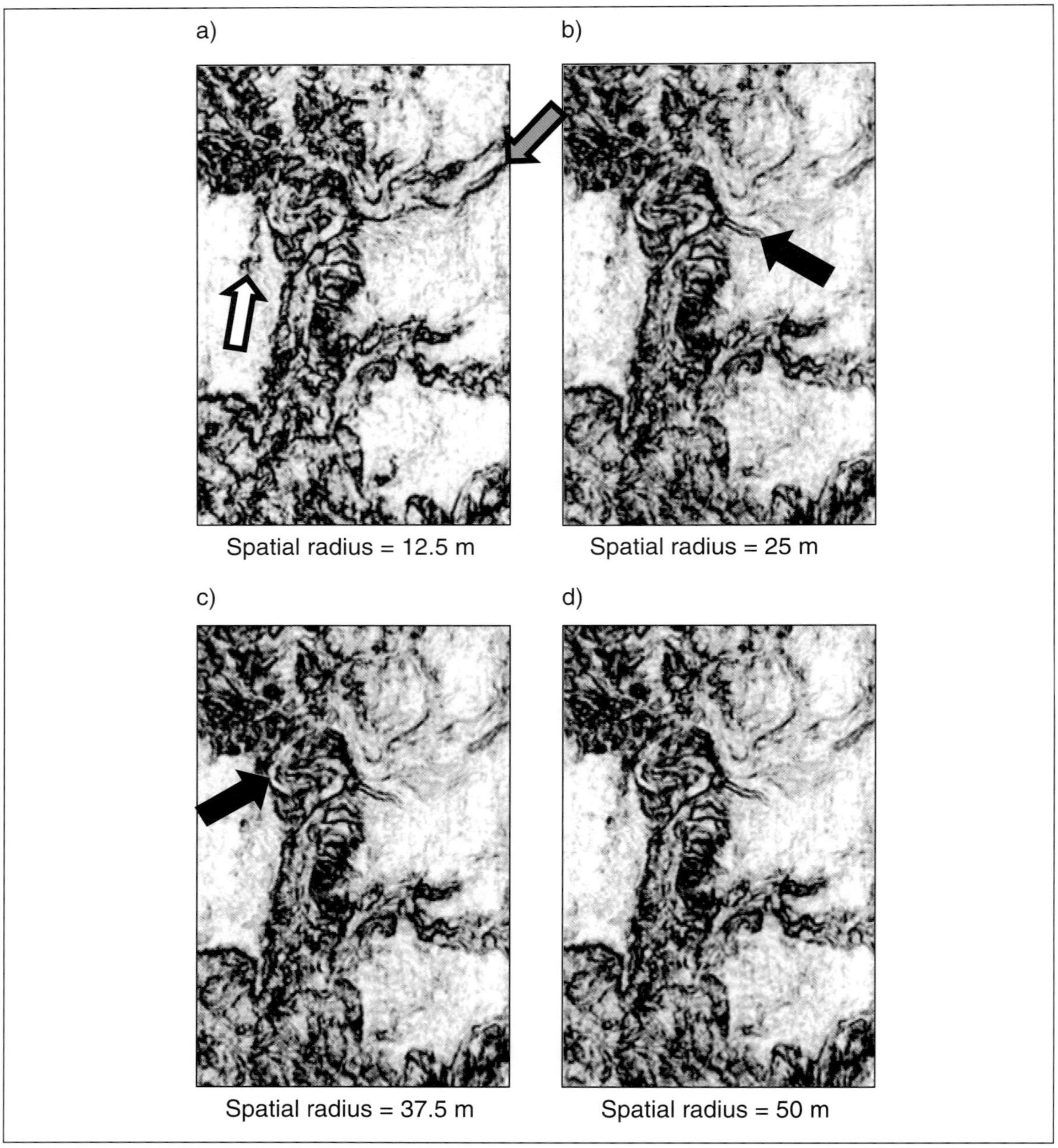

Figure 3.42. The effect of varying the spatial radius on a channel system, using the eigenstructure algorithm with various spatial windows: (a) $r = 12.5$ m, including five traces, (b) $r = 25$ m, including 13 traces, (c) $r = 37.5$ m, including 25 traces, and (d) $r = 50$ m, including 49 traces. The temporal analysis window = 16 ms or five samples for all images. Note that the very thin channel feature (white arrow) and the edges of a broad channel (gray arrow) are illuminated by the 12.5-m window but are washed out in the other images from larger analysis windows. In contrast, a somewhat wider channel (vertically striped arrow) is better illuminated when $r > 12.5$ m. Likewise, the horizontally striped arrow indicates greater detail within a meandering-channel complex.

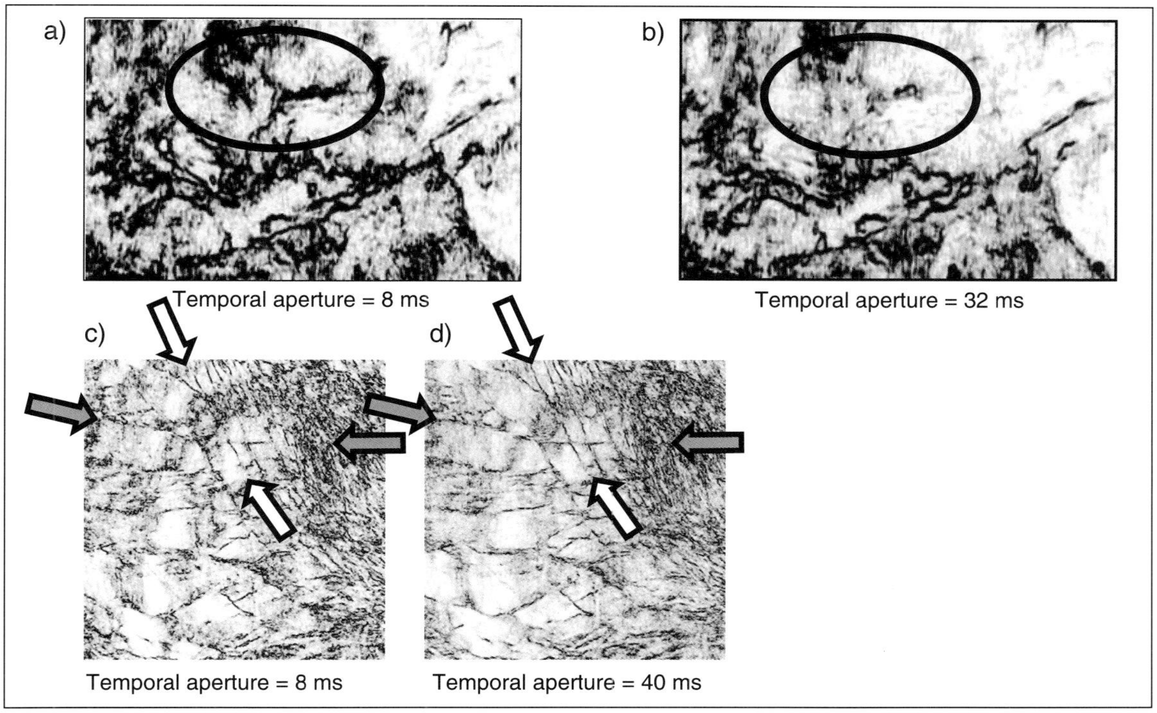

Figure 3.43. A larger temporal analysis window tends to smear stratigraphic features such as channels (compare the circled features in [a] and [b]) that are confined to a given geologic horizon. In contrast, larger temporal analysis windows improve the appearance of vertical faults (compare the faults in [c] and [d] that are identified by white and gray arrows). The latter is because the coherence computation stacks similar discontinuities over a larger vertical window.

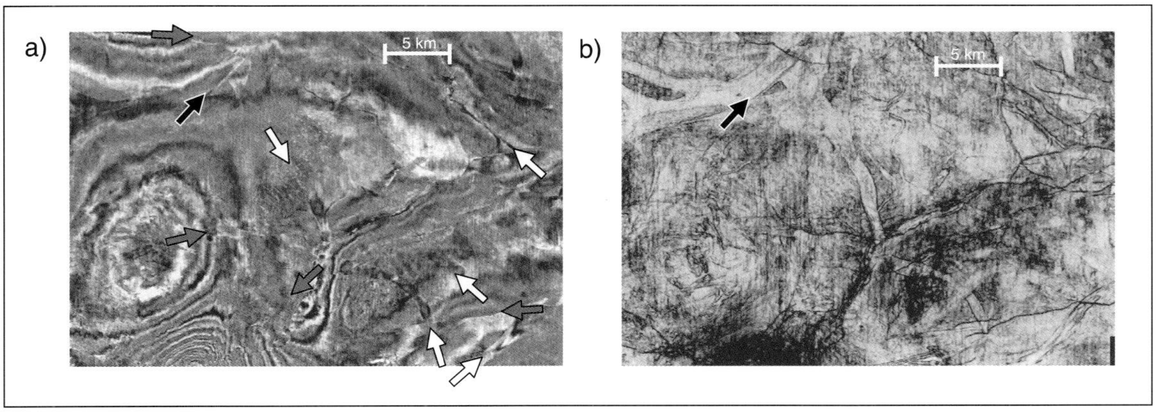

Figure 3.44. Time slices, at $t = 1.200$ s, through (a) a seismic data volume and (b) the corresponding coherence volume calculated using an 11-trace semblance algorithm. Although channels (white arrows) and faults (gray arrows) can be seen on the seismic time slice, lineaments such as that indicated by the black arrow are more ambiguous. The coherence slice allows us to interpret our data more confidently in the time-slice mode. The black arrow clearly indicates the edge of a channel. Previously unrecognizable faults and channels now appear. After Marfurt et al. (1998).

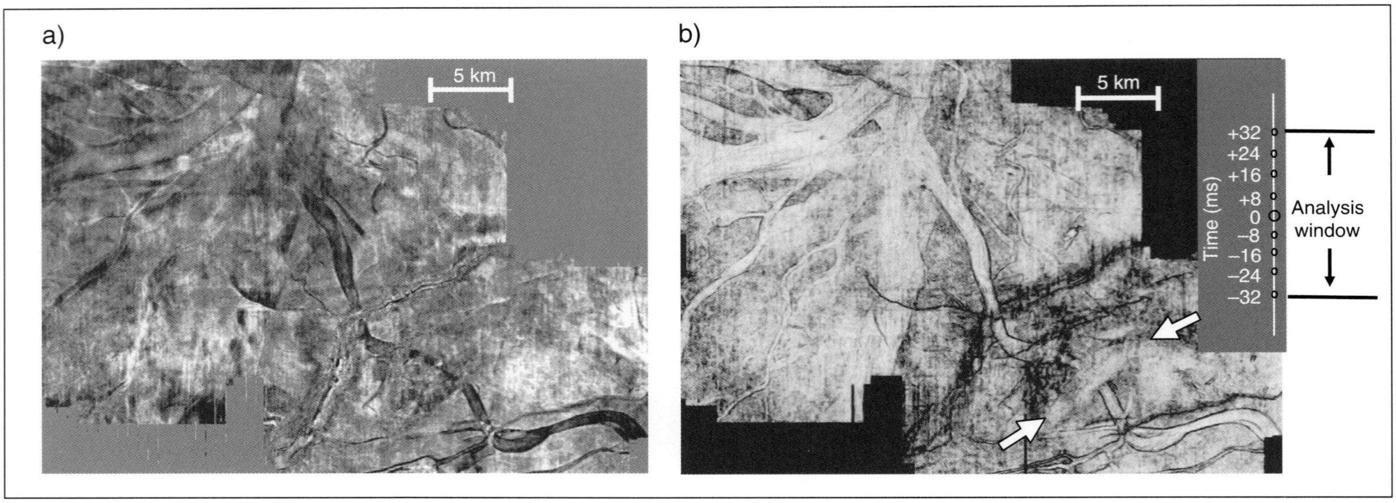

Figure 3.45. Horizon slices through (a) a seismic data volume and (b) a coherence volume along an interpreted Pleistocene horizon corresponding approximately to the time slice shown in the previous figure. The area displayed is smaller than that seen in the previous figure, because the horizon did not exist over a salt diapir in the southwest and was truncated in the northeast. In general, stratigraphic features are seen best on horizon slices, whereas structural features are seen best on time slices. Semblance-based coherence was calculated using 11 traces and a 64-ms vertical analysis window. After Marfurt et al. (1998).

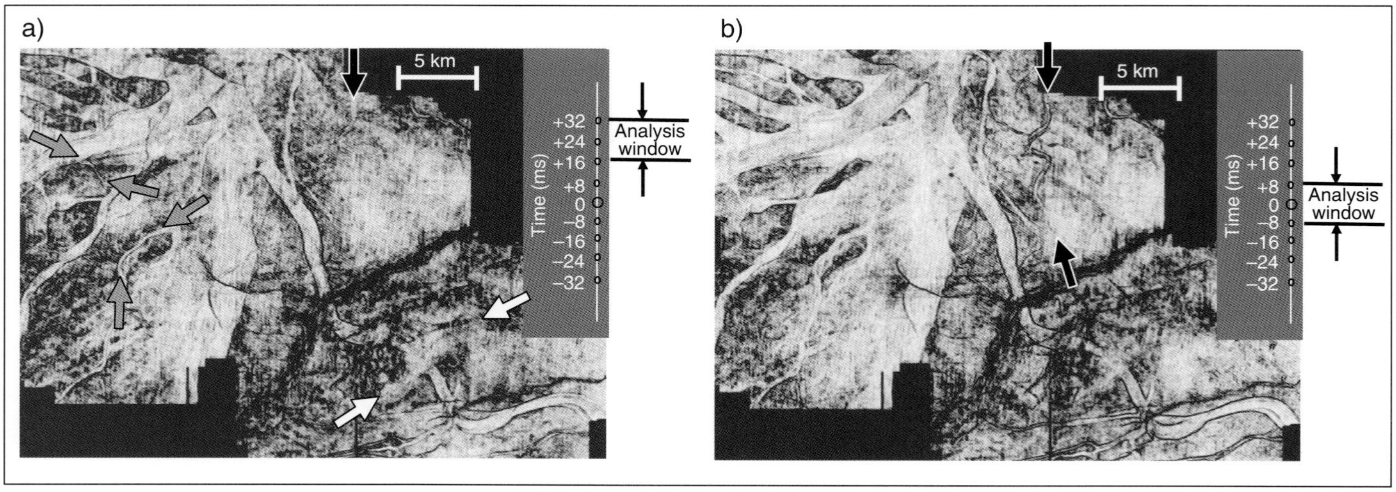

Figure 3.46. The impact on resolving stratigraphic features of using a smaller analysis window. Both images use a three-sample, 16-ms analysis window that is shifted 24 ms higher in the image in (a) than in the image in (b), which is centered about the same horizon shown in part (b) of the previous figure. Note the improved lateral resolution by using a smaller analysis window. Also note that the narrow (and probably thin) channel features indicated by gray arrows are shallower than those indicated by black arrows. The broader channel indicated by white arrows has cut down from a shallower horizon. After Marfurt et al. (1998).

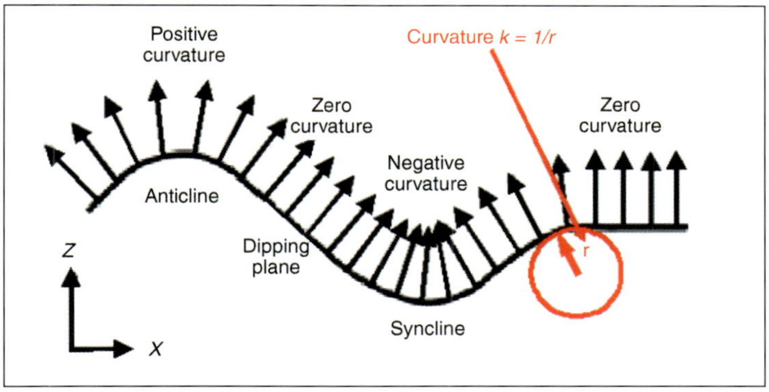

Figure 3.47. An illustrated definition of 2D curvature. Synclinal features have negative curvature, anticlinal features have positive curvature, and planar features have zero curvature. After Roberts (2001).

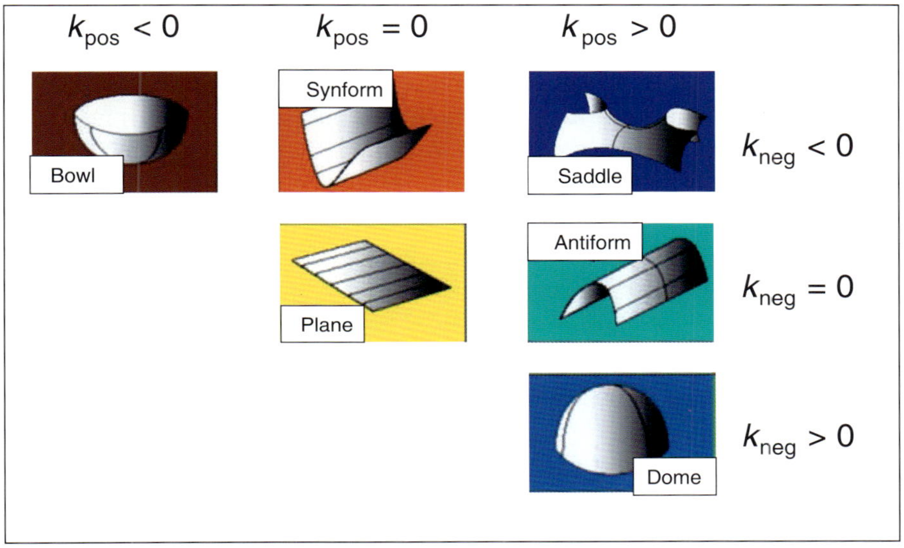

Figure 3.48. The definition of 3D quadratic shapes expressed as a function of the most positive curvature, k_{pos}, and the most negative curvature, k_{neg}. By definition, $k_{neg} \leq k_{pos}$. Thus, if both k_{pos} and k_{neg} are less than zero, we have a bowl; if both are greater than zero, we have a dome; and if both are equal to zero, we have a plane. After Bergbauer et al. (2003). AAPG©2003. Reprinted by permission of the AAPG whose permission is required for further use.

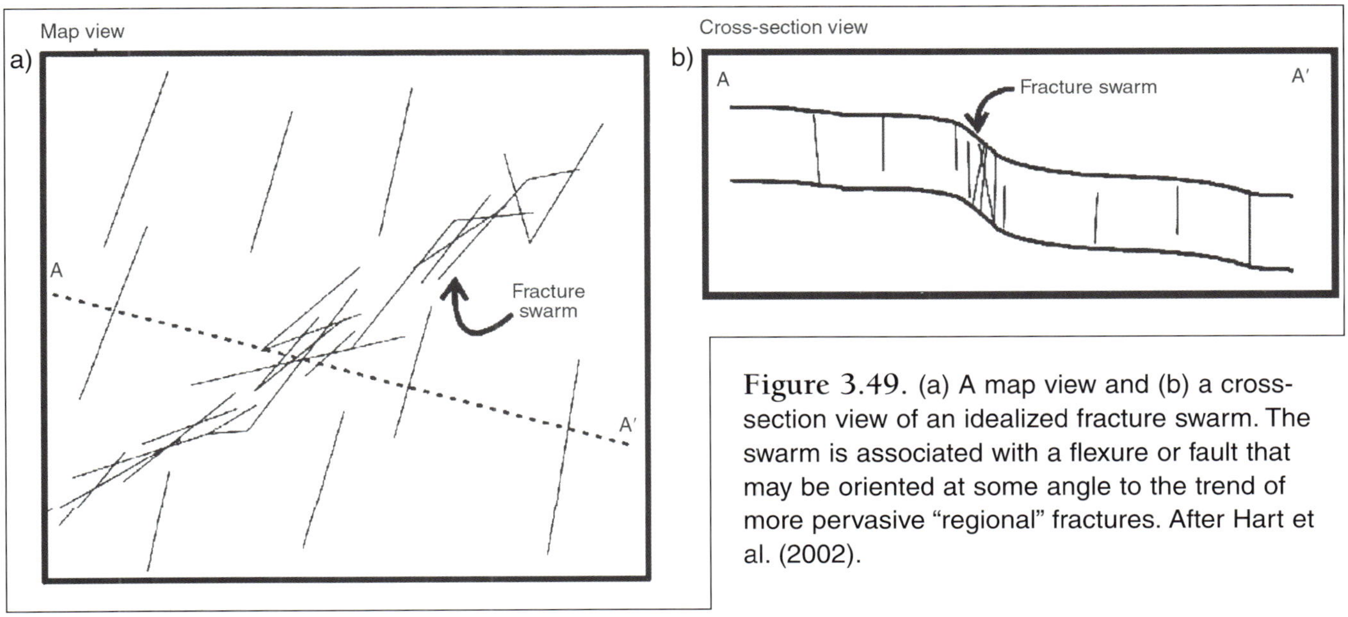

Figure 3.49. (a) A map view and (b) a cross-section view of an idealized fracture swarm. The swarm is associated with a flexure or fault that may be oriented at some angle to the trend of more pervasive "regional" fractures. After Hart et al. (2002).

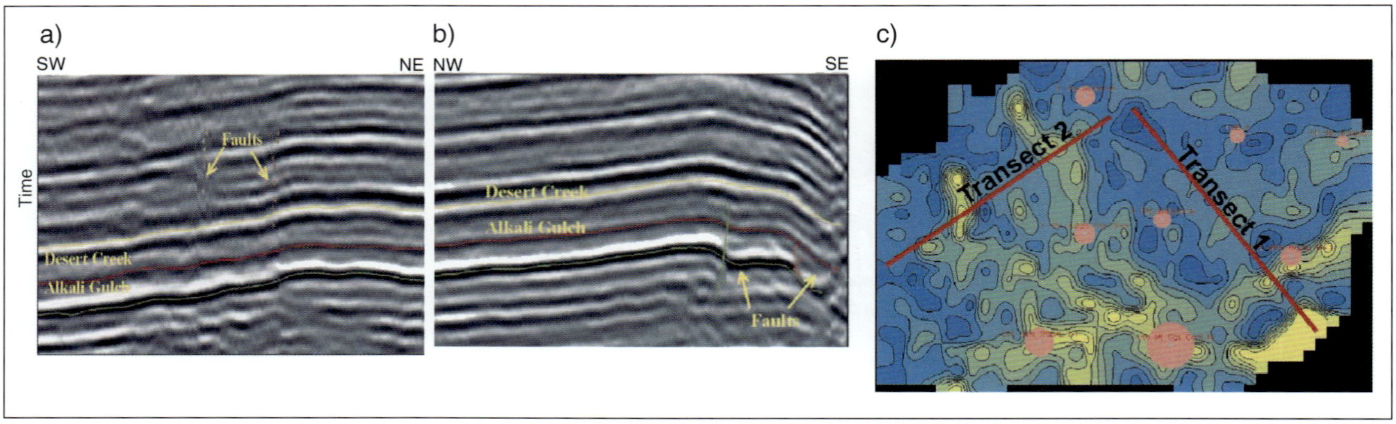

Figure 3.50. Seismic transects (a) through northwest-striking small-offset normal faults (down to the southwest, or left) and (b) through a northeast-striking reverse fault. (c) A dip-magnitude map with high dips shown in yellow. The size of the bubbles is proportional to the best year's production. Note that the larger bubbles fall along intersections of high-dip trends, implying that production is enhanced by the presence of fractures. After Hart et al. (2002).

Figure 3.51. (a) The absolute value of Gaussian curvature calculated from the topographic surface of Goose Egg Dome, Wyoming, U.S.A., and (b) fracture density measured on that outcrop. Note the strong correlation between fracture density and Gaussian curvature. After Lisle (1994). AAPG©1994. Reprinted by permission of the AAPG whose permission is required for further use.

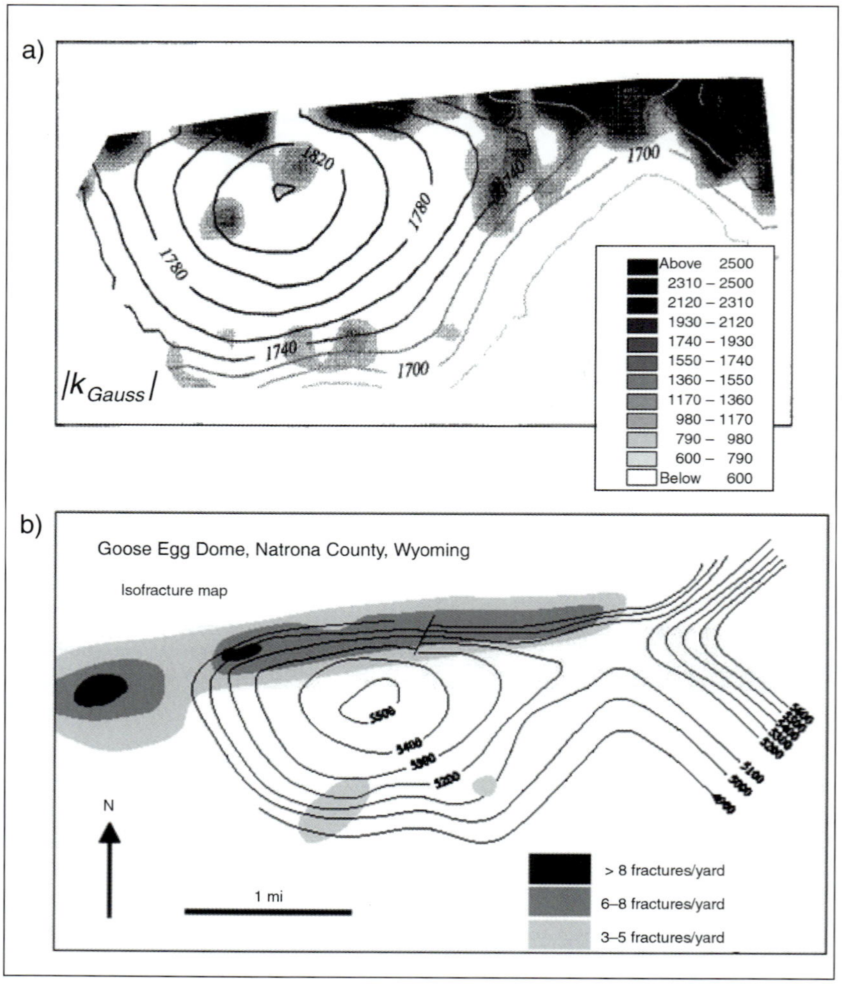

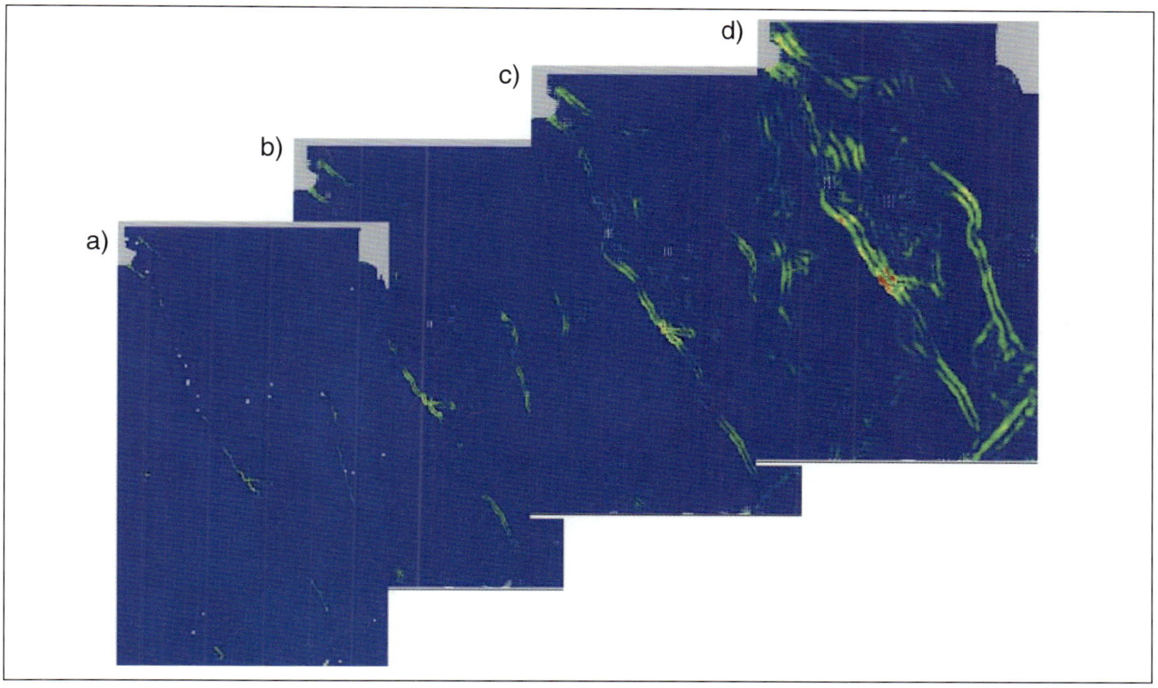

Figure 3.52. Curvedness calculated at different scales, using apertures of (a) 3 × 3, (b) 5 × 5, (c) 7 × 7, and (d) 9 × 9. In this example, the longer-wavelength (larger-aperture) estimates provide a clearer image of the fault-block edges. Recall that the definition of curvedness dictates that it is always greater than or equal to zero. After Sigismondi and Soldo (2003).

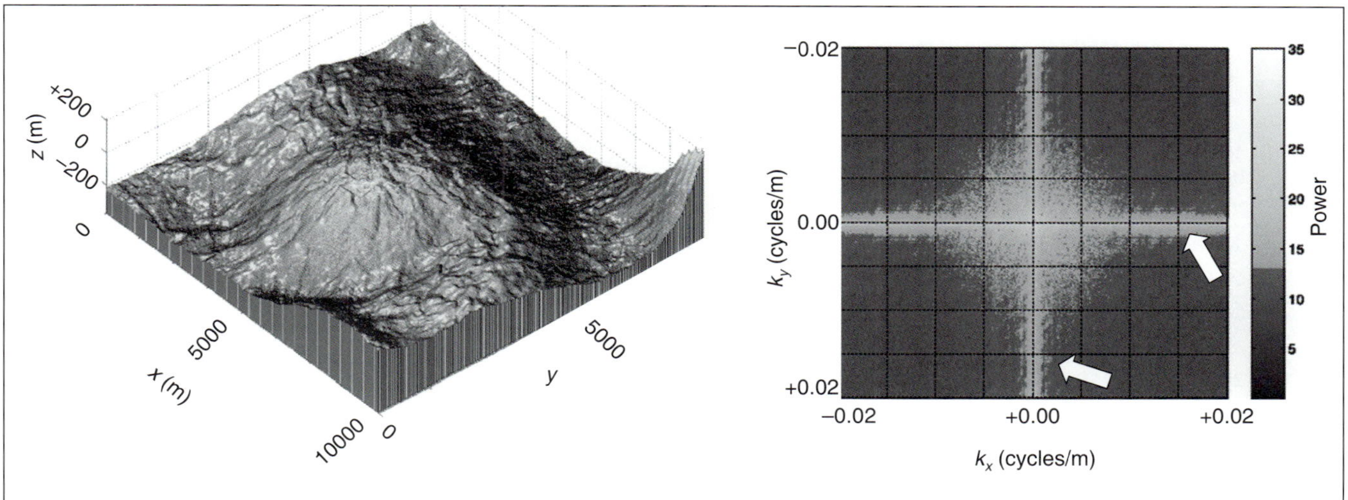

Figure 3.53. (a) A time-structure map and (b) its 2D Fourier transform. Long-wavelength features are represented by small values of k_x and k_y near the center of (b). The higher-amplitude short-wavelength components (indicated by arrows) that appear along the outer reaches of the k_x and k_y axes correspond to vertical and horizontal striping artifacts seen in (a) as a result of acquisition, processing, or autopickers. After Bergbauer et al. (2003). AAPG©2003. Reprinted by permission of the AAPG whose permission is required for further use.

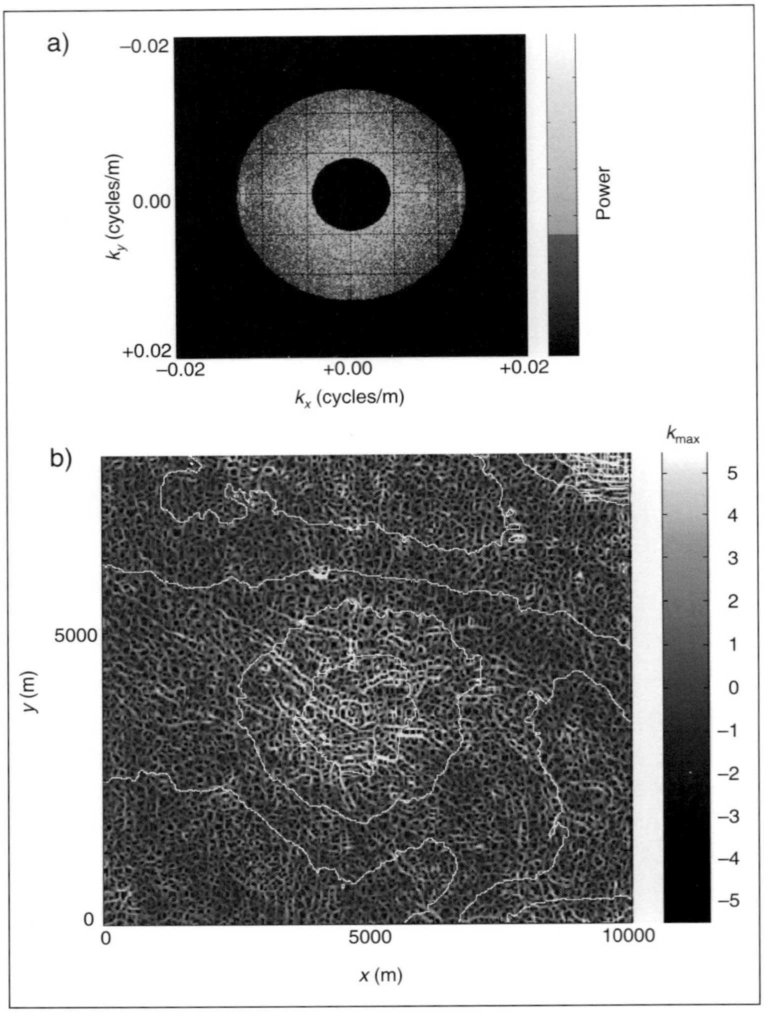

Figure 3.54. (a) A 2D k_x-k_y band-pass-filtered version of the k_x-k_y plane shown in part (b) of the previous figure. Short and long wavelengths have been eliminated. (b) The resulting band-pass-filtered, most-negative-curvature image. Structural contours of the surface shown in part (a) of the previous figure are displayed in white. Note how the spectrally limited (band-pass-filtered) most negative curvature enhances faults. After Bergbauer et al. (2003). AAPG©2003. Reprinted by permission of the AAPG whose permission is required for further use.

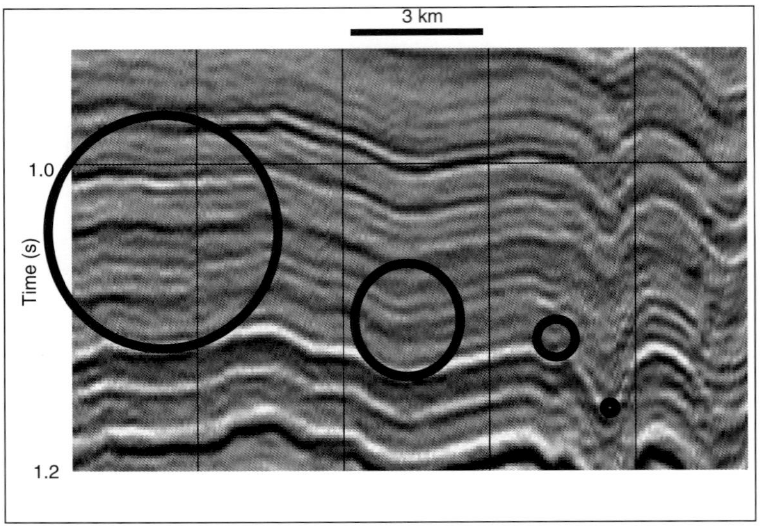

Figure 3.55. A vertical section through a seismic survey from the Forth Worth Basin, Texas, U.S.A., illustrating radii of curvature of approximately 500, 250, 100, and 50 m along the Marble Falls horizon. The seismic data zoom lies along line BB′ in Figure 3.57. Data are courtesy of Devon Energy.

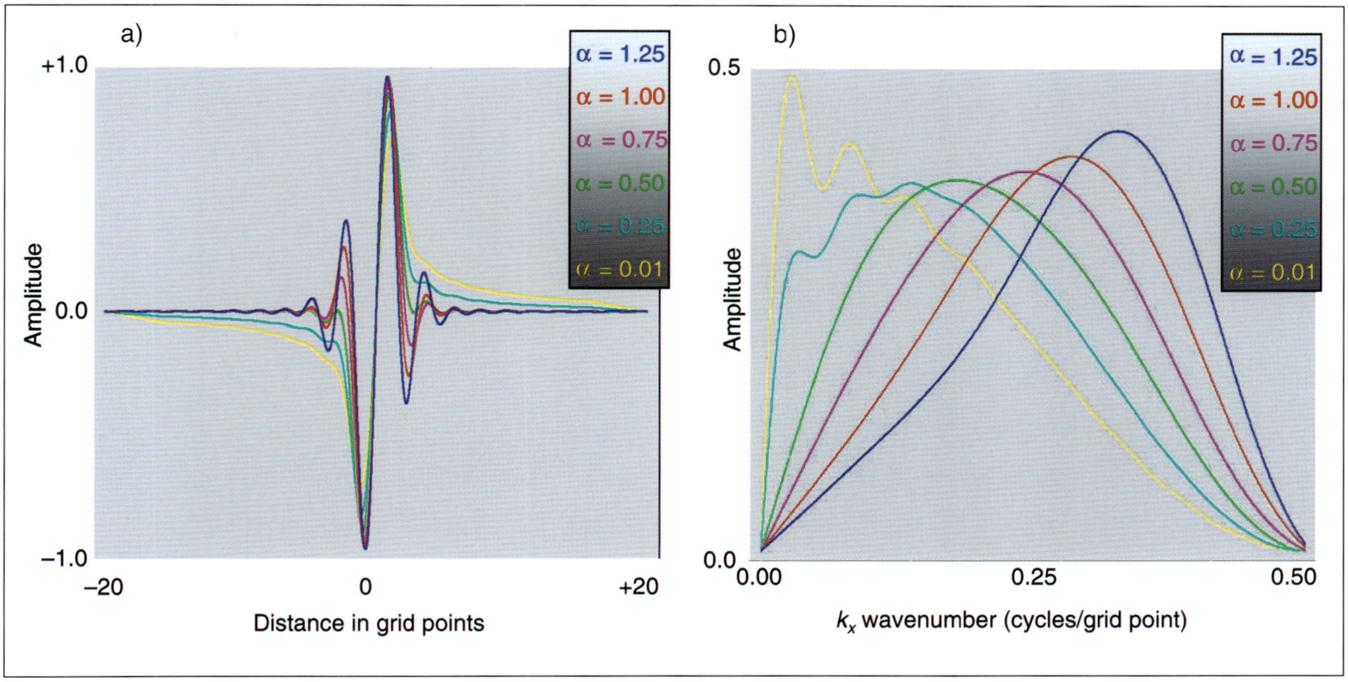

Figure 3.56. Fourier theory allows us to compute a derivative of the inline dip, dp/dx, from its Fourier transform $P(k) \equiv F(p)$, by the following formula: $dp/dx \equiv F^{-1}[ik_x F(p)]$. We generalize this concept by computing a fractional derivative $d^a p/dx^a$ using the formula $d^a p/dx^a \equiv F^{-1}[i(k_x)^a F(p)]$. (a) Spatial operators and (b) their corresponding spectra used in the "fractional-derivative" estimate of curvature. When the value of α = 1.0, it corresponds to the conventional first-derivative operator. As we decrease the value of α, we also decrease the contribution of higher frequencies, thereby shifting the bandwidth toward the longer wavelengths (lower values of k_x). Note that in the limit, as α approaches 0.0, the operator shown in yellow in (b) approximates the Hilbert transform. Thus Luo et al.'s (2002) Hilbert transform edge detector can be interpreted to be an end member of a fractional-derivative edge detector. Courtesy of Saleh al-Dossary.

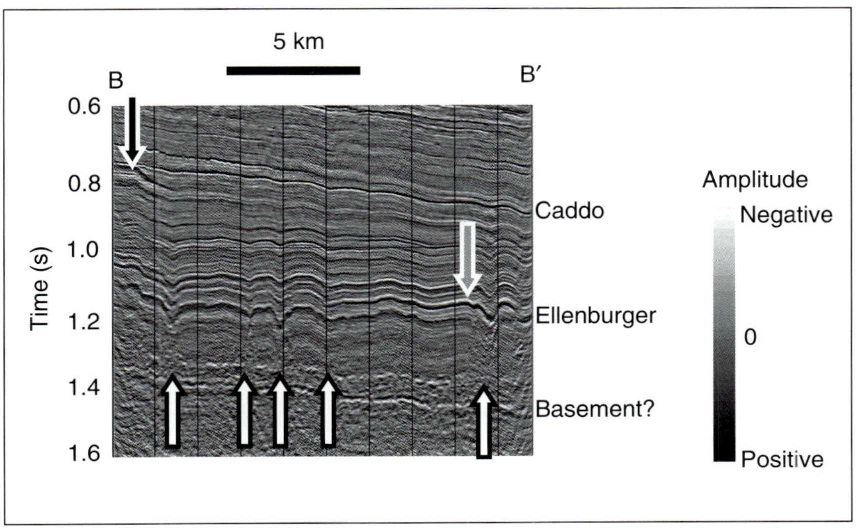

Figure 3.57. Line BB′ through the seismic data cube used in the generation of the coherence and curvature volumes shown in Figures 3.60–3.64. White arrows indicate collapse features that appear to be basement controlled. Black arrow indicates a ridge that can be seen in Figure 3.67b. Gray arrow indicates a dome that can be seen in Figure 3.67c.

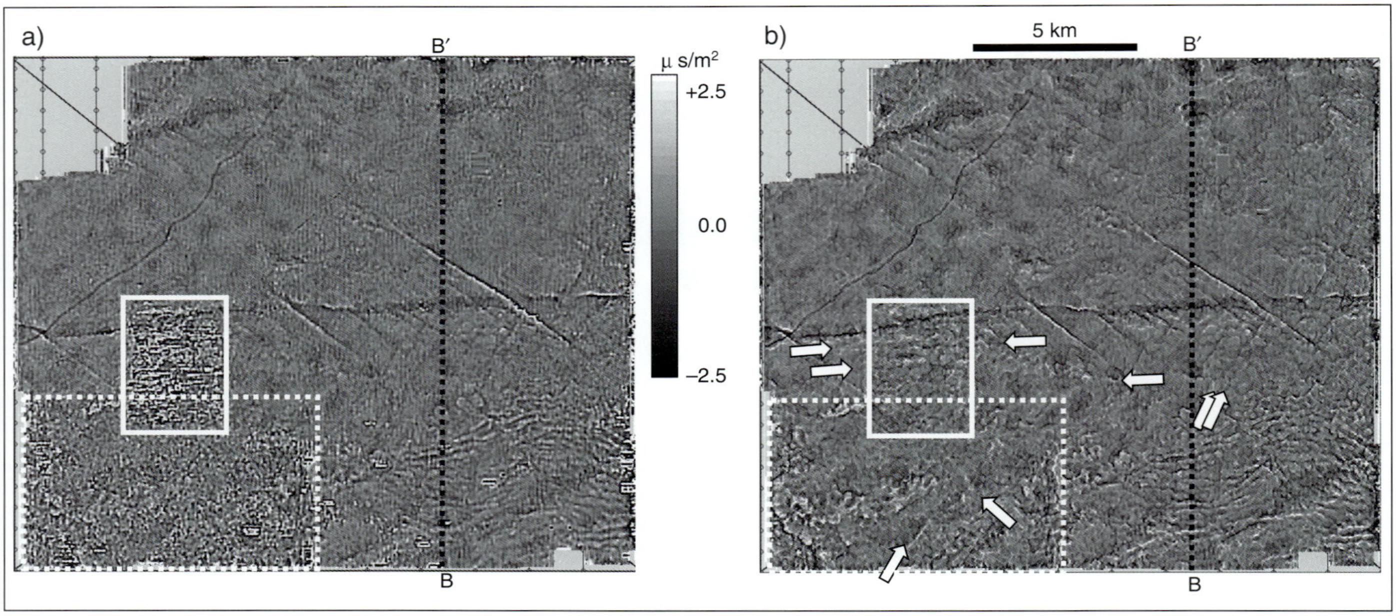

Figure 3.58. Mean curvature calculated (a) from the dip of the picked Caddo horizon shown in Figure 3.14 and (b) from the volumetric estimate of dip shown in Figure 3.15 and sliced along the Caddo horizon. Note the poor quality of the mean curvature estimates derived from the picked horizon in the region of the 1995 survey (indicated by the solid rectangle) and the 1997 survey (indicated by the dotted rectangle). Arrows indicate lineaments that are not seen clearly in (a). Data are courtesy of Devon Energy.

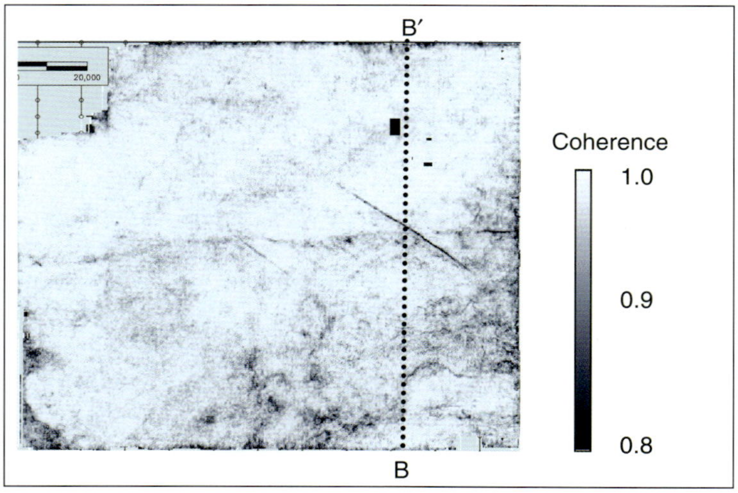

Figure 3.59. A horizon slice along the Caddo horizon, through the coherence volume. This image is relatively featureless when compared with the previous figure, which shows the mean curvature. The reason for this difference is geologic rather than algorithmic — the Caddo horizon is highly deformed but not broken, such that it is represented as an undulating, coherent reflection.

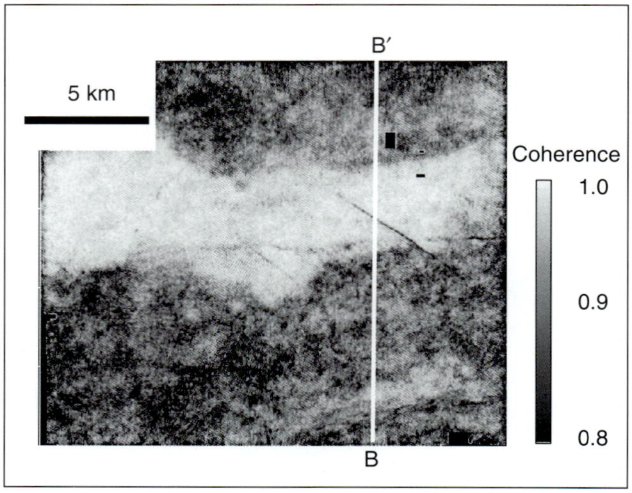

Figure 3.60. A principal-component coherence time slice at $t = 0.800$ s (approximately the Caddo/Atoka level), through a survey from the Fort Worth Basin, Texas, U.S.A. Data are courtesy of Devon Energy.

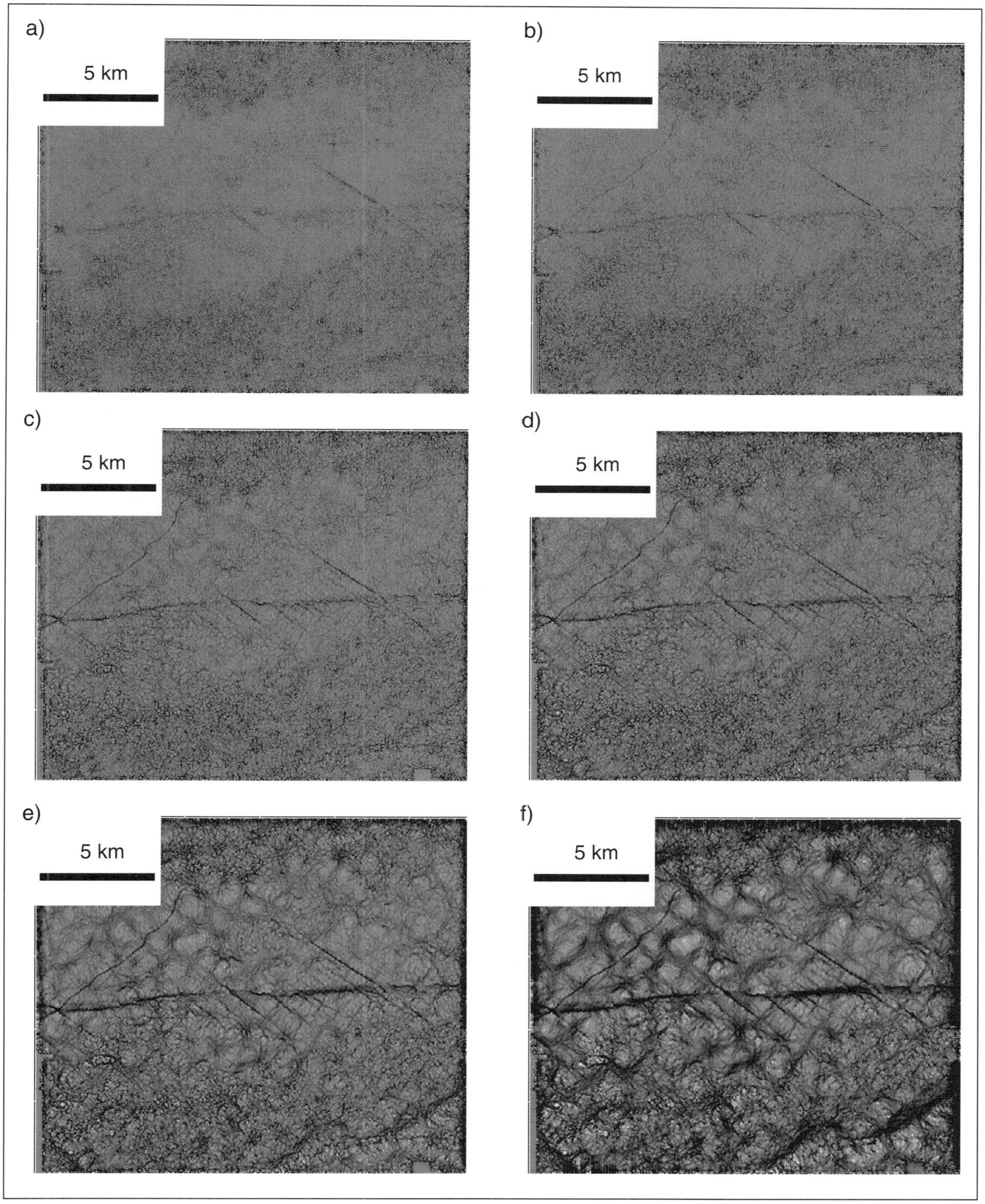

Figure 3.61. A time slice at t = 0.800 s corresponding to Figure 3.60 through the most negative curvature, k_{neg}, computed using the spatial derivative operator shown in Figure 3.56b with α = (a) 2.00, (b) 1.50, (c) 1.00, (d) 0.75, (e) 0.50, and (f) 0.25. Courtesy of Saleh al-Dossary.

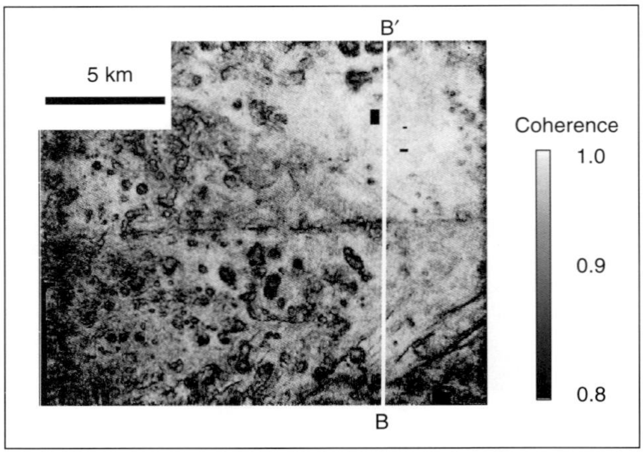

Figure 3.62. A principal-component coherence time slice at $t = 1.200$ s (approximately the Ellenburger level shown in Figure 3.57) through the same data volume shown in Figure 3.60.

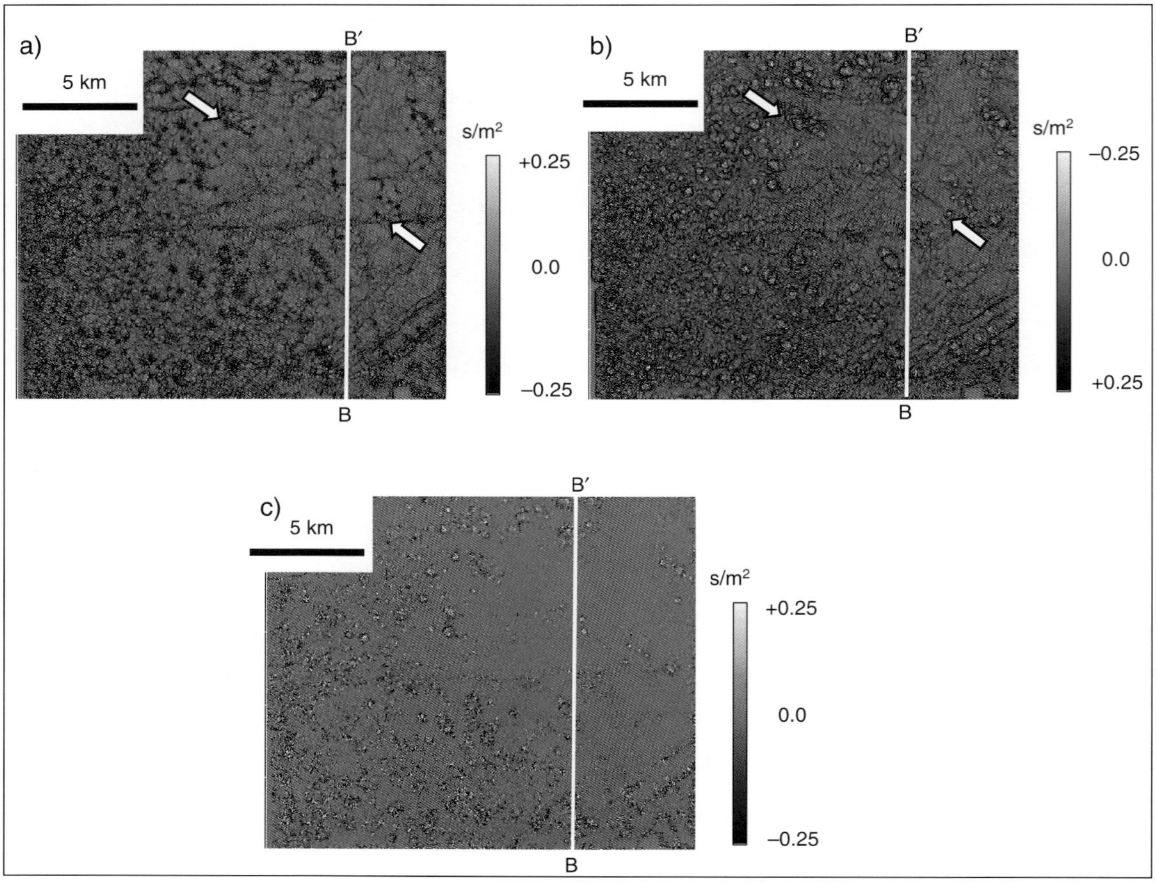

Figure 3.63. Short-wavelength ($\alpha = 1.00$) curvature values at $t = 1.200$ s, corresponding to the previous figure, of (a) most negative curvature k_{neg}, (b) most positive curvature, k_{pos}, and (c) Gaussian curvature, k_{Gauss}. Arrows indicate an elongate collapse feature that appears to be controlled structurally. Courtesy of Saleh al-Dossary.

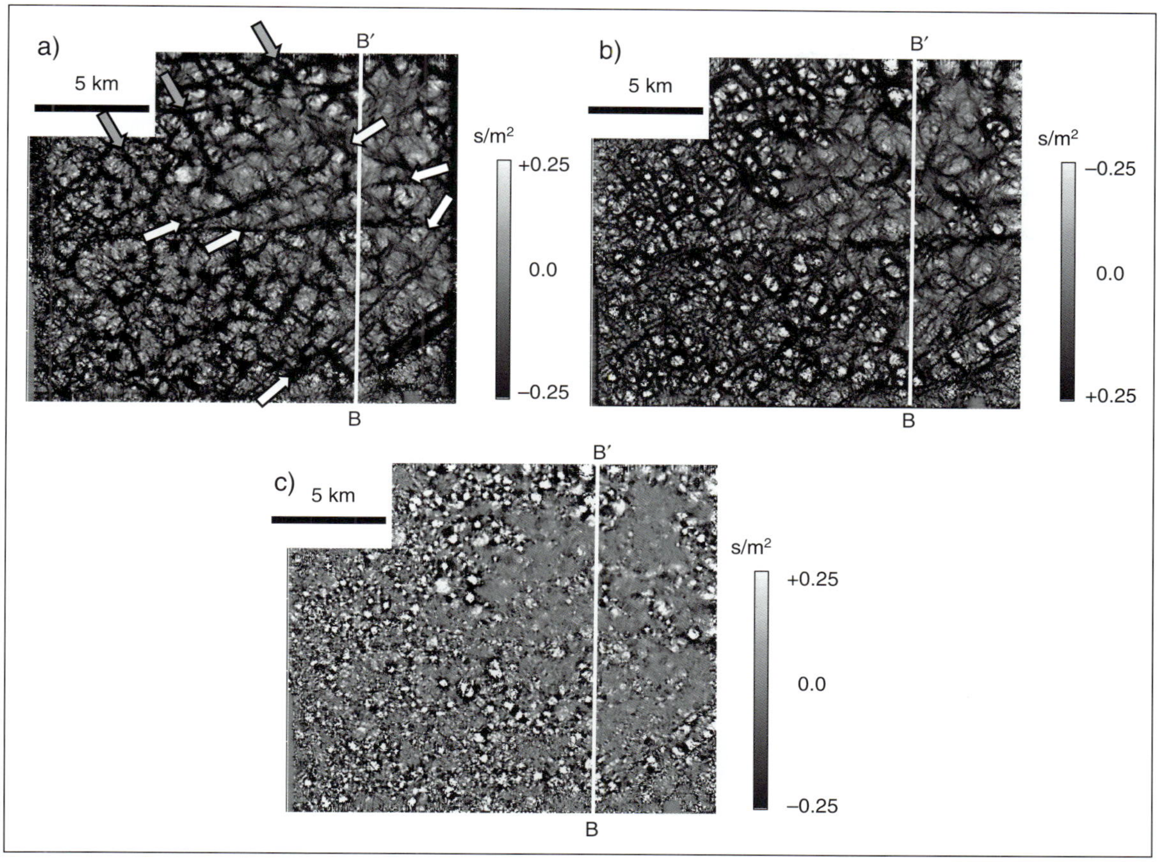

Figure 3.64. Long-wavelength (α = 0.25) curvature values at t = 1.200 s, corresponding to Figure 3.62, of (a) most negative curvature k_{neg}, (b) most positive curvature, k_{pos}, and (c) Gaussian curvature, k_{Gauss}. By examining Figure 3.57, we see that the positive values of k_{neg} in (a) correspond to domes, whereas the negative values of k_{pos} in (b) correspond to bowls. White arrows indicate throughgoing northeast-southwest lineaments that appear to offset the northwest-southeast lineaments indicated by gray arrows. These lineaments are much clearer here than in the previous figure. Courtesy of Saleh al-Dossary.

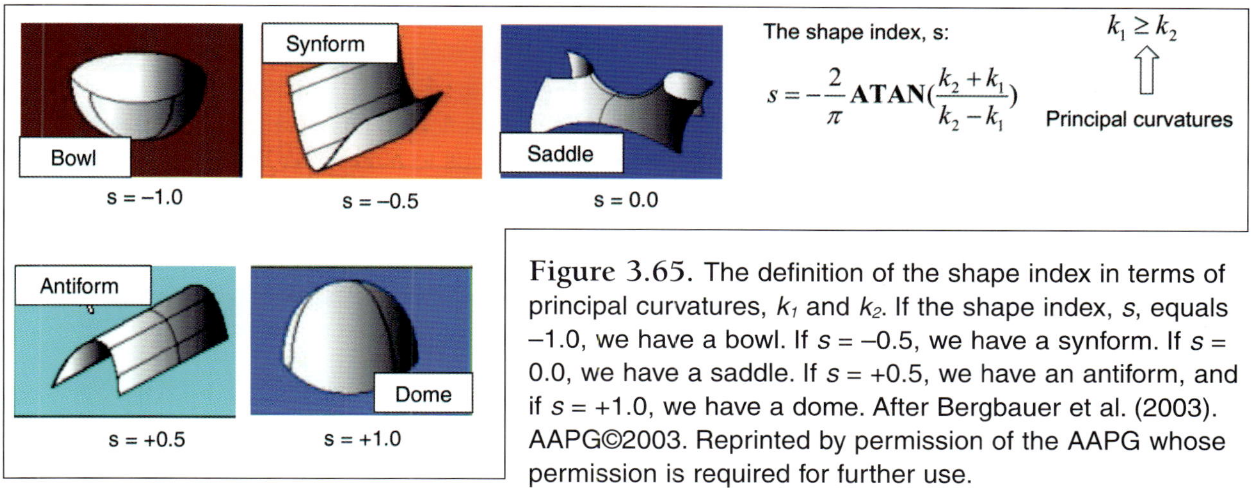

$$s = -\frac{2}{\pi}\mathbf{ATAN}\left(\frac{k_2 + k_1}{k_2 - k_1}\right) \qquad k_1 \geq k_2$$

Figure 3.65. The definition of the shape index in terms of principal curvatures, k_1 and k_2. If the shape index, s, equals −1.0, we have a bowl. If s = −0.5, we have a synform. If s = 0.0, we have a saddle. If s = +0.5, we have an antiform, and if s = +1.0, we have a dome. After Bergbauer et al. (2003). AAPG©2003. Reprinted by permission of the AAPG whose permission is required for further use.

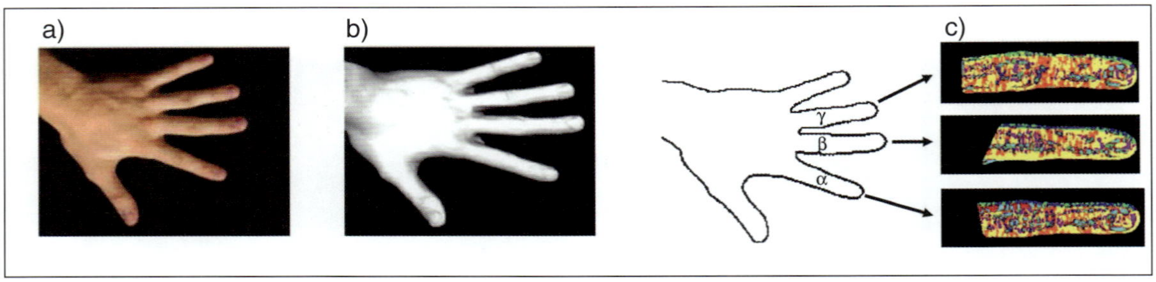

Figure 3.66. (a) An intensity image and (b) a range (what geoscientists would call a "topographic") image of a human hand. (c) Shape indices of the middle three digits, using the equation shown in Figure 3.65. Look at your own fingers and try to match the details of the ridges and valleys around your knuckles to those in the image. After Woodward and Flynn (2004).

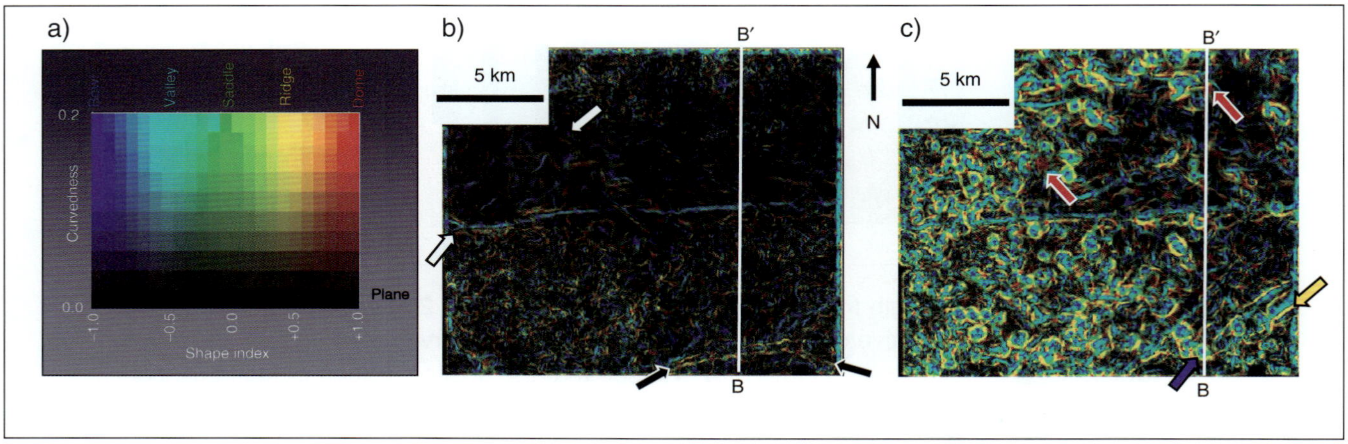

Figure 3.67. (a) A 2D color table that displays the shape index versus curvedness. Planar features will have a curvedness near zero and will be displayed in black. Highly curved features will be plotted as a pure color, thereby allowing us to differentiate between reflector shapes. We plot long-wavelength (α = 0.25) curvedness, *c*, defined as $c = \sqrt{k_1^2 + k_2^2}$, where k_1 and k_2 are the minimum and maximum (or principal) curvatures, and the shape index, *s*, is as defined in Figure 3.65. Using this 2D color table, at (b) t = 0.800 s, and at (c) t = 1.200 s, Compare (b) with Figures 3.60 and 3.61. Note that the shallow section is less contorted and yields a darker image. The cyan valley indicated by white arrows in (b) corresponds to a channel. The yellow ridge indicated by the black arrows corresponds to the ridge seen in Figure 3.57. Other cyan valley and yellow ridge lineaments correspond to deformation about faults and joints. In contrast, (c) is dominated by collapse features in the Ellenburger formation. Collapse features appear as blue circles — such as the one indicated by the blue arrow — ringed by yellow ridges and green saddles. Red arrows indicate two domal features, one of which corresponds to the dome indicated by the gray arrow seen in Figure 3.57. Courtesy of Saleh al-Dossary.

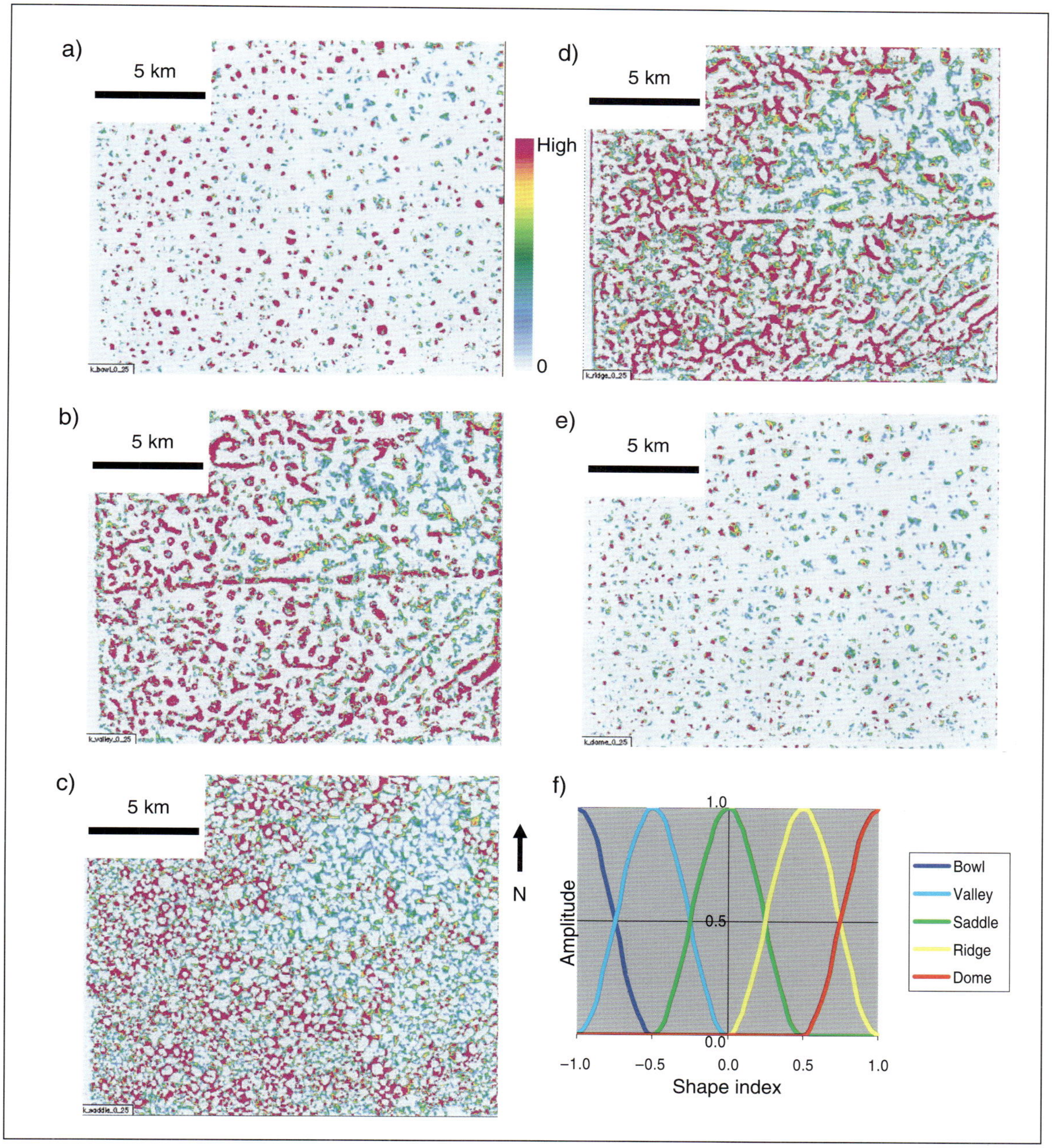

Figure 3.68. Long-wavelength (α = 0.25) curvature anomalies at t = 1.200 s, decomposed according to shape. First we calculate the curvedness defined as $c=\sqrt{k_1^2+k_2^2}$, where k_1 and k_2 are the minimum and maximum (or principal) curvatures. Next we multiply the curvedness by a filter designed to pass the curvedness that has a specific shape: (a) a bowl, (b) a valley, (c) a saddle, (d) a ridge, and (e) a dome. Corresponding filters are shown in (f). In principle, all the information contained in these five images also is contained in part (c) of the previous figure. However, because the sum of a bowl and valley image will include all the features that may correspond to a thicker accommodation space, we expect better performance from statistical transforms that are based on a weighted linear sum of these shape components. Courtesy of Saleh al-Dossary.

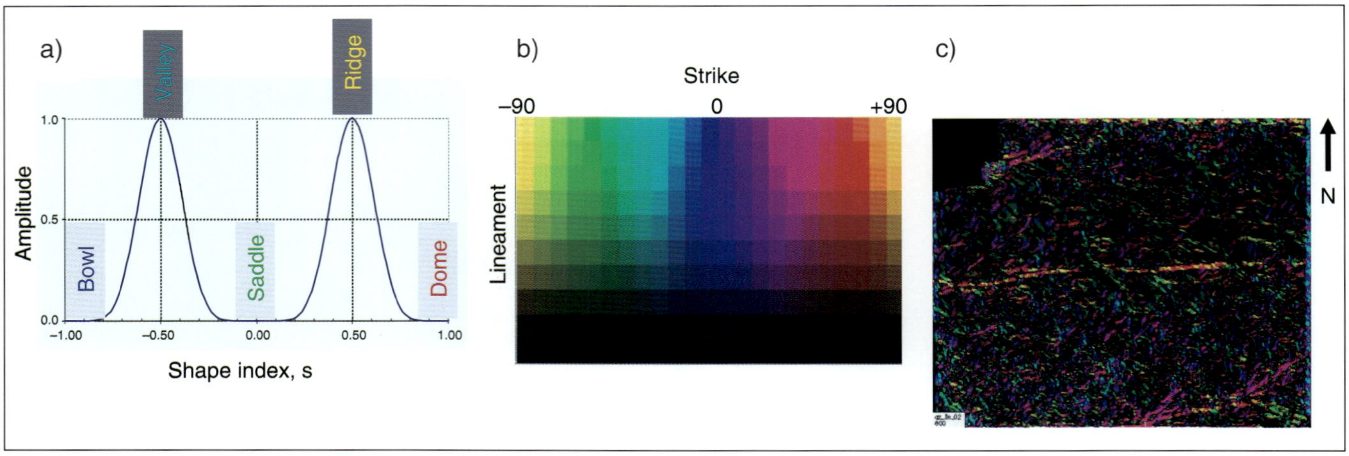

Figure 3.69. (a) Generation of a "lineament" attribute, obtained by applying a filter that passes only those values of curvedness that correspond to ridges and valleys. (b) A 2D color bar to display the lineament attribute against curvedness. (c) A time slice, at $t = 1.200$ s, through the lineament attribute volume. Courtesy of Saleh al-Dossary.

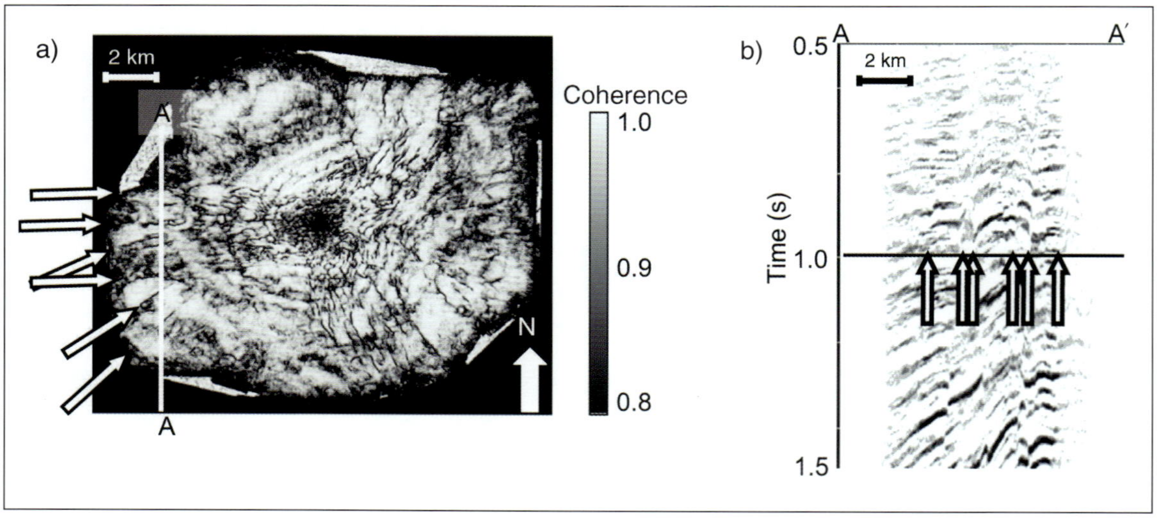

Figure 3.70. (a) A time slice at 1.000 s, through a coherence volume, and (b) a vertical slice through the seismic data, both from a survey acquired over Vinton Dome, Louisiana, U.S.A. The arrows correspond to the same faults shown in Figure 3.71. Note that some of these faults appear to be discrete discontinuities. However, others appear to be folds — as a result either of drag along the fault or of inaccurate migration. Faults that have a folded appearance show up more clearly in the curvature image than in the coherence volumes. Courtesy of Saleh al-Dossary.

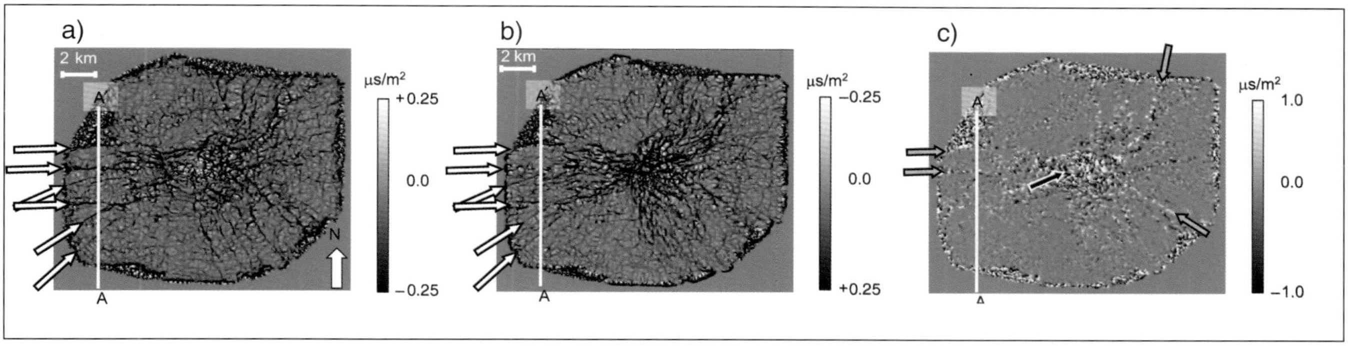

Figure 3.71. Time slices, at t = 1.000 s, through the long-wavelength (α = 0.25) estimates of (a) most negative curvature, k_{neg}, (b) most postive curvature, k_{pos}, and (c) reflector rotation, r_z, corresponding to the same time slice shown in Figure 3.70a. White arrows depict faults that are not seen as clearly on the coherence time slice shown in Figure 3.70b. Gray arrows correspond to faults where reflectors have significant rotation about the fault plane. Such rotations are not represented by the quadratic surface used in curvature calculations. The black arrow indicates noisy, low-fold areas of the data. Courtesy of Saleh al-Dossary.

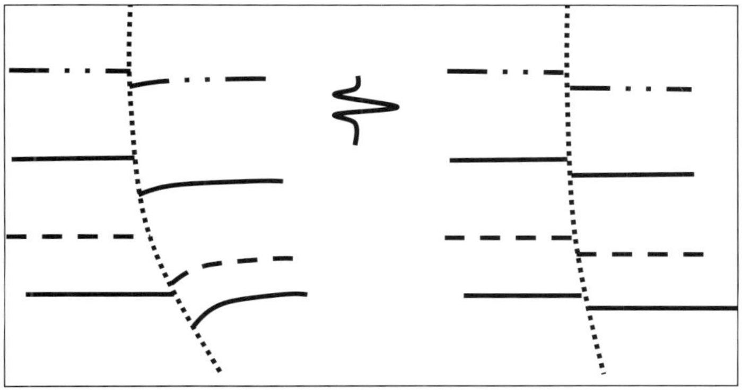

Figure 3.72. Attribute sensitivity to faults, where the amount of fault throw is measured in terms of a seismic wavelet. (a) An idealized growth fault, which will be seen on curvature and coherence-attribute volumes. (b) An idealized strike-slip fault, which will be seen on a coherence volume but not on a curvature-attribute volume.

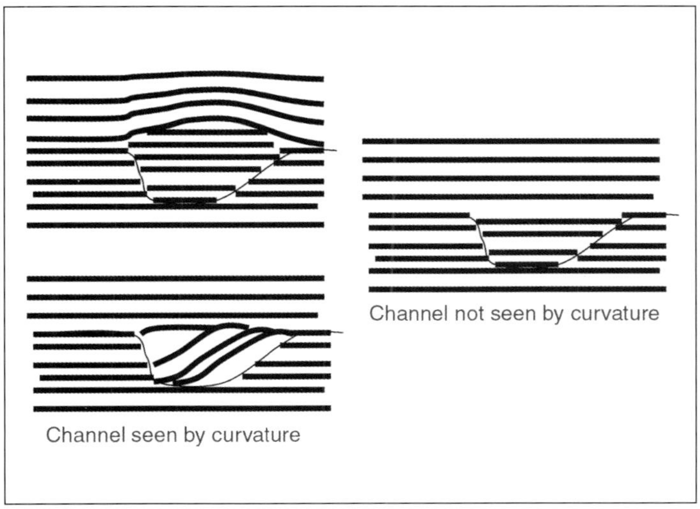

Figure 3.73. The sensitivity of curvature attributes to channels. Channels that have (a) differential compaction or (b) aggradational features generally will be seen by curvature. However, because volumetric curvature calculations are performed on time slices rather than on picked horizons, channels that have little differential compaction or internal dip, such as those in (c), experience no lateral change in dip and thus will not be seen by curvature.

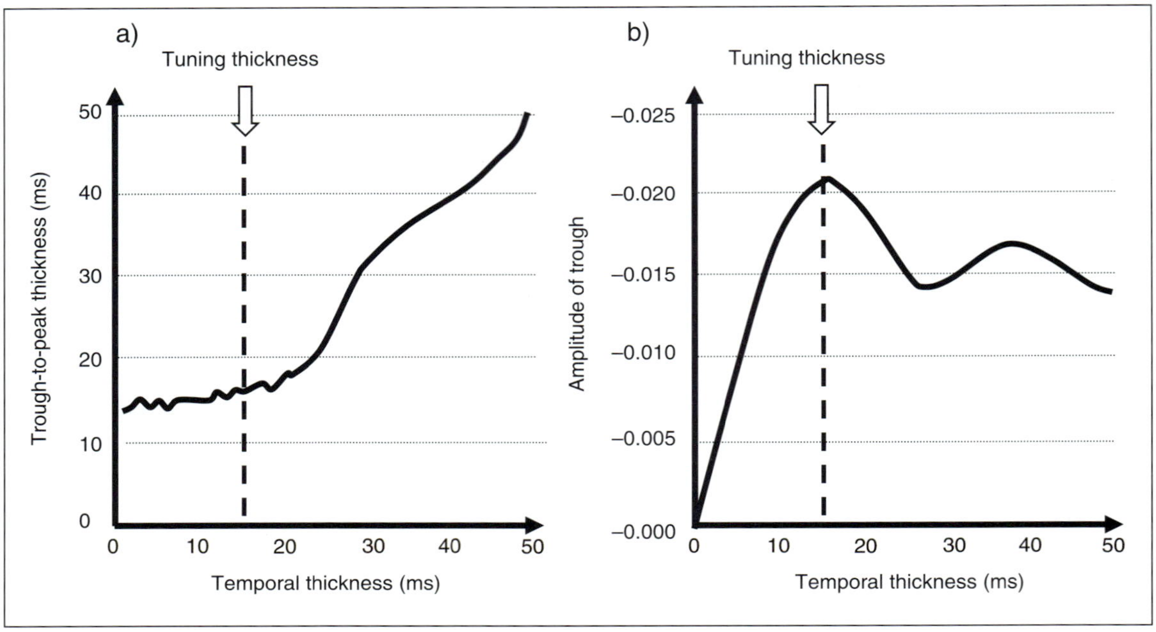

Figure 3.74. Sensitivity of (a) reflector trough-to-peak thickness and (b) trough amplitude, for the simple wedge model shown in Figure 2.1. Note the nearly linear variation of trough amplitude with thickness for small values of thickness (<10 ms). After Partyka (2001), after Widess (1973).

Figure 3.75. (a) An image of the 2006 DISC lecturer. Pixel values, u, range from 0 to 255. (b) The result of applying a Sobel (or edge) filter of the form $[(du/dx)^2 + (du/dy)^2]^{1/2}$ to the pixel values in (a). Figure is courtesy of Saleh al-Dossary.

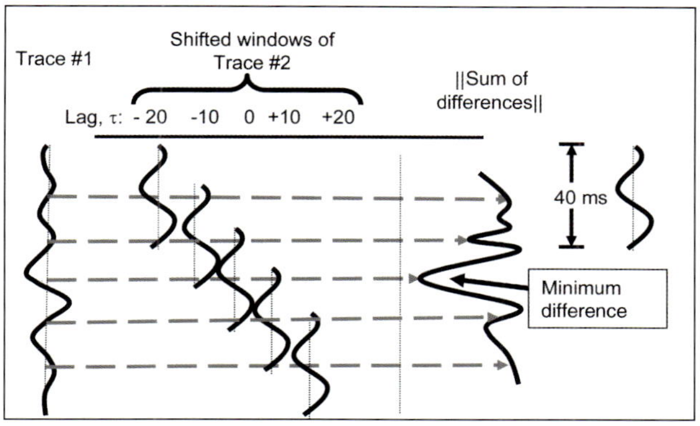

Figure 3.76. Schematic of the original Chevron "edge" algorithm (Luo et al., 1995). As we did with the crosscorrelation algorithm described in Figures 3.26 and 3.27, here we compare two traces in the inline and crossline direction. However, instead of shifting and crosscorrelating the traces to be compared, we shift and compute their differences. The inline and crossline time shifts that produce the minimum squared difference will be combined to generate an edge attribute, *e*, defined by

$$e(t,\tau_x,\tau_y)=\frac{\left[\sum_{k=-K}^{+K}\left\{[u_0(t+k\Delta t)-u_1(t+k\Delta t-\tau_x)]^2+[u_0(t+k\Delta t)-u_2(t+k\Delta t-\tau_y)]^2\right\}\right]^{1/2}}{\left[\sum_{k=-K}^{+K}\left\{2[u_0(t+k\Delta t)]^2+[u_1(t+k\Delta t-\tau_x)]^2+[u_2(t+k\Delta t-\tau_y)]^2\right\}\right]^{1/2}}.$$

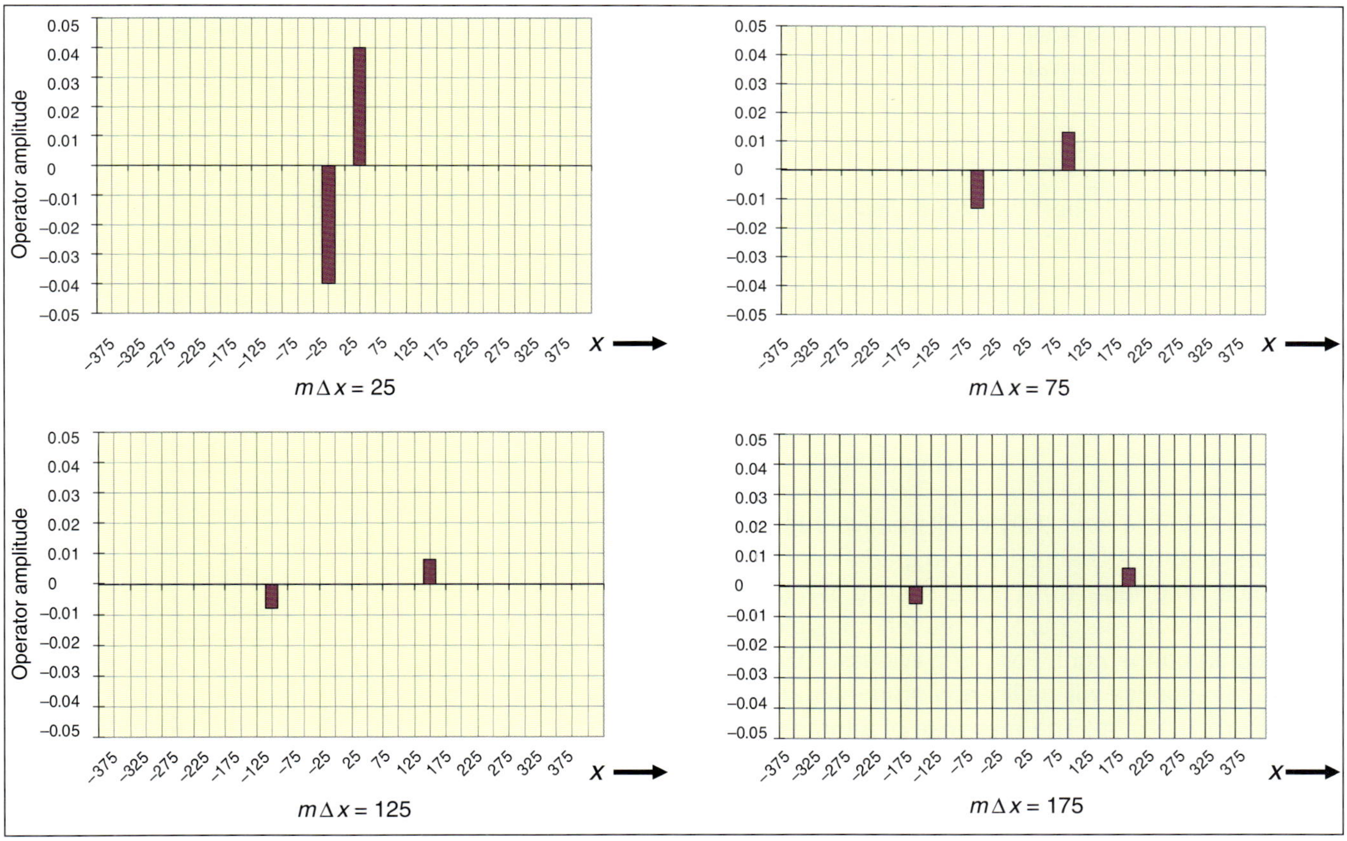

Figure 3.77. Alternative finite-difference approximations to the first derivative defined as $\frac{\partial u}{\partial x}\approx\frac{u(x+m\Delta x)-u(x-m\Delta x)}{2m\Delta x}$, where Δx is the seismic bin size and m is the number of bins away from the center analysis point.

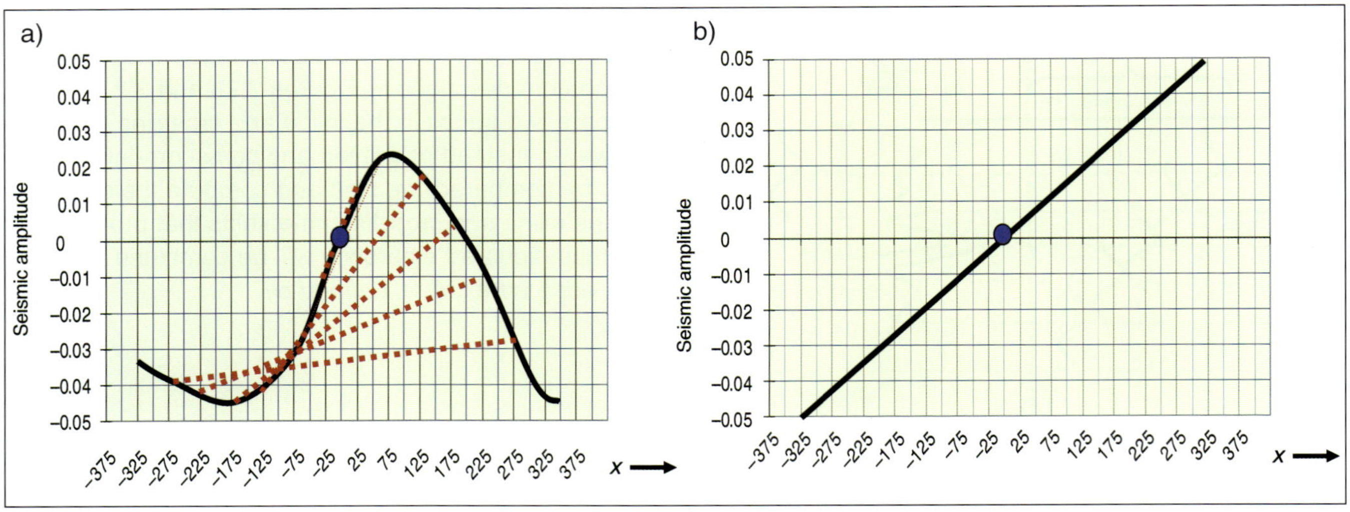

Figure 3.78. A graphical display of the approximations made to the first derivative (or slope) using the operators shown in the previous figure for (a) an amplitude variation that has a moderate wavelength, on the order of 10 seismic bins, and (b) an amplitude variation that has a very long wavelength compared with the seismic bin size Δx. For very long wavelengths, each of the alternative finite-difference approximations provides the same estimate of the first derivative.

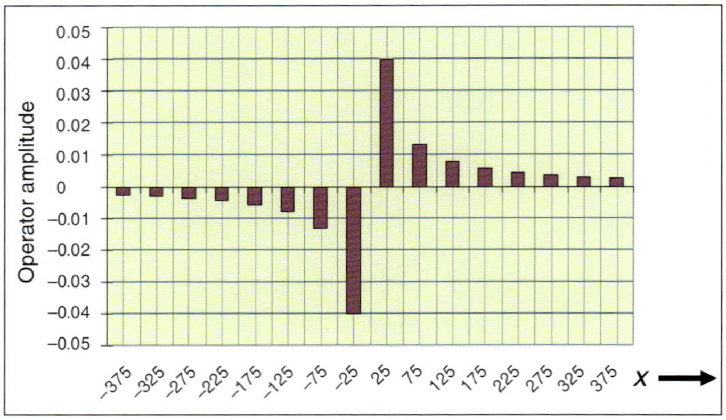

Figure 3.79. The Hilbert transform displayed as a linear combination of the difference operators displayed in Figure 3.77.

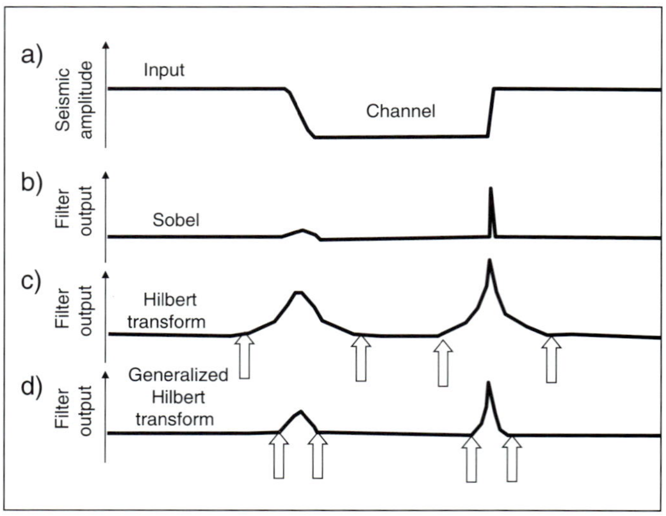

Figure 3.80. The sensitivity of alternative edge detectors to abrupt changes and to gradational lateral changes in amplitude. (a) A cross section of a thin channel, with the inner bank on the left and outer bank (or cutbank) on the right. Following Widess (1974), the amplitude response will be linearly proportional to the channel thickness. (b) The response of a three-point Sobel (or first-derivative) filter. The steep cutbank shows up well, but the gradational inner bank shows up poorly. (c) The response of a 61-point Hilbert transform filter. Both sides of the channel are imaged equally well, but the steep cutbank is overly smeared. (d) The response of a 21-point generalized $r = 2$ Hilbert transform filter. Both sides of the channel are imaged well. After Luo et al. (2003).

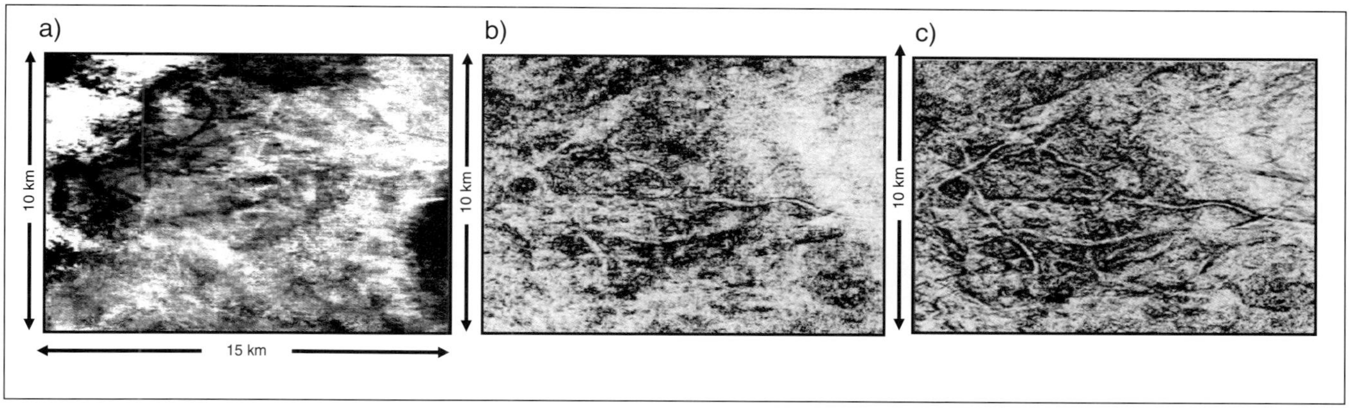

Figure 3.81. (a) A time slice through a seismic data cube, (b) the corresponding eigenstructure coherence volume, and (c) the amplitude-variability volume generated using the generalized $r = 2$ norm Hilbert transform, from a survey acquired on the Arabian Peninsula. Amplitude anomalies associated with channel fill and changes in thickness are clearly present but are somewhat diffuse and difficult to delineate on the seismic-amplitude time slice shown in (a). The eigenstructure estimate of coherence in (b) is insensitive to these changes in amplitude. In contrast, the generalized Hilbert transform in (c) is a direct measure of amplitude variability and delineates the channels clearly. After Luo et al. (2003).

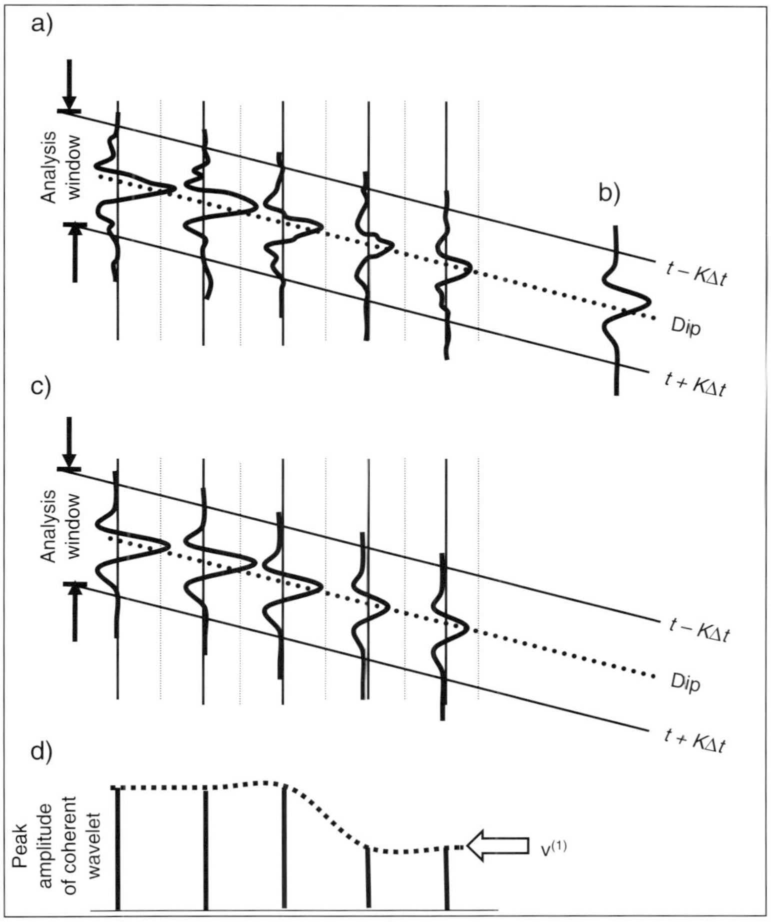

Figure 3.82. A schematic diagram summarizing the steps in the eigenstructure estimation of coherence shown in Figure 3.35. The steps are to (a) calculate the energy of the input traces within an analysis window, (b) calculate the seismic waveform that best approximates the waveform of each input trace, and (c) replace each trace with a scaled version of (b) that best fits the input trace. The amplitudes of the five wavelets in (c) define the components of the five-element-long principal-component eigenvector, $\mathbf{v}^{(1)}$. To calculate energy-weighted coherent-amplitude gradients, we take the derivative of the curve shown by the dotted line in (d) and weight it by the sum of the coherent energy within the analysis window shown in (c).

Figure 3.83. Interpretation of the eigenvector, $\mathbf{v}^1$, corresponding to a nine-trace analysis window, as the representation of a map of coherent amplitude, *v(x,y)*. From this map, we can calculate gradients, $\frac{\partial v}{\partial x}(x,y)$, and $\frac{\partial v}{\partial y}(x,y)$, which will indicate lateral changes in the coherent amplitude. These gradients are particularly useful in mapping lateral changes in layer thickness when the layer is less than one-quarter-wavelength thick. For such thin layers, the amplitude variation is proportional to thickness, and the waveform remains nearly constant.

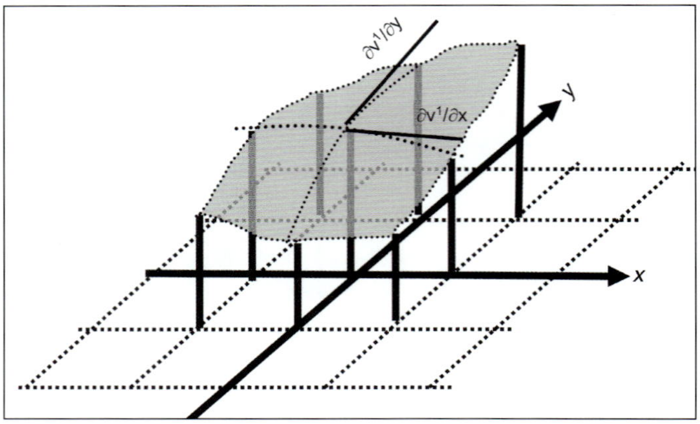

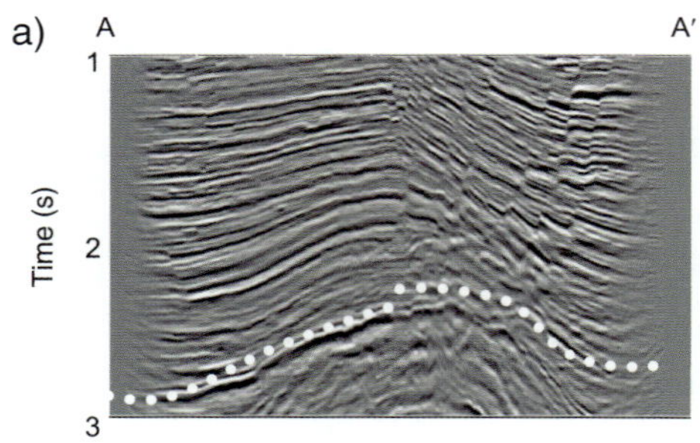

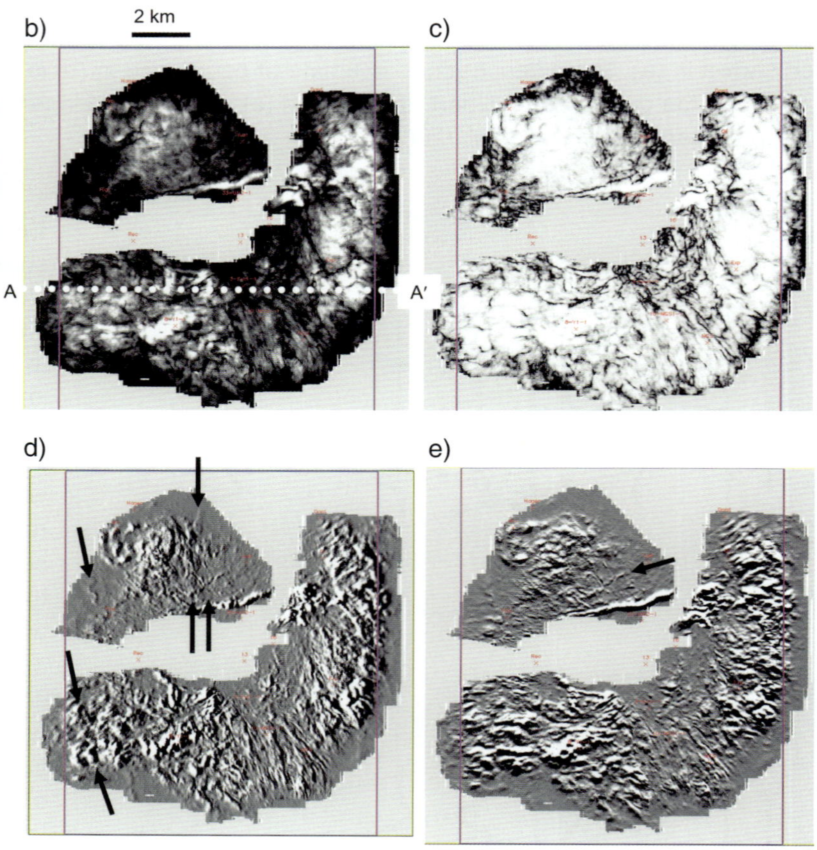

Figure 3.84. (a) A vertical slice through a seismic amplitude volume along a line AA′. The dotted line indicates the Hackberry horizon. (b)-(e). Horizon slices along the Hackberry horizon, through (b) the seismic amplitude, (c) the coherence, (d) the east-west energy-weighted coherent-amplitude gradient, and (e) the north-south energy-weighted coherent-amplitude gradient. Faults have a wormy appearance because the data have been migrated poorly, resulting in interfingering of reflectors. In contrast, the components of the energy-weighted coherent-amplitude gradient away from the faults give accurate images of channels (indicated by arrows) and other stratigraphic features, including areas where the coherence is very high (white). Seismic data are courtesy of OPEX.

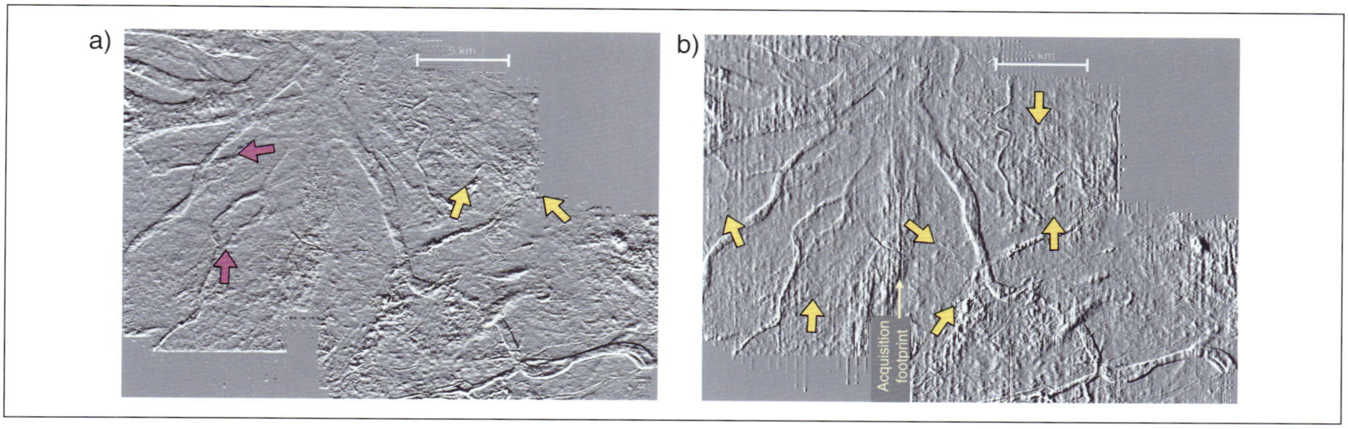

Figure 3.85. Horizon slices corresponding to Figure 3.45b through the (a) north-south and (b) east-west components of the energy-weighted coherent-amplitude gradient described in Figure 3.83. The vertical analysis window was ± 32 ms. Yellow arrows indicate narrow channels not readily seen on the coherence images shown in Figure 3.45b. Statistically, narrow channels also are thin. If they are thinner than the tuning thickness, they will be expressed as a lateral change in amplitude only and not in waveform. Magenta arrows indicate a wider channel that is not seen in the coherence horizon slice in Figure 3.45b. We interpret these channels to have a longer-wavelength gradational change in amplitude, such as that shown in Figure 3.80, so they are not well illuminated by our nine-trace coherence algorithm. After Marfurt and Kirlin (2000).

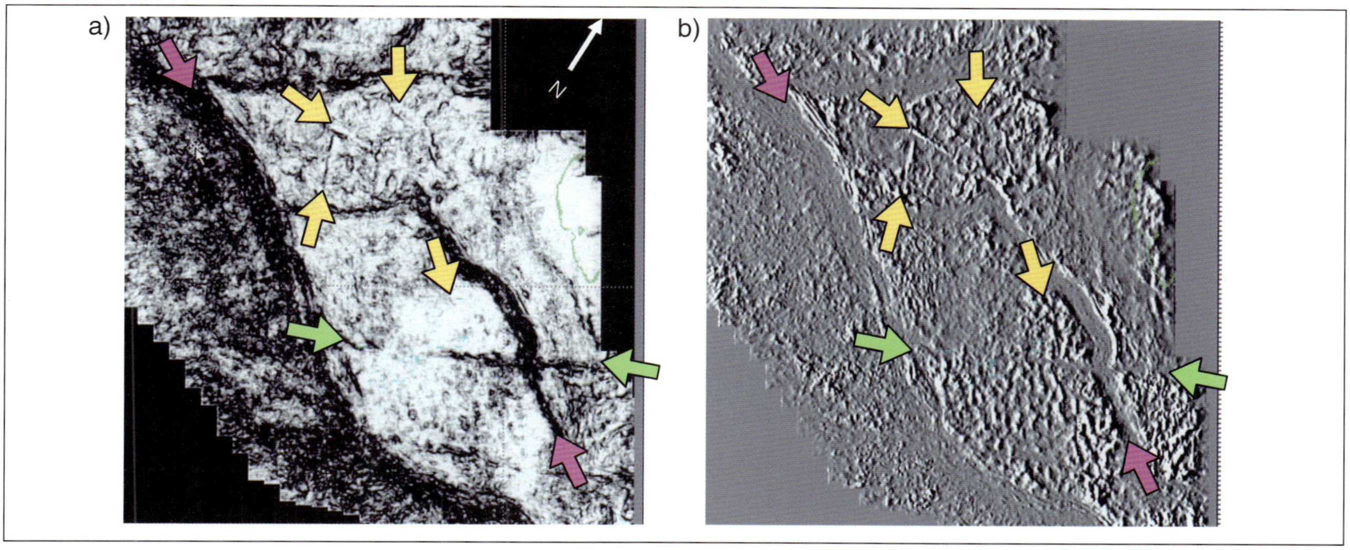

Figure 3.86. A horizon slice along the Devonian Thirtyone formation, from a survey in the Central Basin Platform, Texas, U.S.A., of (a) an eigenstructure coherence and (b) the east-west component of the energy-weighted coherent-amplitude gradient. Note that the reverse faults indicated by yellow arrows show up as low-coherence black zones in the coherence image. Because they are almost totally incoherent, they appear as blank gray zones in the energy-weighted coherent-amplitude-gradient image. We can see channels in the northern one-third of both images. Although the southern one-third of the coherence image is relatively featureless, we still see channel-like features in the southern one-third of the amplitude-gradient image. These "channels" cross an east-west fault indicated by green arrows on the horizon slice, further suggesting that they are sedimentary features. After Blumentritt et al. (2003).

Figure 3.87. Time slices at $t = 0.832$ s, through the (a) east-west and (b) north-south components of energy-weighted coherent-amplitude gradients, and (c) a coherence time slice, all through a survey at approximately Pennsylvanian level from the midcontinent United States. Although the channels can be seen on the coherence time slice, they can be traced farther using the amplitude-gradient images.

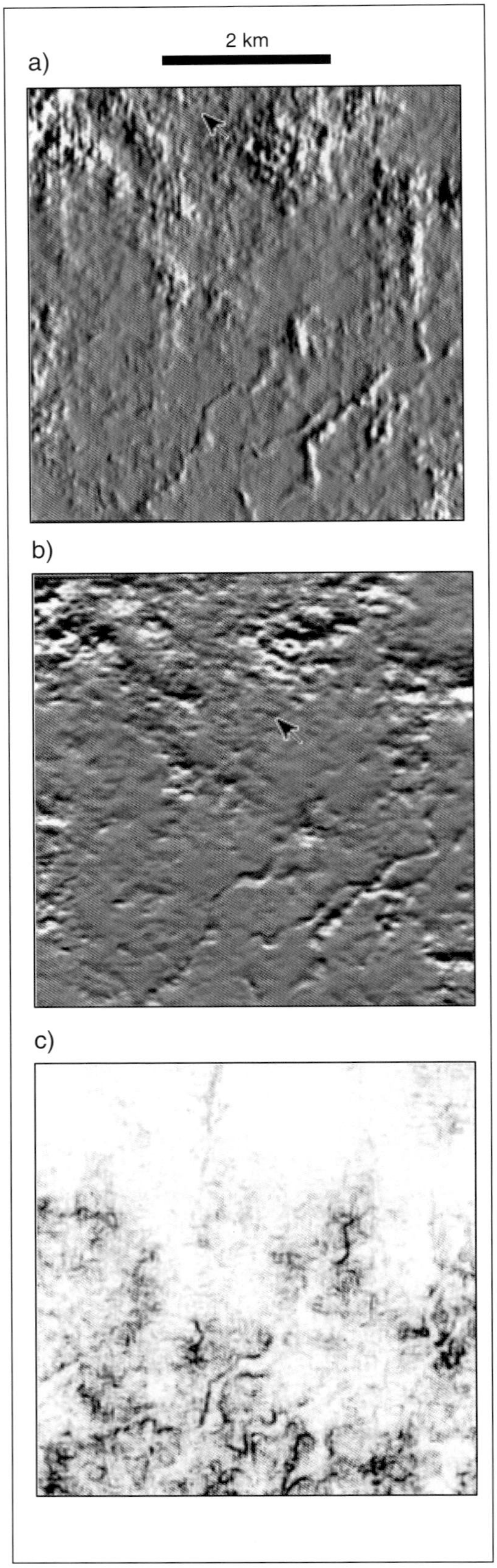

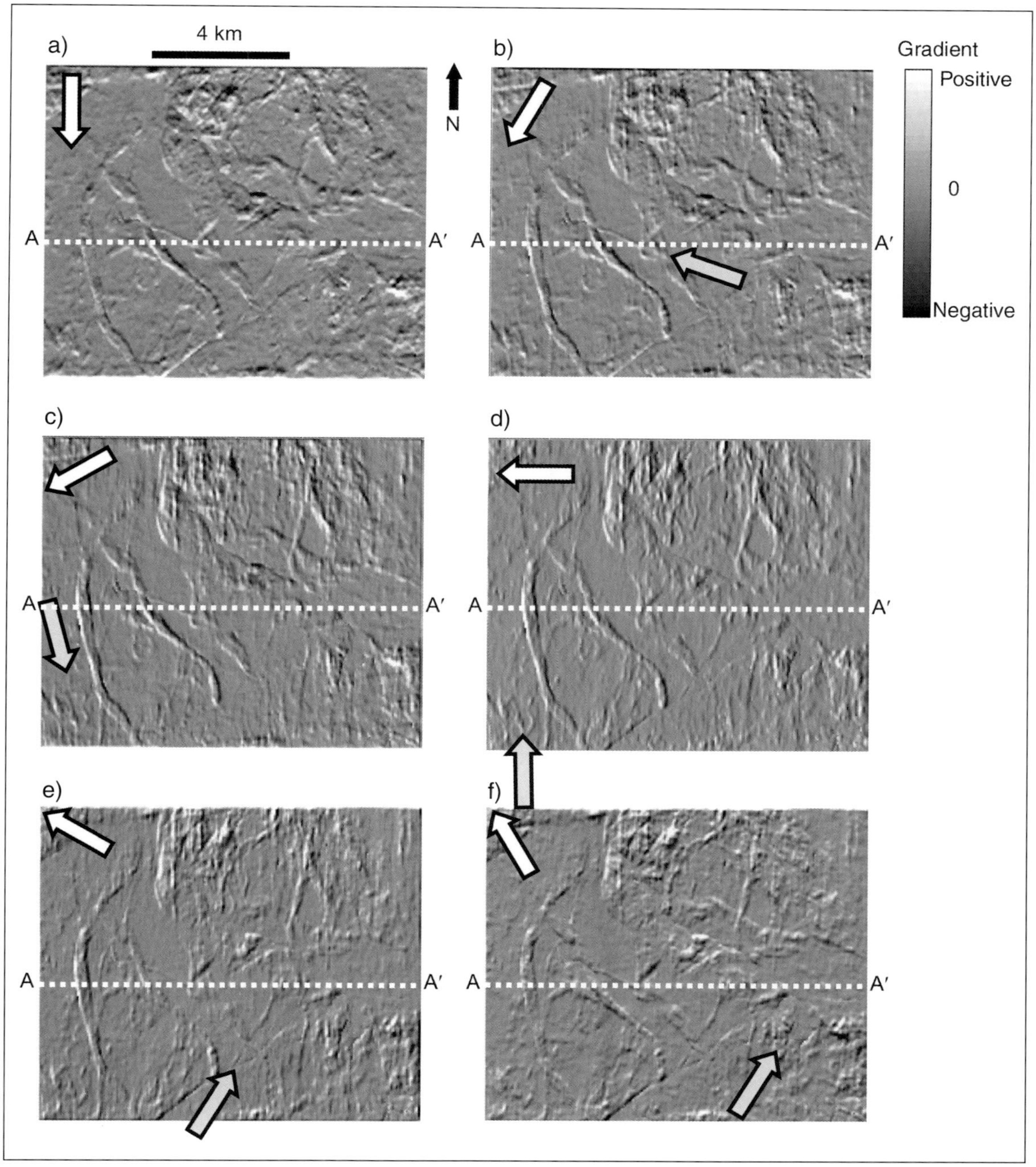

Figure 3.88. Energy-weighted coherent-amplitude gradients at (a) 0°, (b) 30°, (c) 60°, (d) 90°, (e) 120°, and (f) 150° from north (the direction indicated by white arrows), corresponding to the same time slices shown in Figures 4.38 and 4.39. The gray arrows indicate subtle features that show up best at the angle shown in each image. Such subtle features may be important for describing reservoir heterogeneity. Data are courtesy of Fairfield.

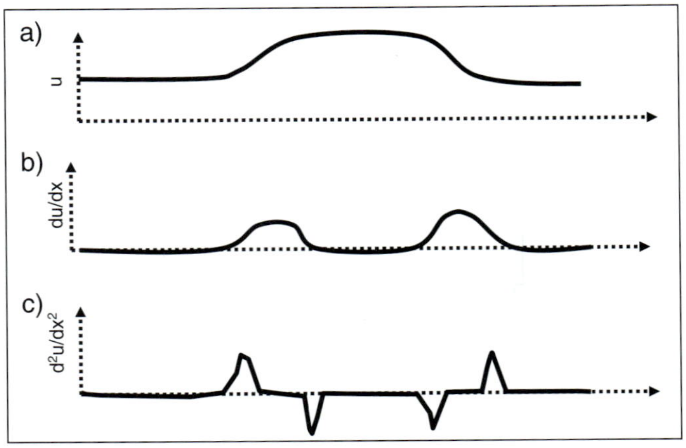

Figure 3.89. A schematic diagram of (a) an amplitude anomaly, (b) its first spatial derivative, and (c) its second spatial derivative. The extrema of (c) demarcate the limits of the anomaly.

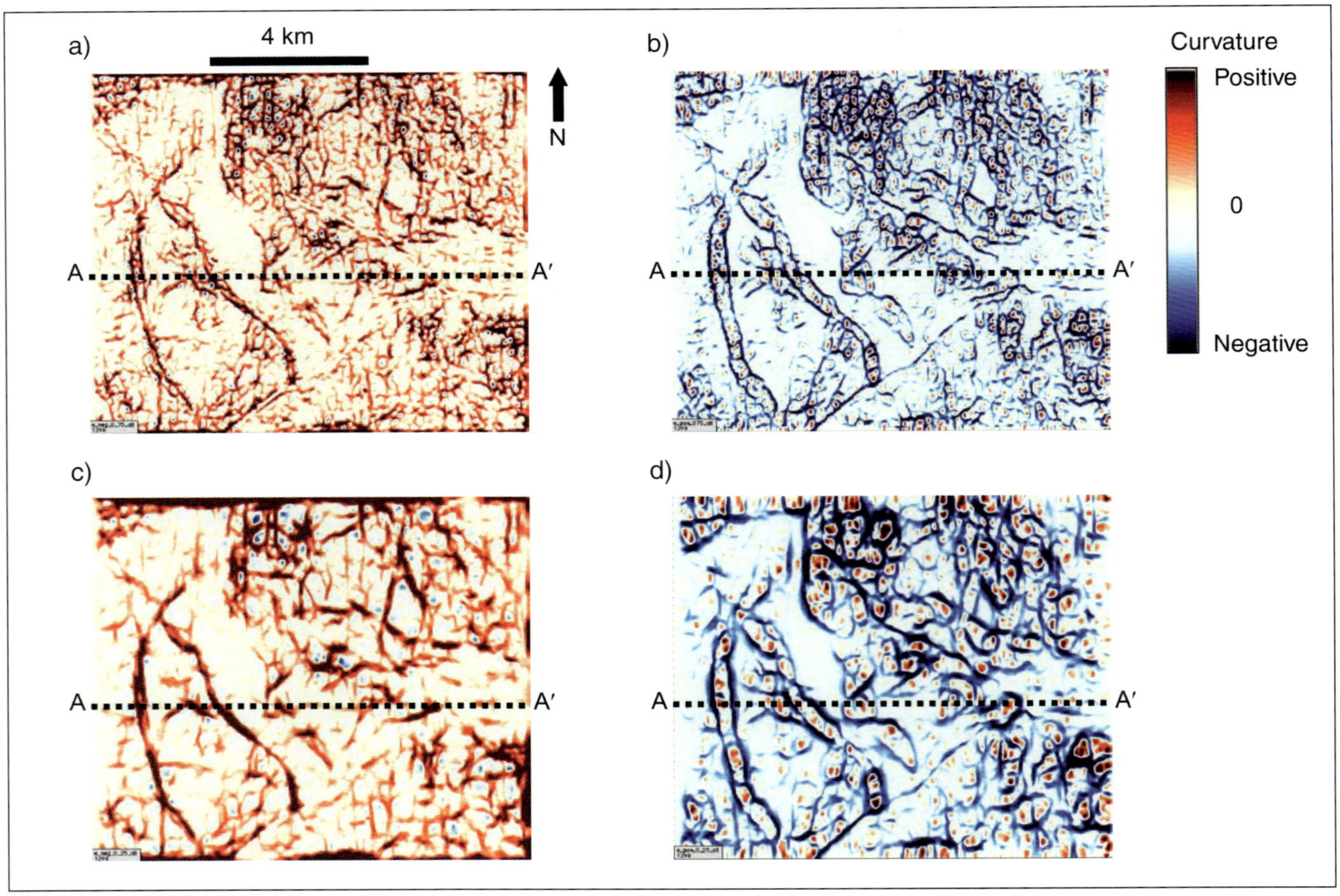

Figure 3.90. Time slices, corresponding to the time slice at t = 1.300 s shown in Figures 4.38, 4.39, and 3.88. Images (a)-(b) are of moderate wavelength (α = 0.75) — (a) most positive curvature and (b) most negative curvature of the coherent amplitude. Images (c) and (d) are longer-wavelength (α = 0.25) images, again, of most positive curvature and most negative curvature, respectively. Strong red lineaments in (a) and (c) correspond to maximum constructive interference and, in this image, indicate the thalweg of the two channels. Strong blue lineaments in (b) and (d) indicate strong destructive interference and correspond to the edges of the channel. The vertical analysis window was ± 20 ms. Data are courtesy of Fairfield.

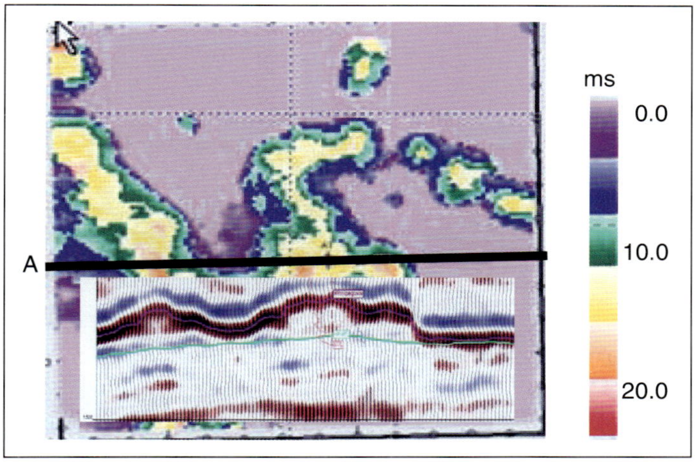

Figure 3.91. A time-thickness map and a vertical slice through a Winnepegosas reef from southeastern Saskatchewan, Canada. The thickness is calculated between the blue pick (the reef's top) and the green pick (the reef's base). Note that here, the maximum amplitude corresponds to the thinnest part of the reef complex. Data are courtesy of Talisman.

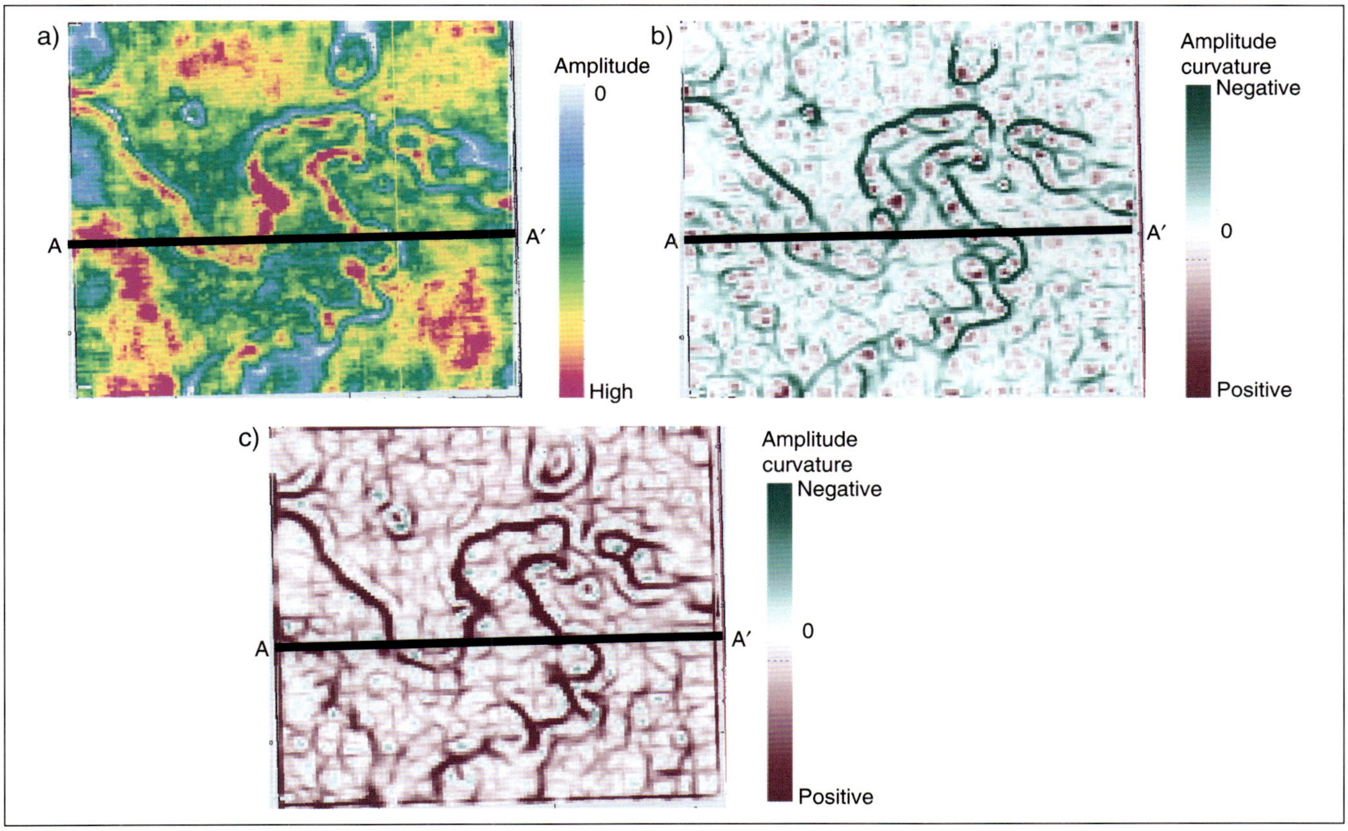

Figure 3.92. (a) An amplitude horizon slice, (b) a long-wavelength most-negative-amplitude curvature volume, and (c) a long-wavelength most-positive-amplitude curvature volume, along the top of the reef (blue pick) shown in the previous figure. Here, the strong negative values of the most negative curvature, seen as dark green, track the low-amplitude light-blue trends in the amplitude horizon slice. In contrast, the strong positive values of the most positive curvature, seen as dark red, track the high-amplitude red trends in the amplitude horizon slice. These latter anomalies delineate the edges of the reef. The rectilinear features seen in both (b) and (c) are the result of an acquisition footprint. Two hundred traces were used in the curvature calculations. Data are courtesy of Talisman.

Figure 3.93. (a) Computation of GLCMs for different reflection characteristics selected by polygons marked 1, 2, and 3. The size of the matrix used for computation is (a) 16 × 16, (b) 32 × 32, and (c) 64 × 64. Notice that with a matrix size of 16, fewer elements are seen populating the GLCM, compared with the 32-element GLCM. The highest density of elements is seen with a 64-element GLCM.

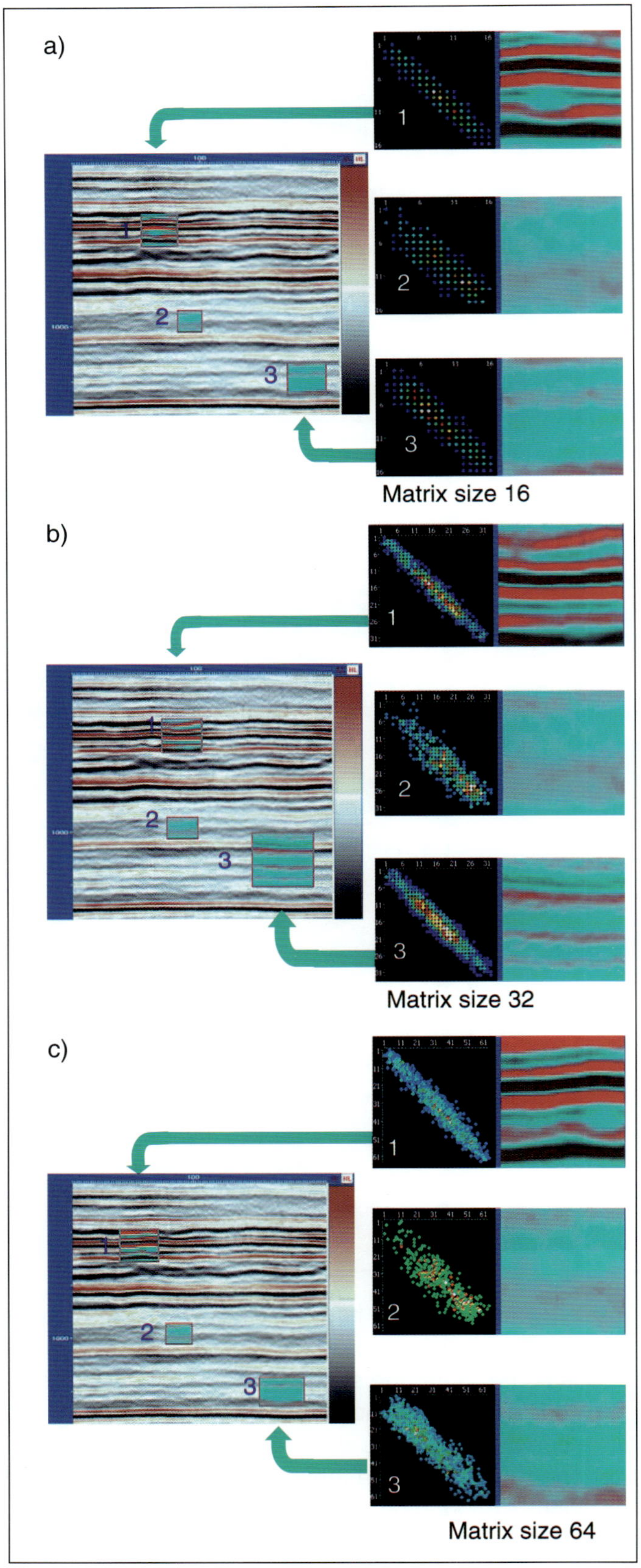

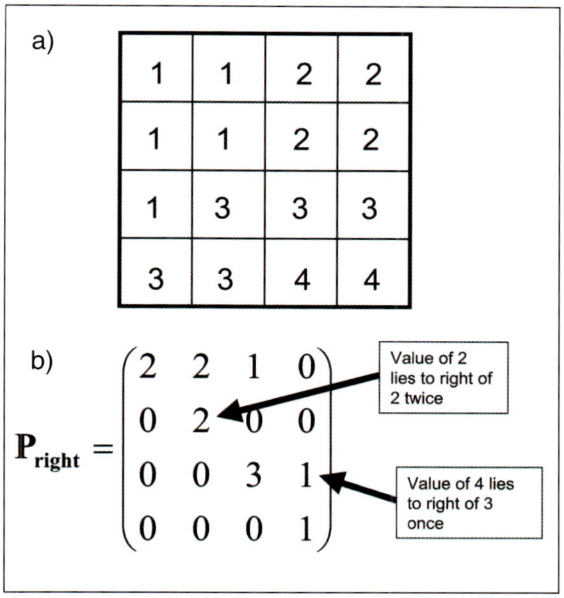

Figure 3.94. (a) A representative 4 × 4 patch of seismic data that have been scaled and biased so that their values fall from 1 through 4. (b) Each row *i*, column *j* element of the gray-level co-occurrence matrix, $\mathbf{P}_{\mathbf{right}}$, of the data shown in (a) indicates how many times the occurrence of a value *j* lies to the right of the value *i*. For 3D data, generally we compute the gray-level co-occurrence matrix in six or 26 directions and sum the results. Texture attributes summarize the properties of the gray-level co-occurrence matrices, the most popular being energy = $\Sigma\Sigma P_{ij}^2$, entropy = $\Sigma\Sigma P_{ij}(\log P_{ij})$, contrast = $\Sigma\Sigma(i-j)^2 P_{ij}$, and homogeneity = $\Sigma\Sigma P_{ij}/[1-(i-j)^2]$.

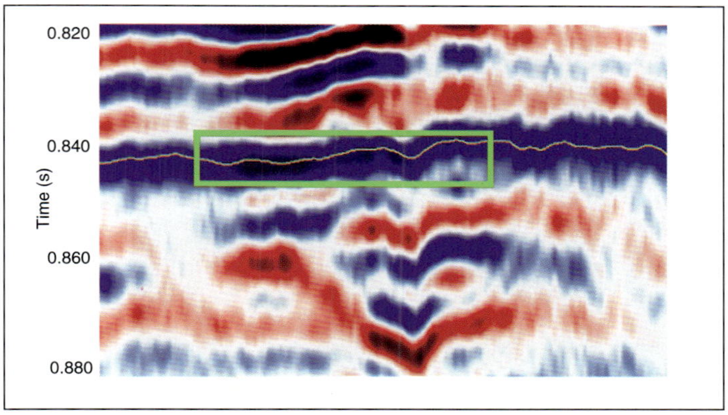

Figure 3.95. A segment of a seismic section showing Lower Cretaceous glauconite fluvial deposits, seen tracked in the highlighted zone.

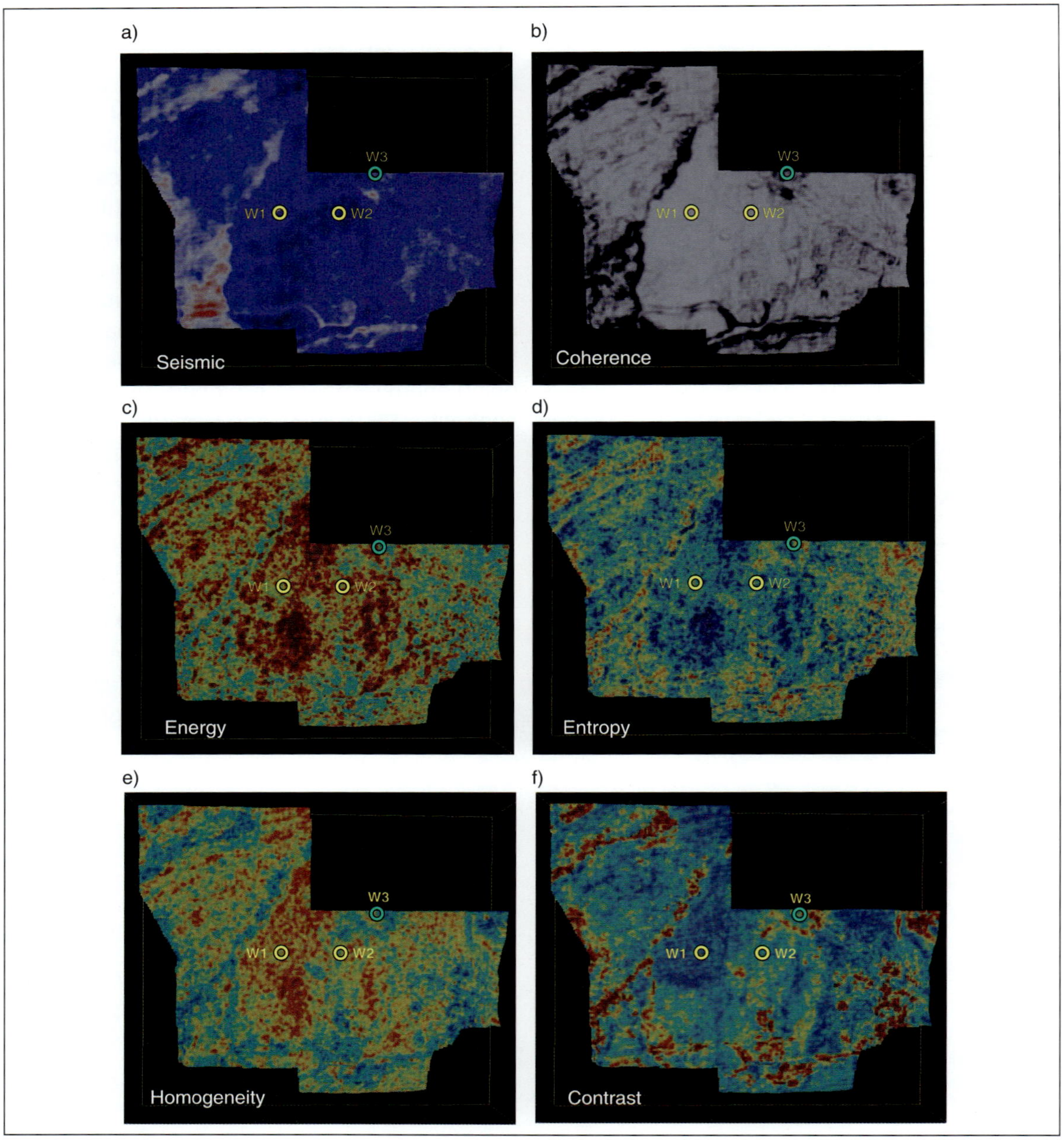

Figure 3.96. Stratal slices through (a) a seismic amplitude volume and (b) a corresponding coherence volume, at the level of a producing sandstone. The seismic amplitudes indicate the sandstone distribution. The coherence slice indicates the channel edges clearly, and channel sands are indicated in the top left section. However, nothing on the images corresponds to the productive sands. (c)-(f). Stratal slices from (c) energy, (d) entropy, (e) homogeneity, and (f) contrast attributes at the same zone of interest. The energy attribute indicates high values that are interpreted to be associated with the same productive sands seen in wells W1 and W2. Corresponding to these high values of energy, we see low values of entropy and high values of homogeneity, which we would expect of fluvial deposits. The low-contrast feature seen around well W3 indicates a separate block, which is confirmed by a different pressure observed in that well.

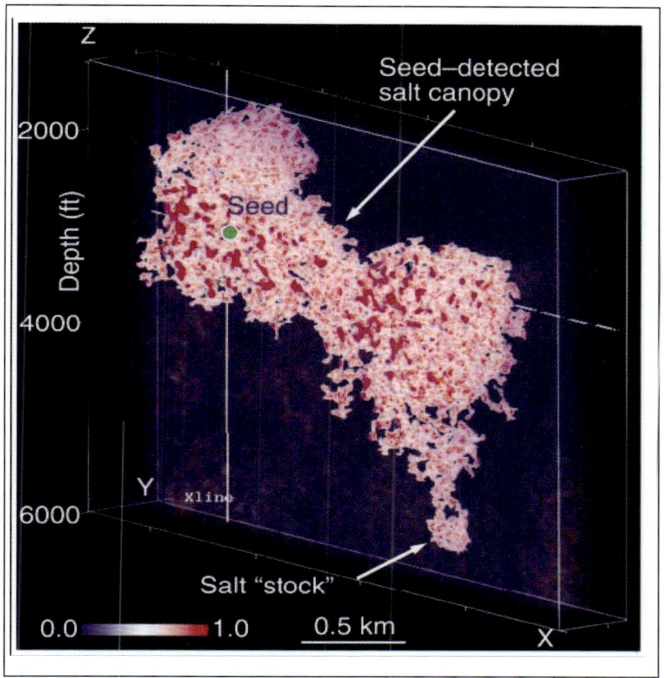

Figure 3.97. A salt canopy detected by using a "seed" in a texture energy volume. The energy values corresponding to the salt are higher than the surrounding areas, and the salt body can be detected more effectively from the energy volume than from using a seismic volume. Use of a seismic volume would result in "bleeding" across the boundaries and would prevent a crisper definition of the salt canopy. After Gao (2003).

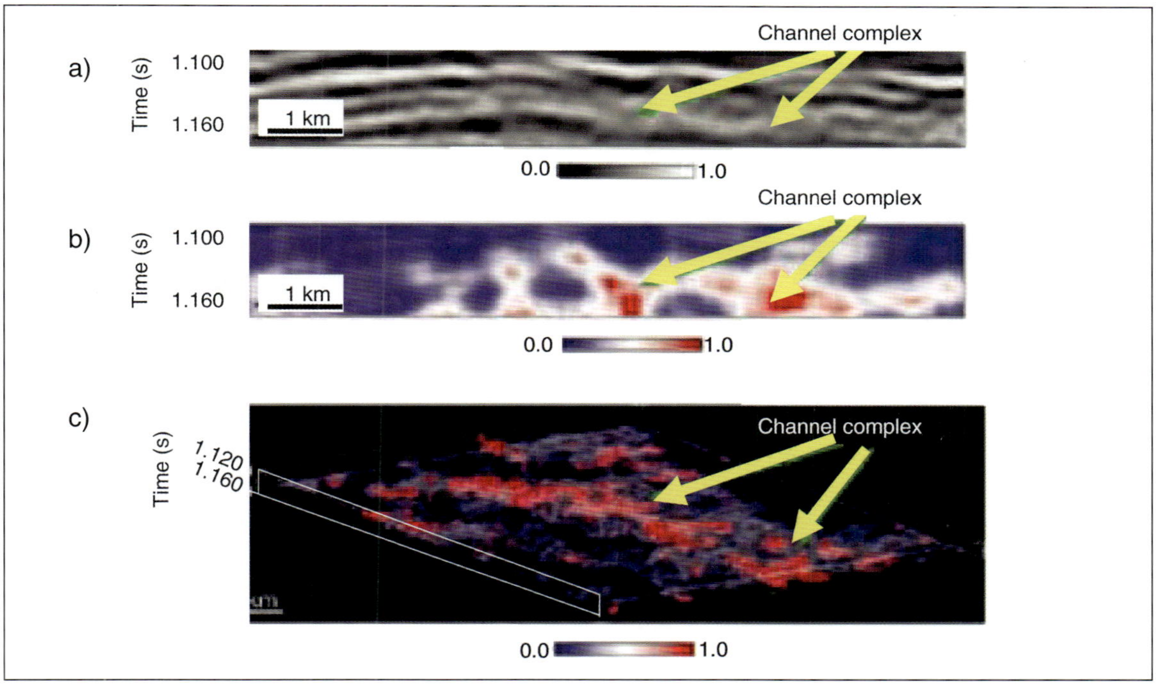

Figure 3.98. (a) A segment of an original seismic section. (b) The corresponding texture energy section. (c) The texture energy subcube, with an opacity filter applied. The high-energy feature (red) along a channel system is revealed clearly by rendering the low-energy amplitudes transparent. The same feature would be difficult to visualize and isolate from the original amplitude volume. After Gao (2003).

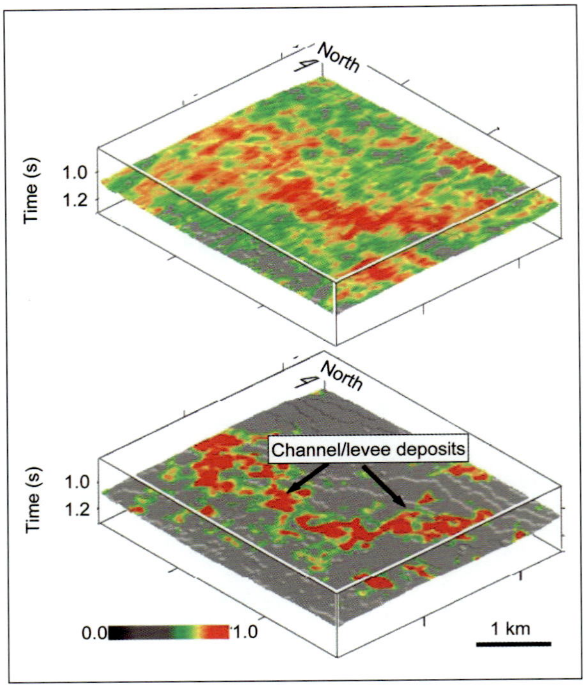

Figure 3.99. A horizon-slice comparison at the same stratigraphic level, from (a) a texture energy volume and (b) a seismic amplitude volume. Notice how the channel and levee deposits can be recognized, mapped, and detected more effectively from the texture energy volume than from the seismic amplitude volume. After Gao (2003).

Unit 4 Attributes and the Seismic Interpreter

Overview

By identifying reflector terminations and subtle changes in dip and azimuth, we can infer coherent progradational and transgressive packages, as well as more chaotic slumps, fans, and braided-stream complexes, infill of karsted terrains, gas seeps, faults, and angular unconformities.

Geometric attributes such as coherence allow interpreters to quickly visualize and map complex fault systems, salt and shale diapirs and, in some cases, incoherent overpressured shales. Volumetric curvature provides images of folds, domes, and collapse features and help workers map faults that have a vertical throw below seismic resolution. Color displays of dip and azimuth on vertical sections and time slices allow interpreters to visualize the rotation of fault blocks within the 3D seismic volume. Therefore, such displays can be an important aid in risking four-way closure. Used together, spectral components, dip and azimuth, coherence, curvature, and lateral changes in amplitude support a better understanding of deformational and depositional processes, thereby allowing the interpreter to link a fault system via connecting folds and to visualize structural control of channels, diapirs, and karst.

We begin by illustrating the sensitivity of volumetric attributes to structural deformation by presenting images of faulting, folding, and diapirism. Then we illustrate the sensitivity of volumetric attributes to clastic depositional environments, using images of channels, levees, turbidites, and angular unconformities. Our next topic is carbonate depositional environments, with an emphasis on pinnacle reefs and karst. That topic is followed by several examples of shallow depositional environments and drilling hazards, including mass-transport complexes, shale-dewatering features, and gas pockmarks. Finally, we demonstrate how we can use volumetric attributes to better define reservoir heterogeneity. Along the way, we introduce several work flows that show how the use of attributes can accelerate a conventional interpretation.

Summary

Volumetric attributes possess several advantages over attributes computed from horizons. Those advantages include avoiding the time needed to pick the horizon, avoiding interpreter error and/or bias, and using a vertical analysis window. Given sufficient time and skill, a human interpreter can avoid the first two problems. However, volumetric attributes not only are calculated for an entire volume, they also are computed within a small vertical-analysis window, which provides higher signal-to-noise estimates of reflector dip and azimuth.

Geometric attributes enable us to quickly define and name a coarse fault network. Whereas coherence clearly defines well-imaged faults that appear as discrete discontinuities, volumetric curvature allows us to map subtle folds and flexures associated with tectonic deformation. Volumetric curvature also illuminates faults that are imaged inaccurately or that have very small vertical throw. Geometric attributes allow us to visualize plastic deformation in ductile shales and brittle deformation in more competent carbonates and sandstones.

In general, time slices through seismic attributes offer better fault images (with less interpreter bias) than do horizon slices. Similarly, time slices show better images of salt and shale diapirs, gas clouds, pinnacle reefs, karst, and other geologic features that cut through multiple geologic strata. In contrast, slices that are chosen to follow or run parallel to an interpreter-picked horizon provide better images of clastic and carbonate depositional environments, including channels, levees, fans, turbidites, and mass-transport complexes.

Because of their long, narrow, often meandering nature, channels are particularly easy to identify on seismic-attribute volumes. Other features of geologic interest, such as dunes, bars, and fans, are more subtle and must be interpreted in the context of related features seen on a horizon slice and linked through models of depositional processes.

The interpretation work flow of carbonate shoals, turbidites, and debris fans resembles that used in siliciclastic environments. However, pinnacle reefs and karst features present a more challenging interpretation problem. Such features tend to have limited lateral extent, and they either grow with (in the case of pinnacle reefs) or cut through (in the case of karst and collapse features) surrounding lithologies that may have radically different lithologic properties. For that reason, conventional time-structure maps of pinnacle reefs and karst features are difficult to generate and have limited value. In contrast, a sequence of attribute time slices (or of phantom-horizon slices from a nearby undisturbed stratigraphic horizon) through reef and karst features offer an unbiased image of the lateral extent of these features and of their relationship to neighboring features. A sequence of attribute time slices also can reveal the internal character of reef and karst features. A 3D visualization is particularly effective.

Not all geologic features of interest can be represented by edges. Salt and shale diapirs usually appear as low-amplitude, chaotic zones that show up as low-coherence anomalies. Karst and other collapse features often appear as moderate-amplitude chaotic features. Likewise, mass-transport complexes, amalgamated channels, and turbidites have a distinct internal seismic texture that can be recognized on seismic attributes. Preliminary results show that hydrothermally altered dolomite produces similar "muddy" textures. Structure-oriented filtering (SOF), which can greatly enhance our ability to visualize faults and channel edges, often suppresses these more subtle volumetric textures.

Many geologic hazards are easier to see on attribute volumes than on conventional seismic data. The 3D volumetric attributes, coupled with 3D graphics tools, allow us to quickly visualize, rather than manually pick, features such as mass-transport complexes, shale dewatering, amalgamated channels, plow marks, gas chimneys, and pockmarks. Differential compaction over deeper channels can generate shallow faults which can be drilling hazards.

Together, 3D geometric attributes and spectral decomposition allow us to extend geomorphologic concepts to 3D data volumes, thereby helping us use our understanding of geologic processes to infer lithology and porosity. In combination with well logs, image logs, production, and tracer data, volumetric attributes can help us to better define reservoir heterogeneity and compartmentalization. Production history, chemical tracers, and microseismic emissions can be used to determine which fault sets are conduits and which are seals. Statistically estimating permeability that results from fractures requires correlating production data with the distance from faults and other lineaments of azimuthally limited orientations seen on attribute volumes. Statistical correlations often can be improved by including very simple geologic concepts, such as the presence of an oil-water contact and the amount of throw on nearby faults.

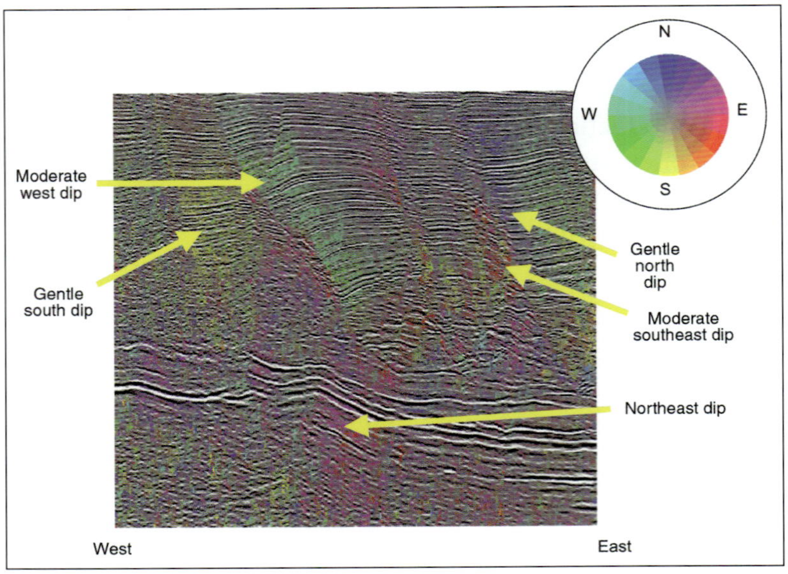

Figure 4.1. Complex growth faults from the Gulf of Mexico, displayed as a composite of dip and azimuth combined with seismic data, using the method described in Unit 1. Although this seismic line is aligned along the east-west axis, the dip and azimuth allow us to see dip in and out of the plane of the vertical section, as well as within that plane. In this example, green indicates moderate dip to the west, yellow indicates gentle dip to the south, blue indicates gentle dip to the north, red indicates moderate dip to the southeast, and magenta indicates moderate dip to the northeast. This color coding allows us to see how the fault blocks are rotated.

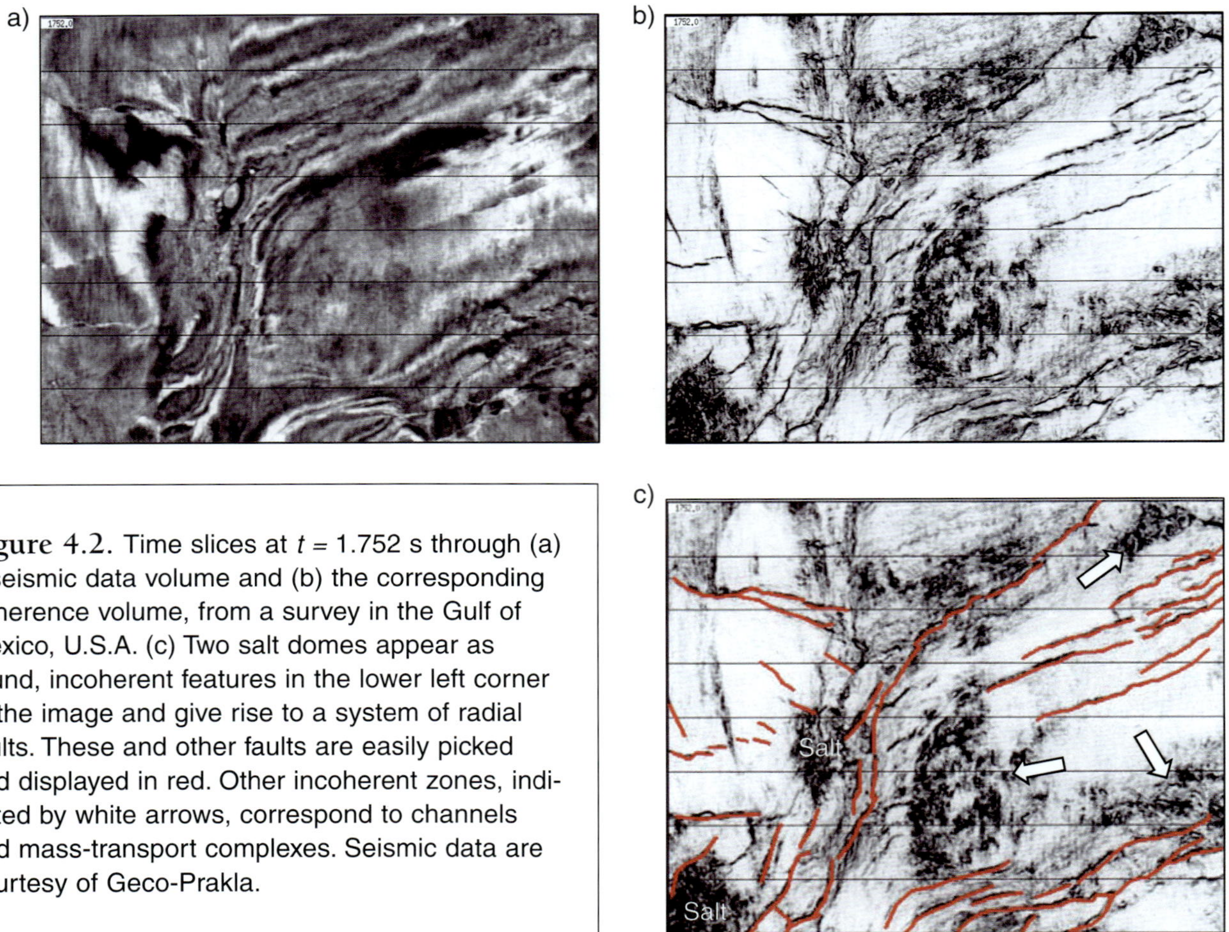

Figure 4.2. Time slices at $t = 1.752$ s through (a) a seismic data volume and (b) the corresponding coherence volume, from a survey in the Gulf of Mexico, U.S.A. (c) Two salt domes appear as round, incoherent features in the lower left corner of the image and give rise to a system of radial faults. These and other faults are easily picked and displayed in red. Other incoherent zones, indicated by white arrows, correspond to channels and mass-transport complexes. Seismic data are courtesy of Geco-Prakla.

Figure 4.3. Time slices at t = 1.200 s through (a) a seismic data volume, (b) a coherence volume, and (c) a coherence volume overlaid on a seismic data volume, from a survey in southern Alberta, Canada. The arrow indicates a zone where the faults appear to horsetail, which could endanger the quality of the seal in a reservoir.

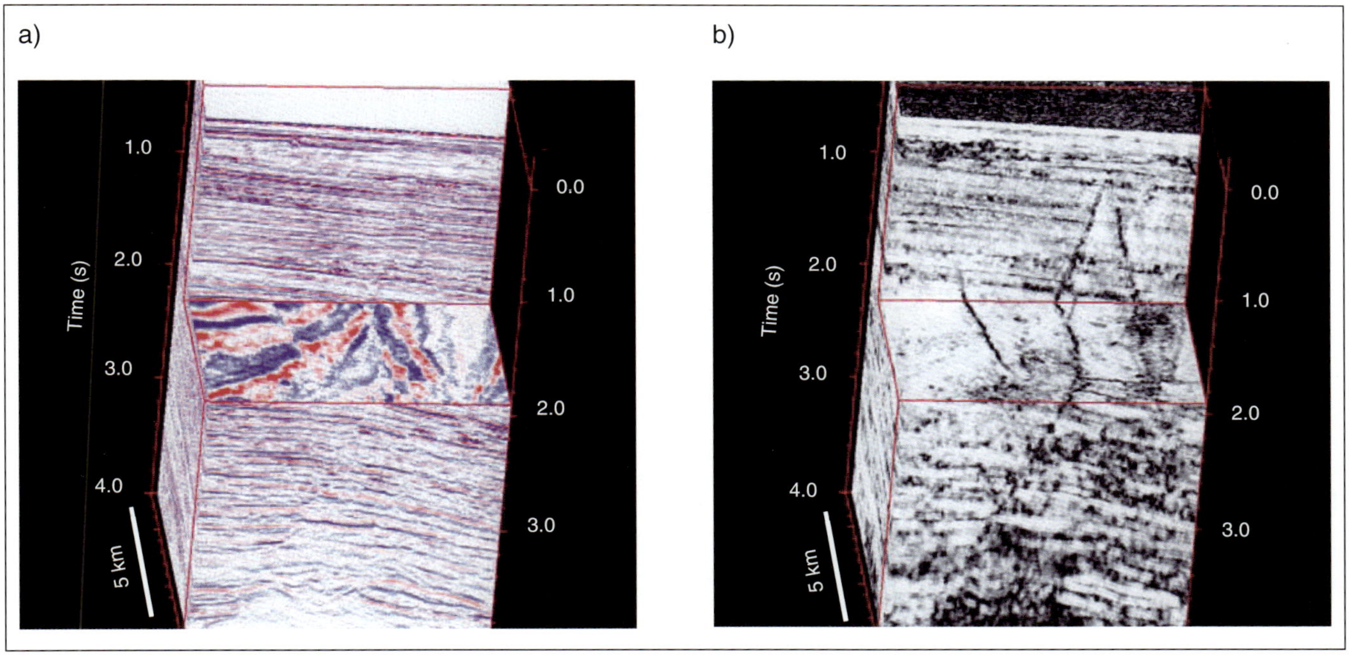

Figure 4.4. Chair diagrams through (a) a seismic data volume and (b) a coherence volume, over a salt dome from the Gulf of Mexico, U.S.A. Faults are seen clearly on the coherence volume. The low-coherence banding following stratigraphy has two major causes in clastic terrains. The more common cause is low reflectivity associated with shale-on-shale lithologies within a fixed (higher) background of incoherent seismic noise. A less common, but not infrequent, cause is truly chaotic geology, associated with slumping, channel cannibalization, and laterally reworked tidal complexes.

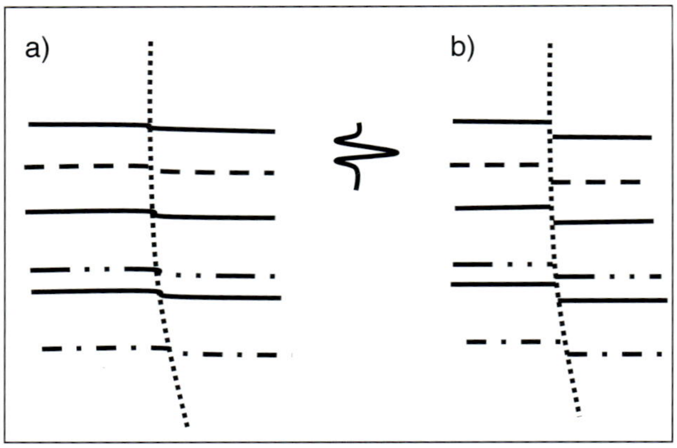

Figure 4.5. Attribute sensitivity to faults, where the amount of fault throw is measured in terms of a seismic wavelet. (a) An idealized fault with an offset much less than one-quarter the size of the seismic wavelet. This fault will be seen on curvature attribute volumes but not on coherence attribute volumes. (b) An idealized fault with finite offset and no reflector rotation. This fault will be seen on coherence attribute volumes but not on curvature attribute volumes.

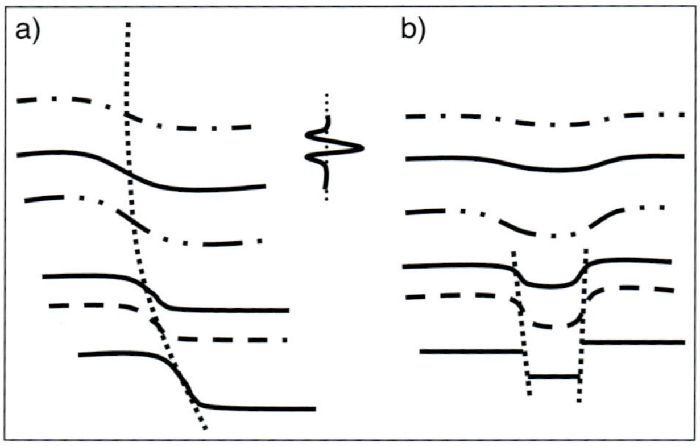

Figure 4.6. (a) Attribute sensitivity to folds and flexures. This "fault" will be seen on curvature attribute volumes but not on coherence attribute volumes. (b) Attribute sensitivity of infill/collapse features. The deeper part of the fault will be seen on coherence attribute volumes, whereas the shallower part of the fault and the overlying infill will be seen on curvature attribute volumes.

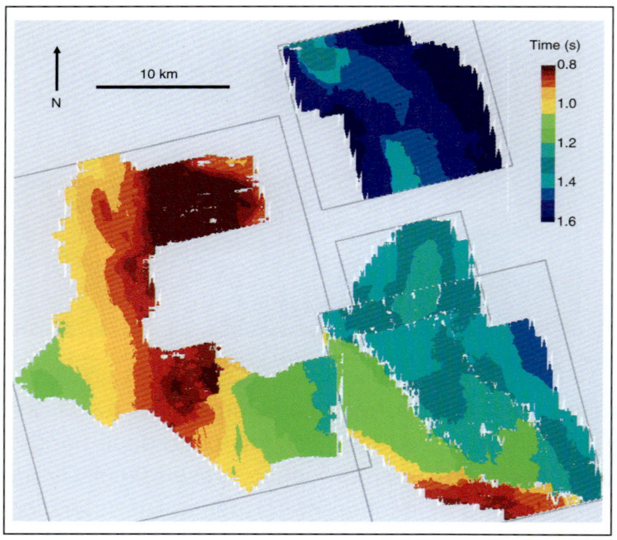

Figure 4.7. A time-structure map of the Ellenburger horizon for the four surveys shown in Figure 4.8. Note the structural misties among the surveys. The interpretation is courtesy of Chuck Blumentritt, University of Houston.

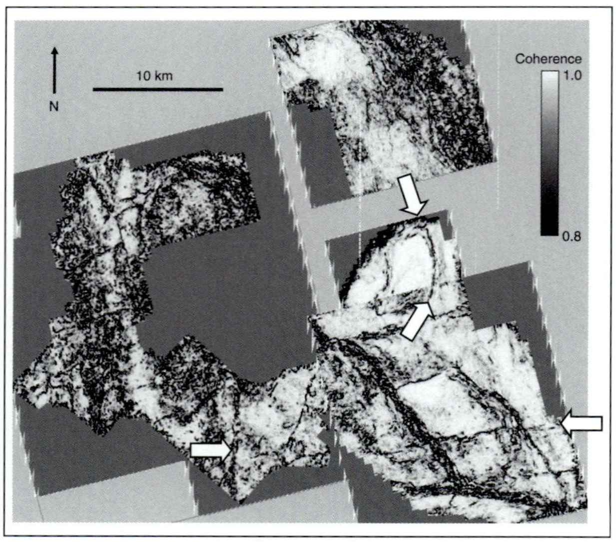

Figure 4.8. Time slices at $t = 1.000$ s through coherence volumes generated for four surveys acquired by different operators during a 15-year period on the Central Basin Platform, Texas, U.S.A. Arrows indicate faults that can be carried across surveys. Seismic data are courtesy of Oxy Permian, Devon, Burlington, and BP.

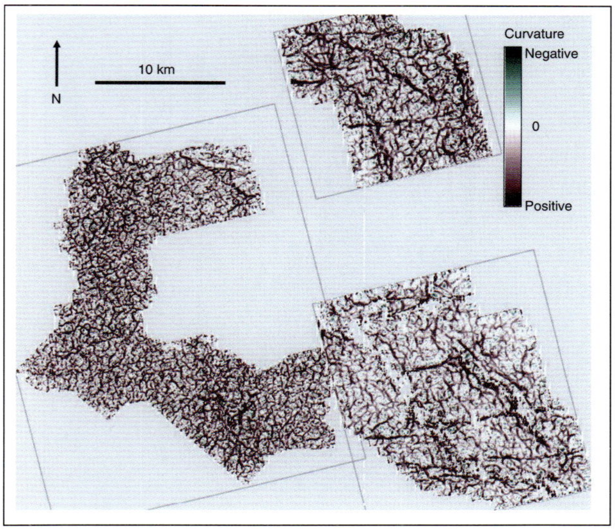

Figure 4.9. Horizon slices along the Ellenburger horizon shown in Figure 4.7, displayed here through the most-positive-curvature volumes. Interpretation is courtesy of Chuck Blumentritt.

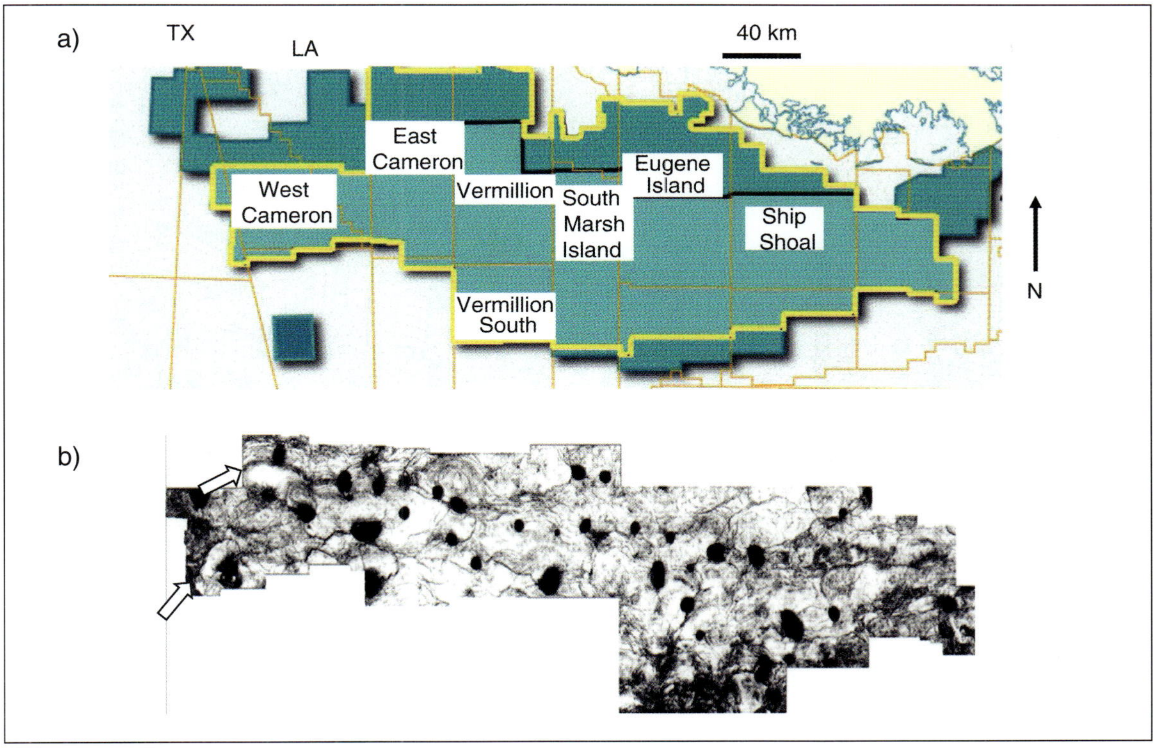

Figure 4.10. (a) The area of a very large regional study over the northern Gulf of Mexico, U.S.A. (b) A time slice at $t = 2.000$ s through a variance volume for the area bordered in yellow. This time slice shows more than 40 salt domes, many of which are linked by faults. After Biles et al. (2003).

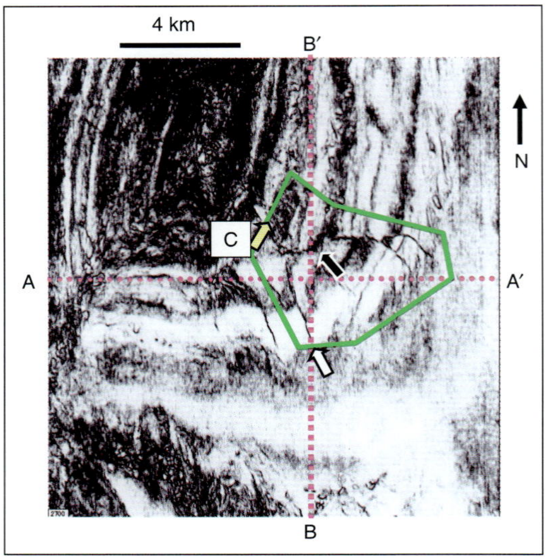

Figure 4.11. Using coherence to help pick faults and horizons: Lines AA′ (Figure 4.12a) and BB′ (Figure 4.12b) will cut through the fault indicated by the yellow and black arrows, almost perpendicularly to the dip. We will choose not to interpret the fault indicated by the white arrow, because we cut it at a steep angle. The green traverse that begins and ends at C (Figure 4.12c) was chosen to avoid cutting all but the main fault, thereby allowing us to correlate horizon picks by running around their ends. The time slice was at $t = 2.700$ s. Seismic data are courtesy of BP.

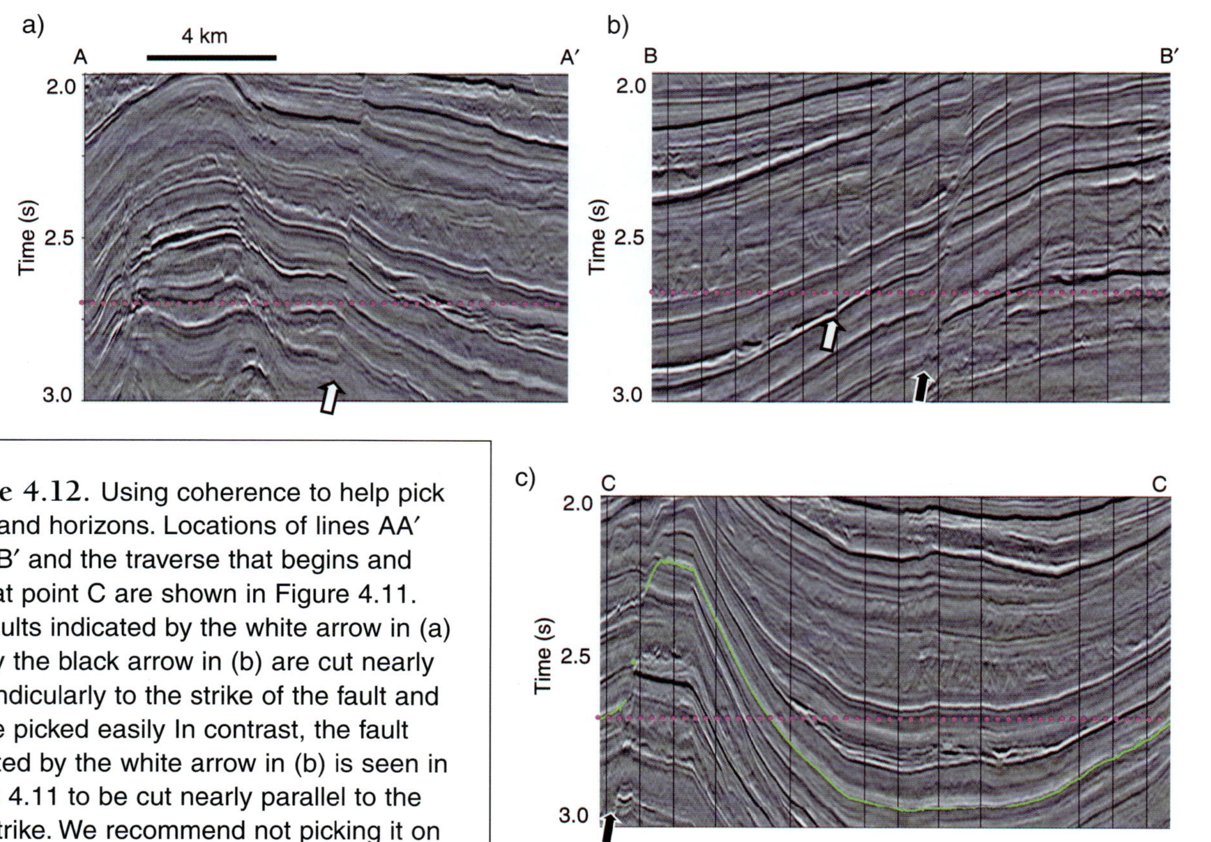

Figure 4.12. Using coherence to help pick faults and horizons. Locations of lines AA′ and BB′ and the traverse that begins and ends at point C are shown in Figure 4.11. The faults indicated by the white arrow in (a) and by the black arrow in (b) are cut nearly perpendicularly to the strike of the fault and can be picked easily In contrast, the fault indicated by the white arrow in (b) is seen in Figure 4.11 to be cut nearly parallel to the fault strike. We recommend not picking it on this line. Correlating horizons across faults can be difficult, particularly if there is a strike-slip component along the fault. The traverse displayed in (c) cuts the major fault indicated by the black arrow but avoids all other faults, thereby allowing the interpreter to tie the green horizon pick rapidly and unambiguously. The time slice was at $t = 2.700$ s and is indicated by the dotted magenta line. Seismic data are courtesy of BP.

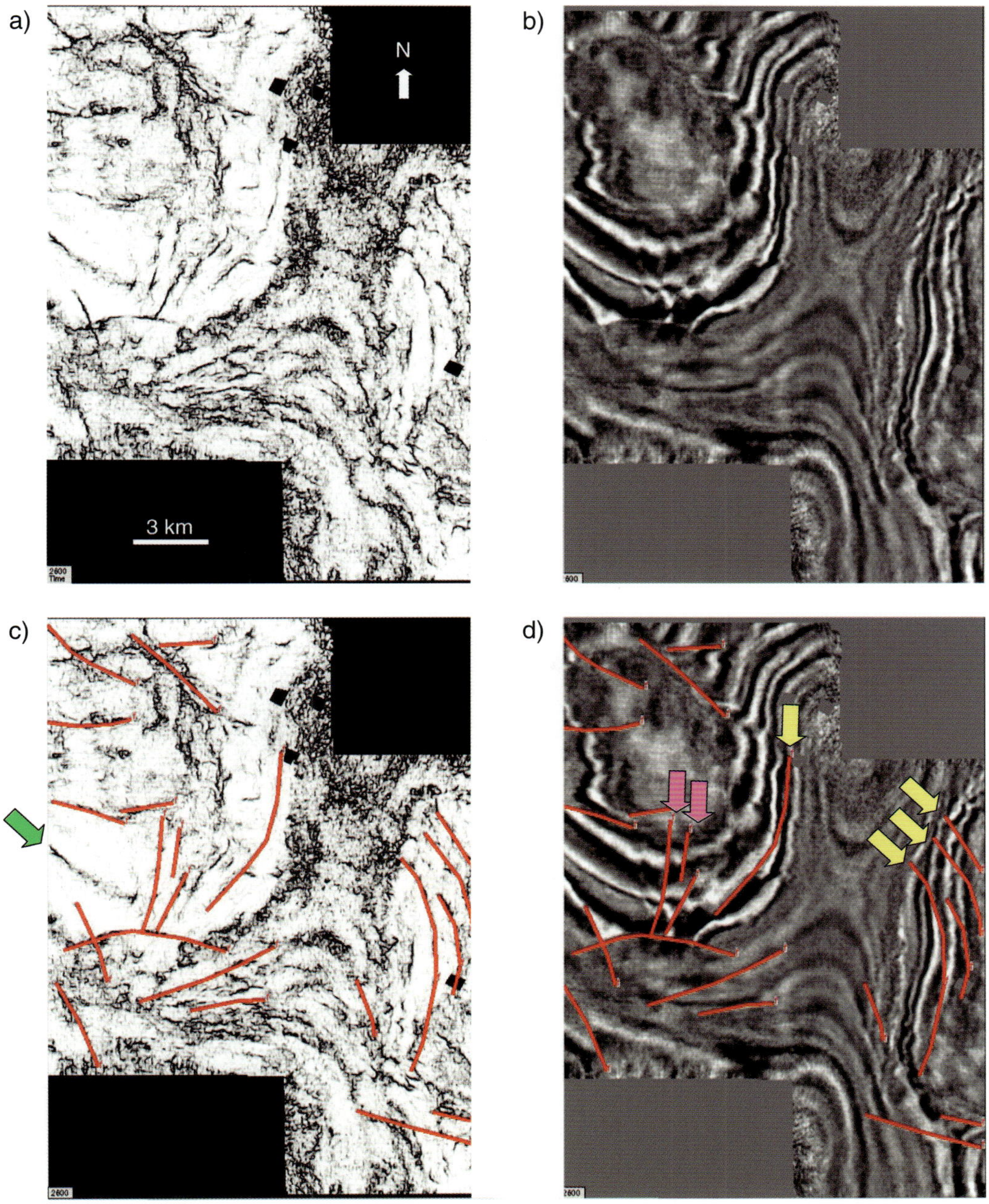

Figure 4.13. Time slices at $t = 2.600$ s through (a) an uninterpreted coherence volume and (b) an uninterpreted seismic volume and through (c) an interpreted coherence volume and (d) an interpreted seismic volume. Interpretation begins by displaying both slices at the same time and picking faults that will be assigned subsequently. Faults that cut across the seismic fabric (magenta arrows) are easy to pick on the seismic time slice, whereas those that cut subparallel to the seismic fabric (yellow arrows) are more visible by using coherence. In contrast, the feature indicated by the green arrow in (c) does not cut across the seismic fabric. We delay picking it for the moment. Seismic data are courtesy of Seitel.

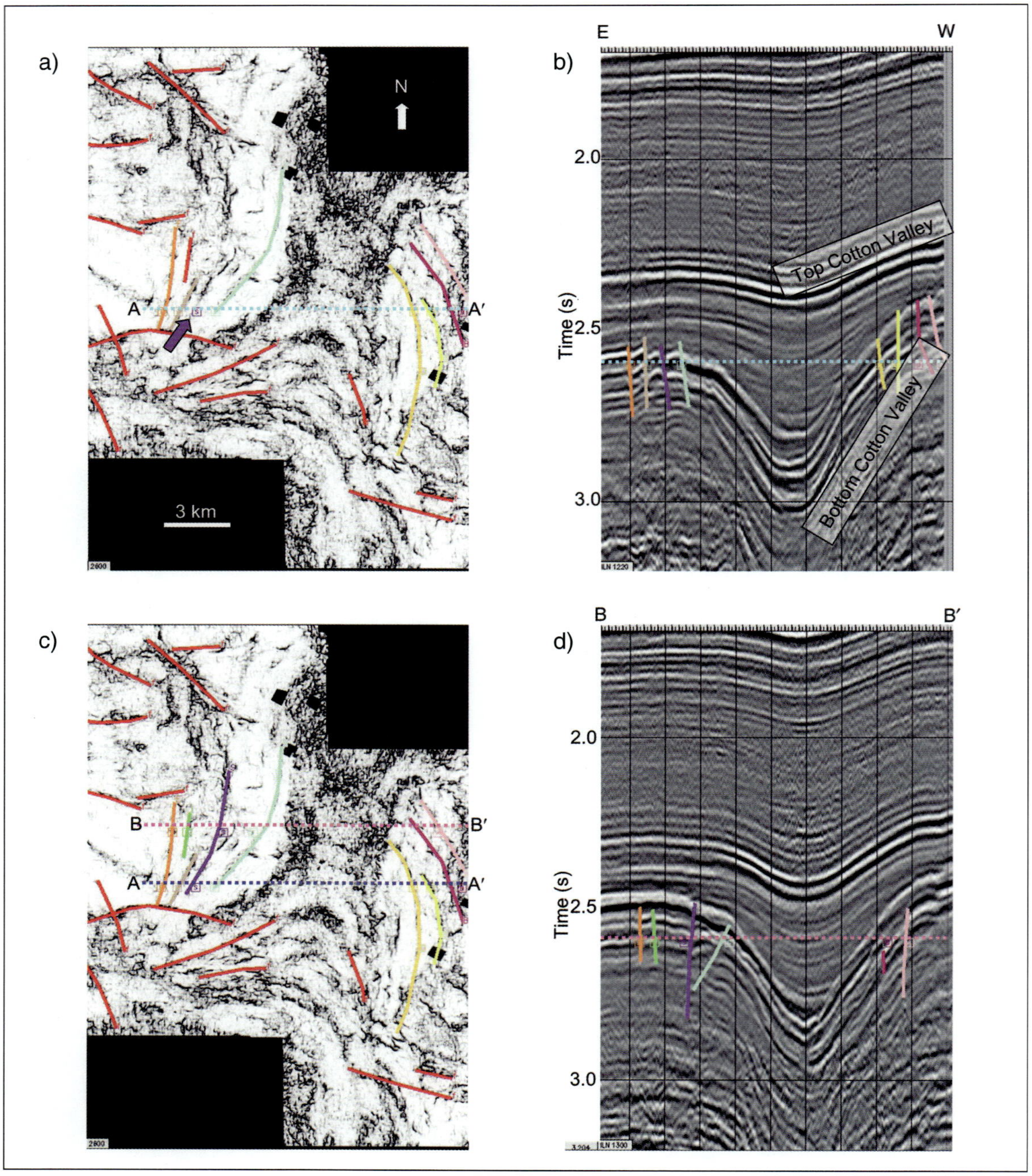

Figure 4.14. Fault assignment using coherence and seismic volumes. In the vertical seismic slice in (b), we have chosen an east-west section roughly perpendicular to several suspected faults seen in the coherence volume in (a). We pick and assign a unique name and color to these faults, one at a time. We add an additional purple fault on the seismic section in (b), note its location (purple arrow) in (a), and pick it in (c). Next, in (d), we display a seismic section 100 lines to the north and note the posted faults.

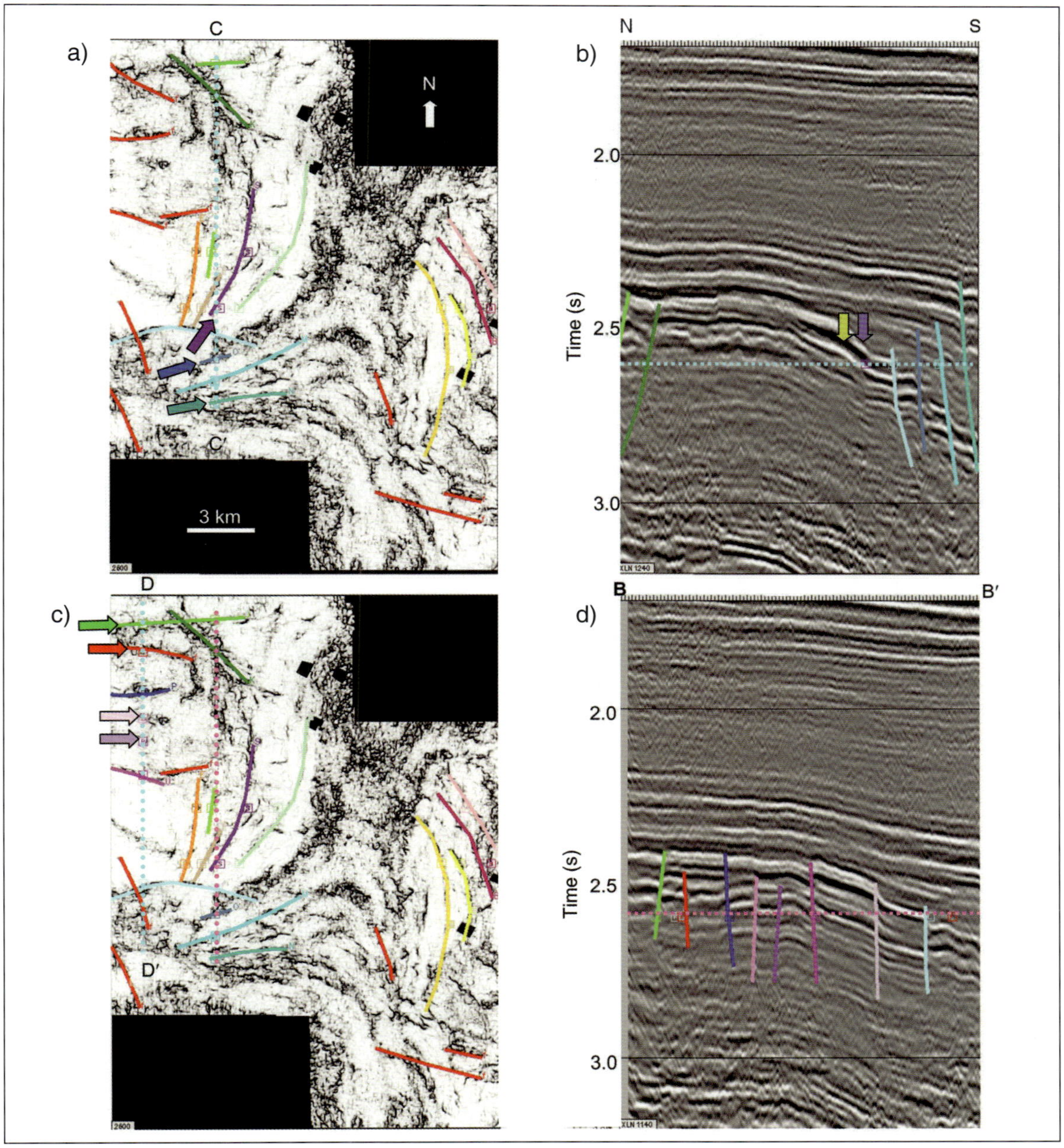

Figure 4.15. Fault assignment using coherence and seismic volumes. The coherence volume (a), from Figures 4.13 and 4.14, is shown here as we continue our interpretation. (b) In this vertical seismic slice, we display the cyan dotted north-south seismic line shown in (a), on which we assign unique names and colors to faults trending roughly east-west. Note that we do not pick the purple and tan faults (indicated by arrows) because we are cutting them closer to their strike direction. We add a blue fault and extend the previously unassigned red fault. (c) The magenta-dotted north-south seismic section 100 lines to the west, on which we (d) extend a previously assigned red fault and assign it a teal color. We note that the green fault extends farther to the west, and we modify both the green and red picks on the coherence slice in (c). We also add a pink fault and a violet fault.

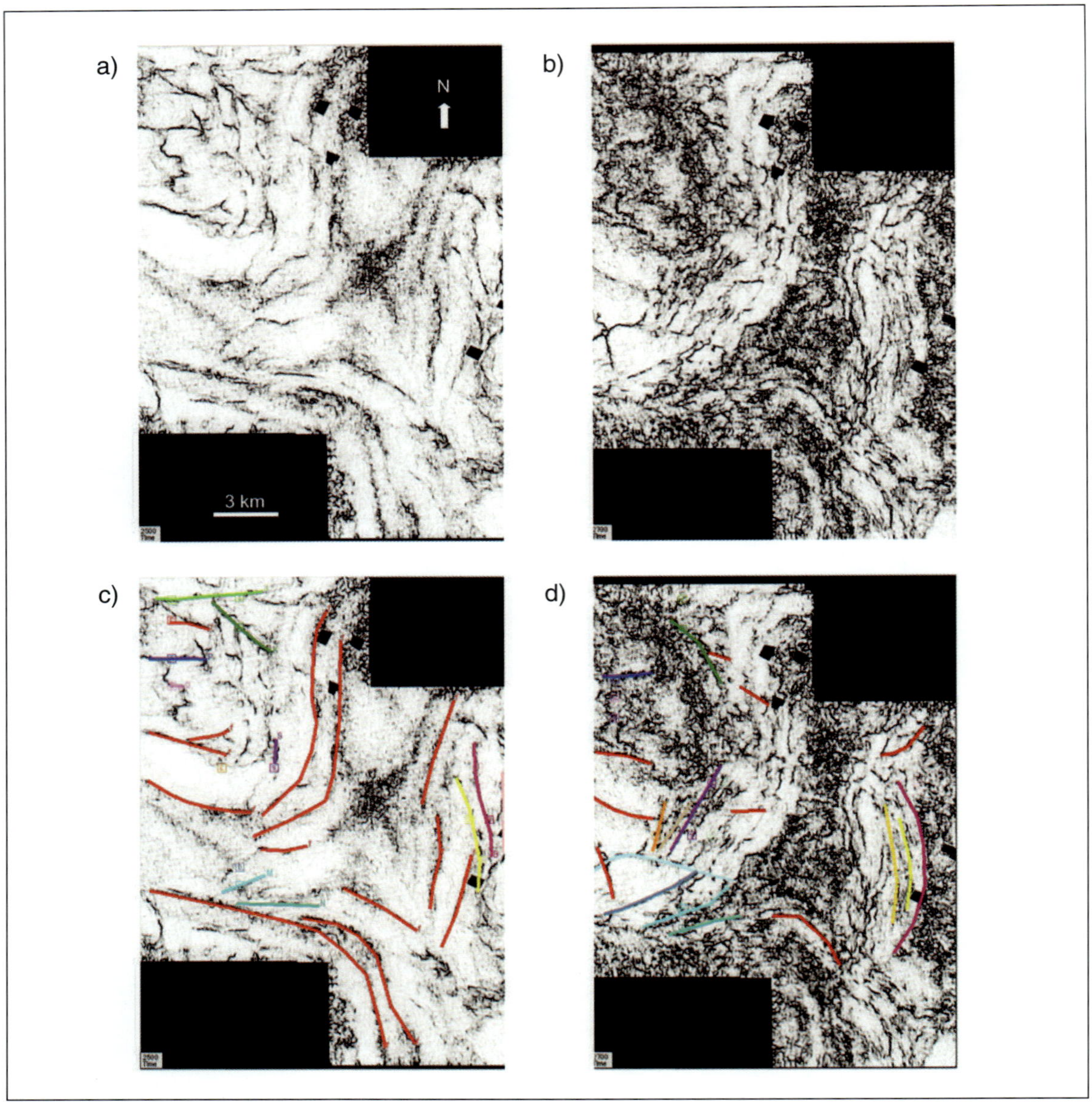

Figure 4.16. Fault assignment using coherence and seismic volumes. Coherence time slices at (a) $t = 2.500$ s and (b) $t = 2.700$ s, which are respectively 0.100 s above and below the image shown in Figure 4.13a. In (c) and (d), we follow the same process as before, checking the quality of previous interpretation picks and identifying new faults seen on coherence and seismic volumes until we have generated a consistent fault map.

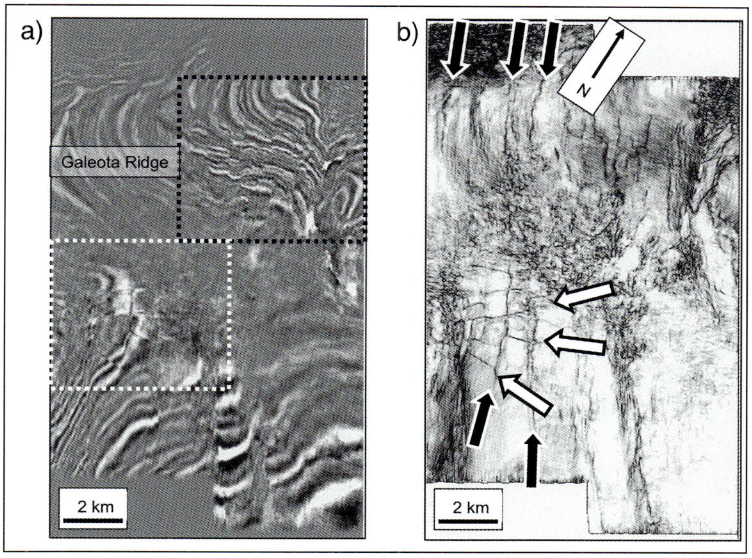

Figure 4.17. Time slices at $t = 1.200$ s at the level of the El-Diablo-1 discoveries offshore Trinidad, through (a) a seismic data volume and (b) a coherence volume. The discoveries were found in blocks bounded by strike-slip and antithetic faults such as those indicated by arrows. Complex faulting is difficult to detect on the seismic time slice but is clear on the coherence slice. After Gersztenkorn et al. (1999).

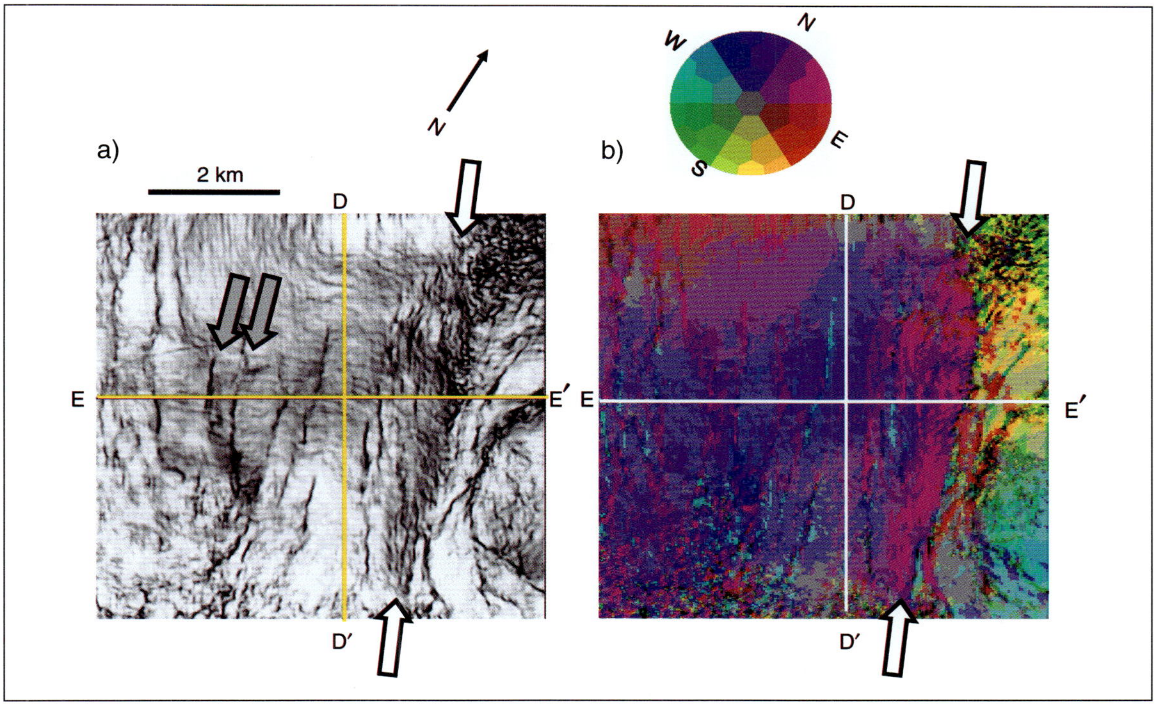

Figure 4.18. Time slices at $t = 1.100$ s through (a) a coherence volume and (b) a dip and azimuth volume, corresponding to the area of the dotted black rectangle in Figure 4.17. In the western (left) half of the figures, the faulting is rather simple and the dip is gentle and toward the north and northeast (magenta and blue). To the east of the zone indicated by white arrows, deformation becomes much more intense, which results in lower (blacker) coherence and reflectors that dip toward the south, southeast, and west. After Gersztenkorn et al. (1999).

Figure 4.19. Seismic lines DD′ along strike and EE′ perpendicular to strike, as indicated on the time slices shown in Figure 4.18. Line EE′ shows a simple monocline. The reflectors do not rotate along the normal faults (indicated by gray arrows) on the western half of line DD′, so the colors in Figure 4.18b vary smoothly. In contrast, a significant change in dip occurs across the fault indicated by white arrows. Figure 4.18 shows that this rotation is complicated. Given these images and knowledge of the regional tectonics, we conclude that this is a wrench fault. After Gersztenkorn et al. (1999).

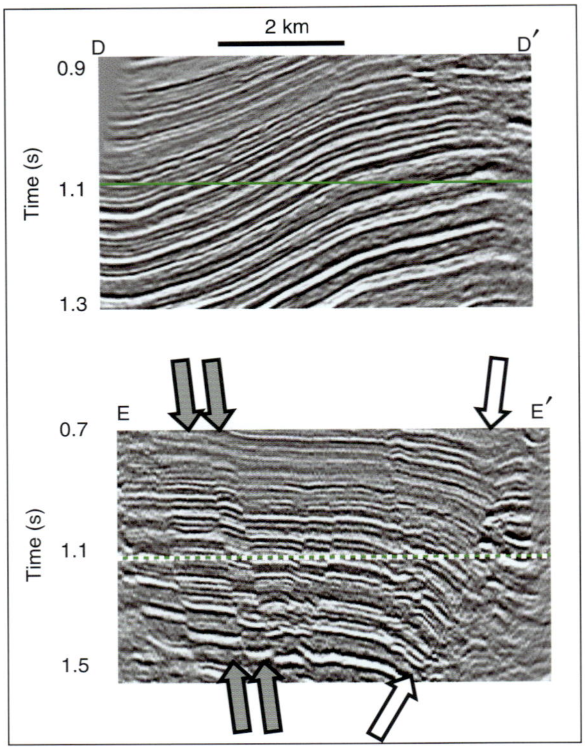

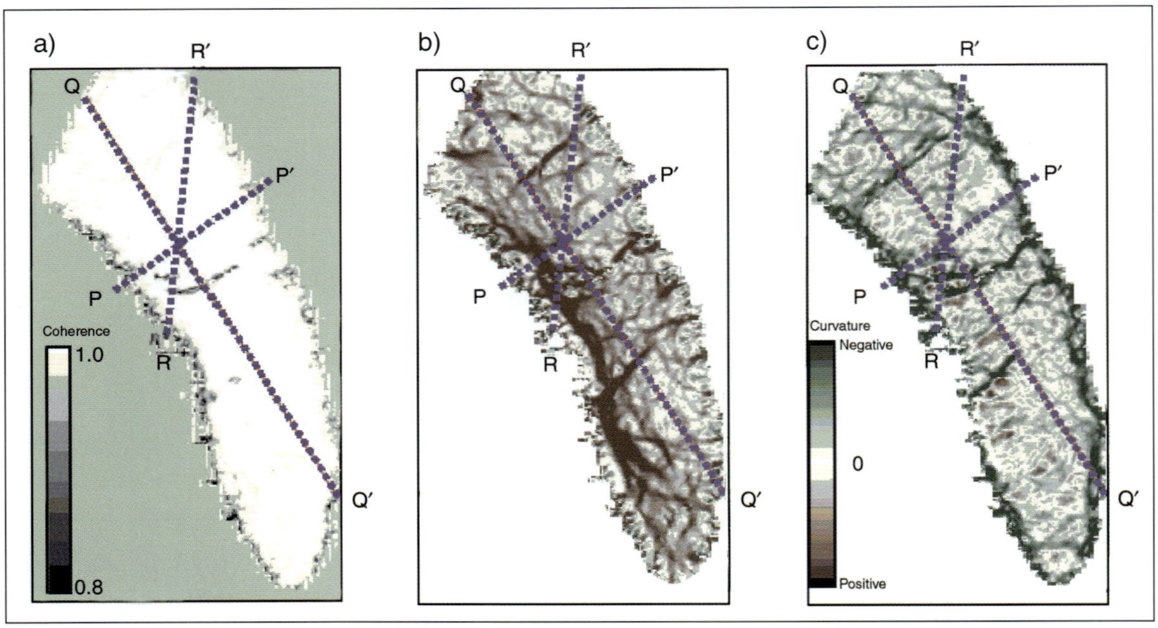

Figure 4.20. Reverse faulting that gives rise to strike-slip faulting and pop-up blocks, seen through seismic attributes, from a survey conducted in the Teapot Dome area, Wyoming, U.S.A. Horizon slices along the purple horizon shown in Figure 4.21 cut though (a) a coherence volume, (b) a most-positive-curvature volume, and (c) a most-negative-curvature volume. Seismic data are courtesy of RMOTC.

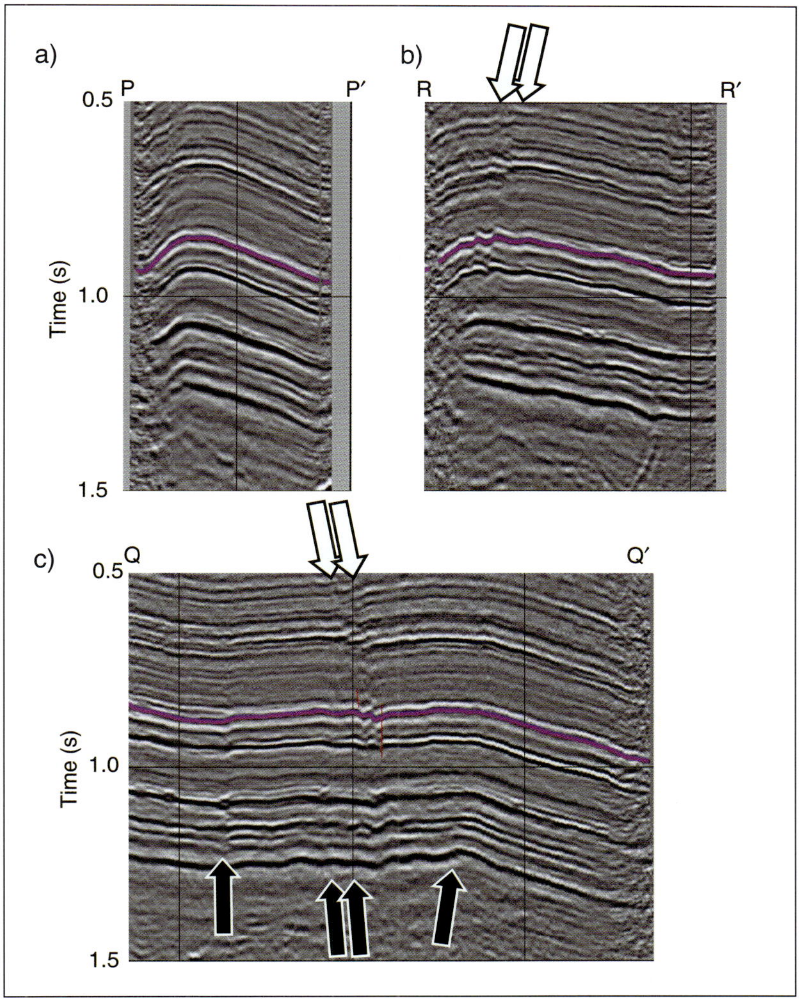

Figure 4.21. Seismic cross sections through faults seen in the curvature volumes in Figure 4.20: (a) dip line PP′; (b) RR′, a line perpendicular to the third curvature lineaments; and (c) strike line QQ′. The strike-slip fault indicated by white arrows is a distinct discontinuity at this level and therefore is seen in the coherence slice. The subtler faults indicated by black arrows are not seen in the coherence slice, but they do appear on the curvature horizon slices. Seismic data are courtesy of RMOTC.

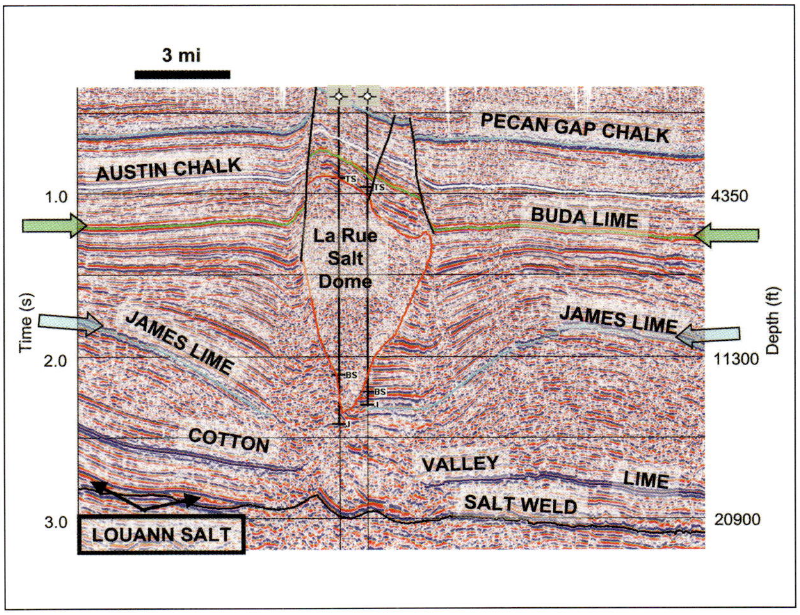

Figure 4.22. A vertical section through the La Rue salt dome, east Texas, U.S.A. The geologic section is composed of competent carbonate limestones and chalks interbedded with more plastic sandstones and shales. The Louann Salt has been withdrawn from below, giving rise to a salt-withdrawal basin and turtle structures. After Maione (2001).

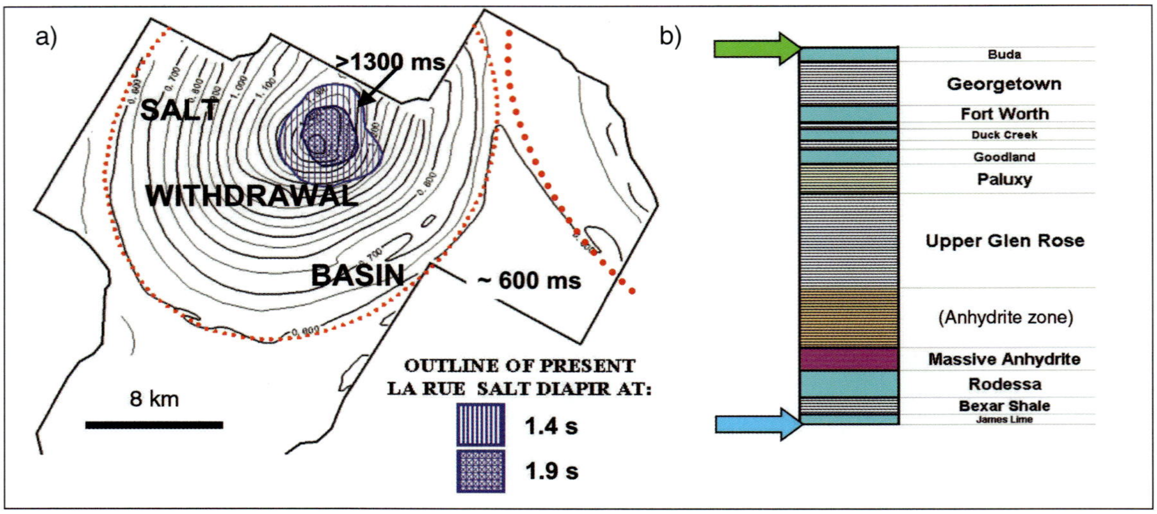

Figure 4.23. Isochron map between the Buda Limestone and the James Limestone, outlining the salt-withdrawal basin shown in Figure 4.22. After Maione (2001).

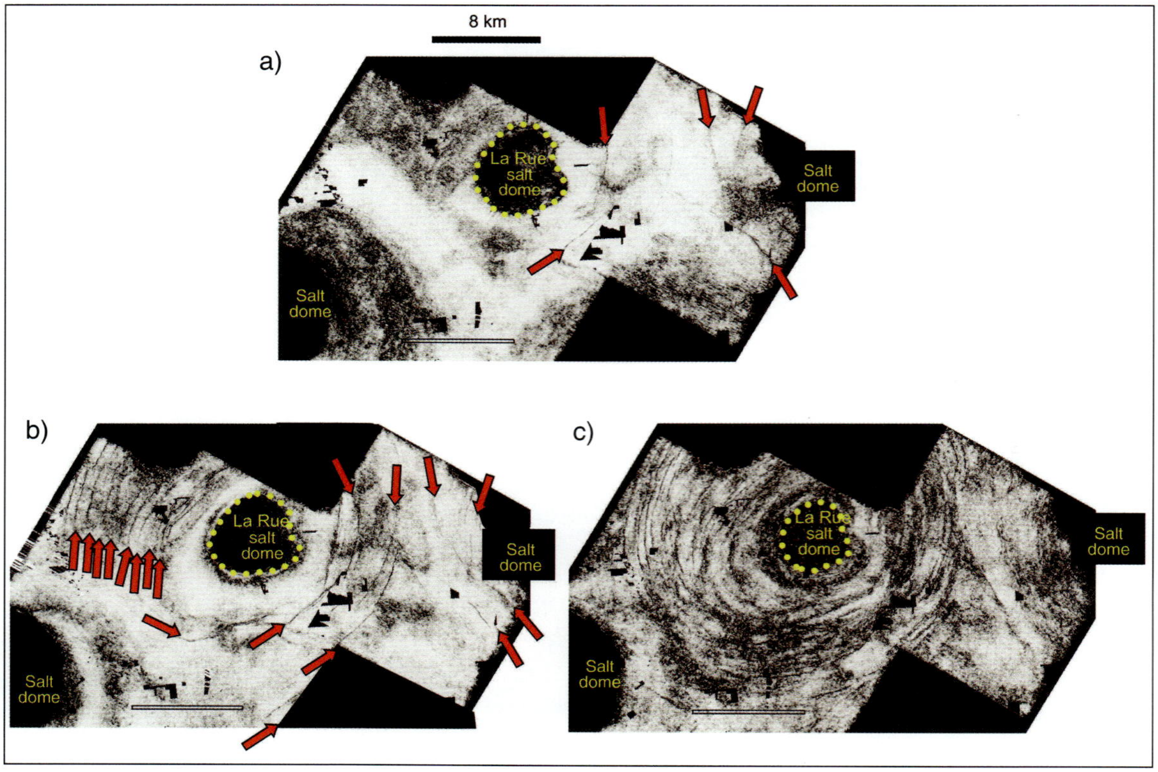

Figure 4.24. Time slices through the coherence volume computed over the La Rue salt dome at t = (a) 1.232, (b) 1.400, and (c) 1.636 s. Arrows indicate ring faults associated with salt withdrawal in (a) and (b). In (c), there are too many ring faults to indicate by arrows. Note several ring faults at the far eastern edge of the survey associated with a salt dome to the east. After Maione (2001).

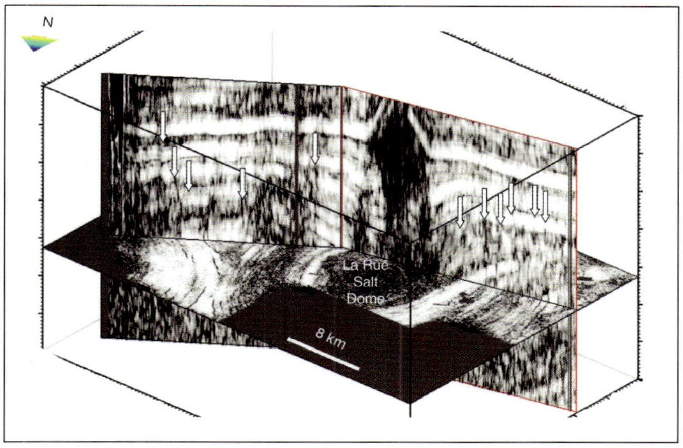

Figure 4.25. A 3D view from the south of a coherence volume over the La Rue salt diapir, east Texas, U.S.A. Arrows denote ring faults associated with salt withdrawal. After Maione (2001).

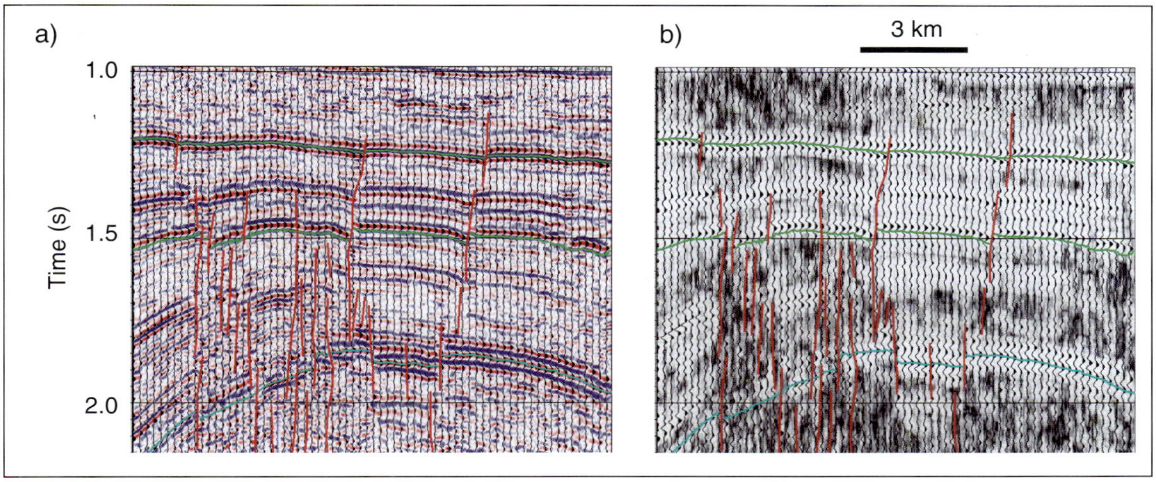

Figure 4.26. Vertical sections from along the northern part of a survey between two salt-withdrawal basins, cutting through (a) a seismic data volume and (b) a seismic volume overlaid on coherence. Ring faults (red lines) were interpreted first on the coherence image and then transferred to the seismic image. After Maione (2001).

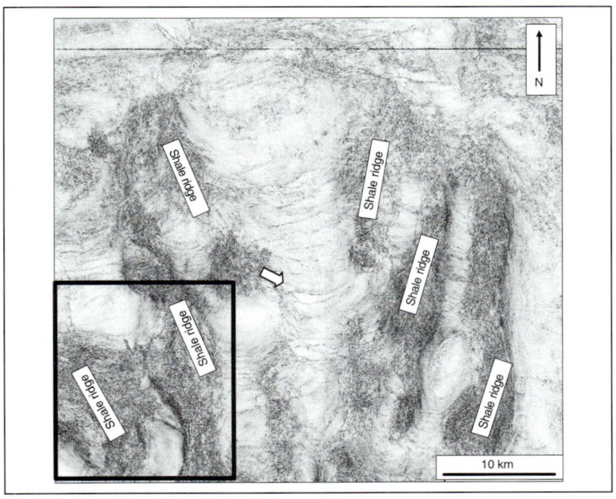

Figure 4.27. A phantom horizon slice from a coherence volume 1.175 s below the water bottom, from a large survey offshore Nigeria, showing multiple shale ridges. Ridges are linked by faults in the long minibasins between ridges (arrow). The area indicated by the box is redisplayed in Figure 4.28. After Haskell et al. (1999).

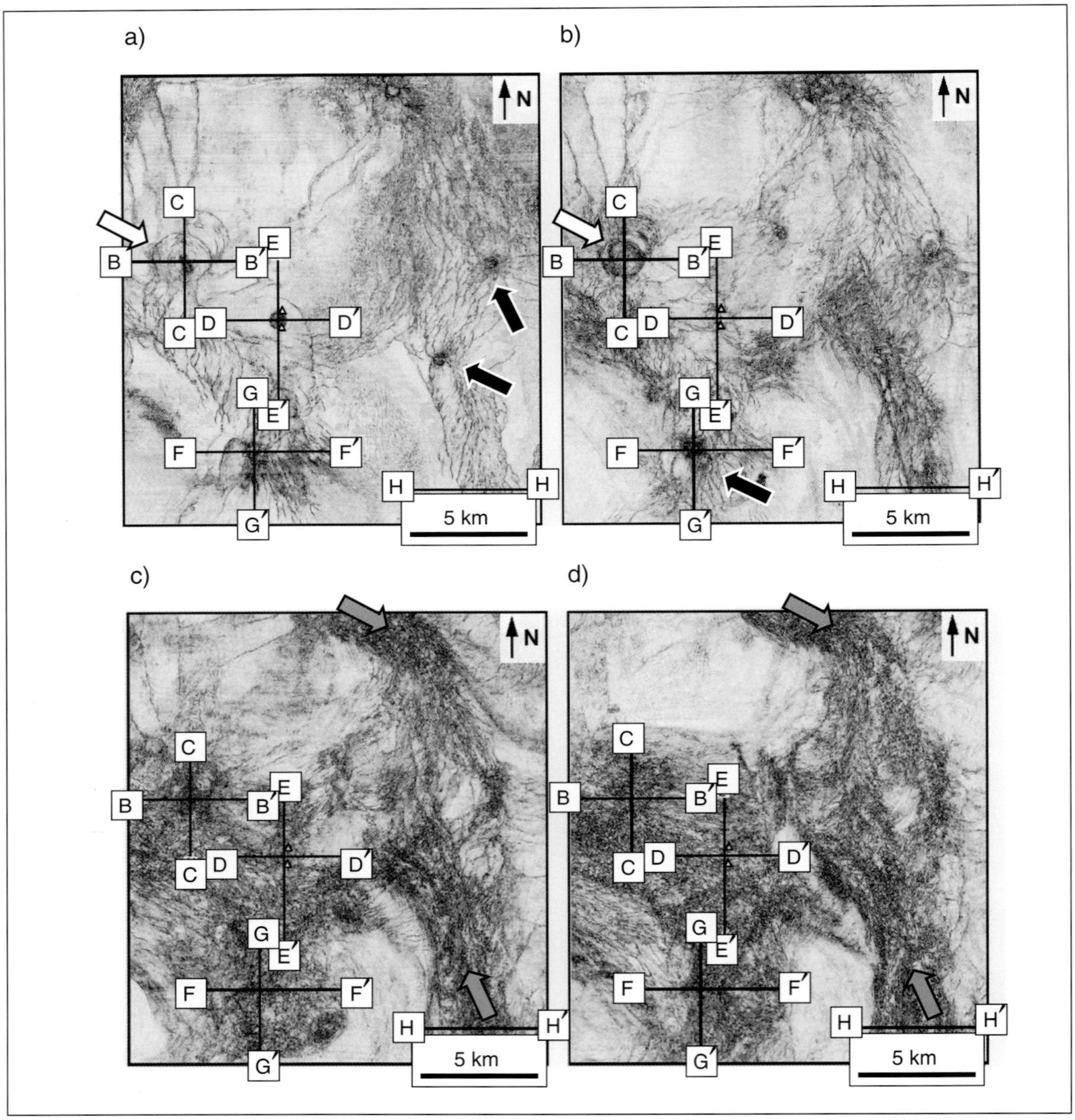

Figure 4.28. Time slices through a coherence volume at (a) 0.225, (b) 0.725, (c) 1.225, and (d) 1.725 s below the seafloor, cutting a shale ridge and associated diapirs and corresponding to the area shown by the black box in Figure 4.27. Gray arrows in (c) and (d) indicate shale ridges. Black and white arrows in (a) and (b) indicate radial and ring faults, respectively. After Haskell et al. (1999).

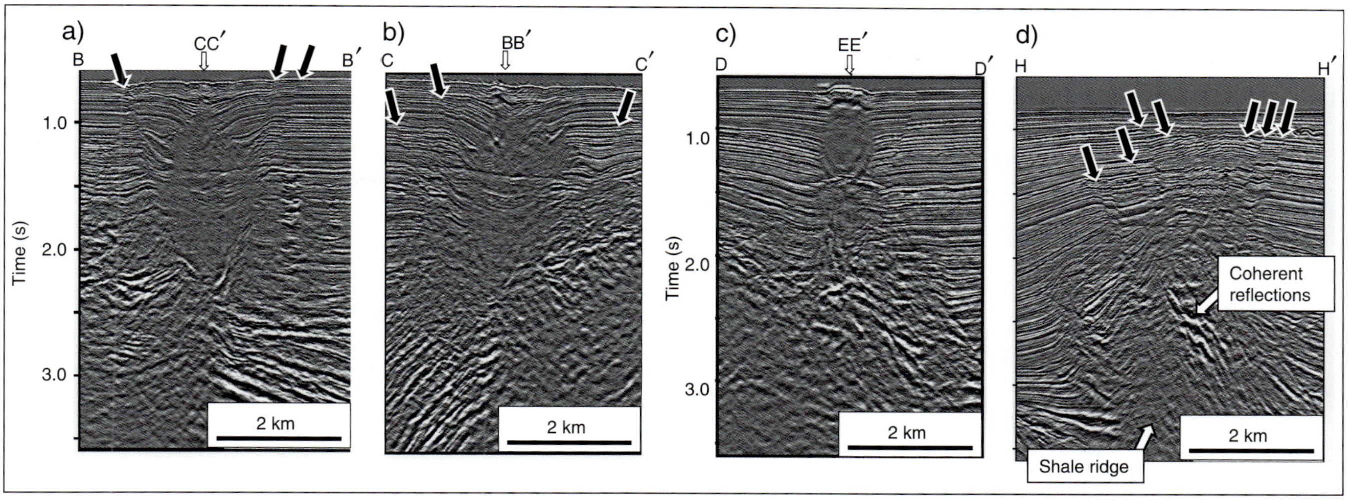

Figure 4.29. Vertical slices through a seismic survey acquired offshore Nigeria. Line locations are shown in Figure 4.28. The diapirs are rooted in a deeper shale ridge, such as that shown in (d) in line HH′. The faults (indicated by black arrows) in (d) are linked to radial faults emanating from a shale diapir to the north. The faults indicated by black arrows in (a) and (b) are ring faults associated with a collapse feature. In contrast, line DD' in (c) shows doming. Other cross sections are displayed in the original paper. After Haskell et al. (1999).

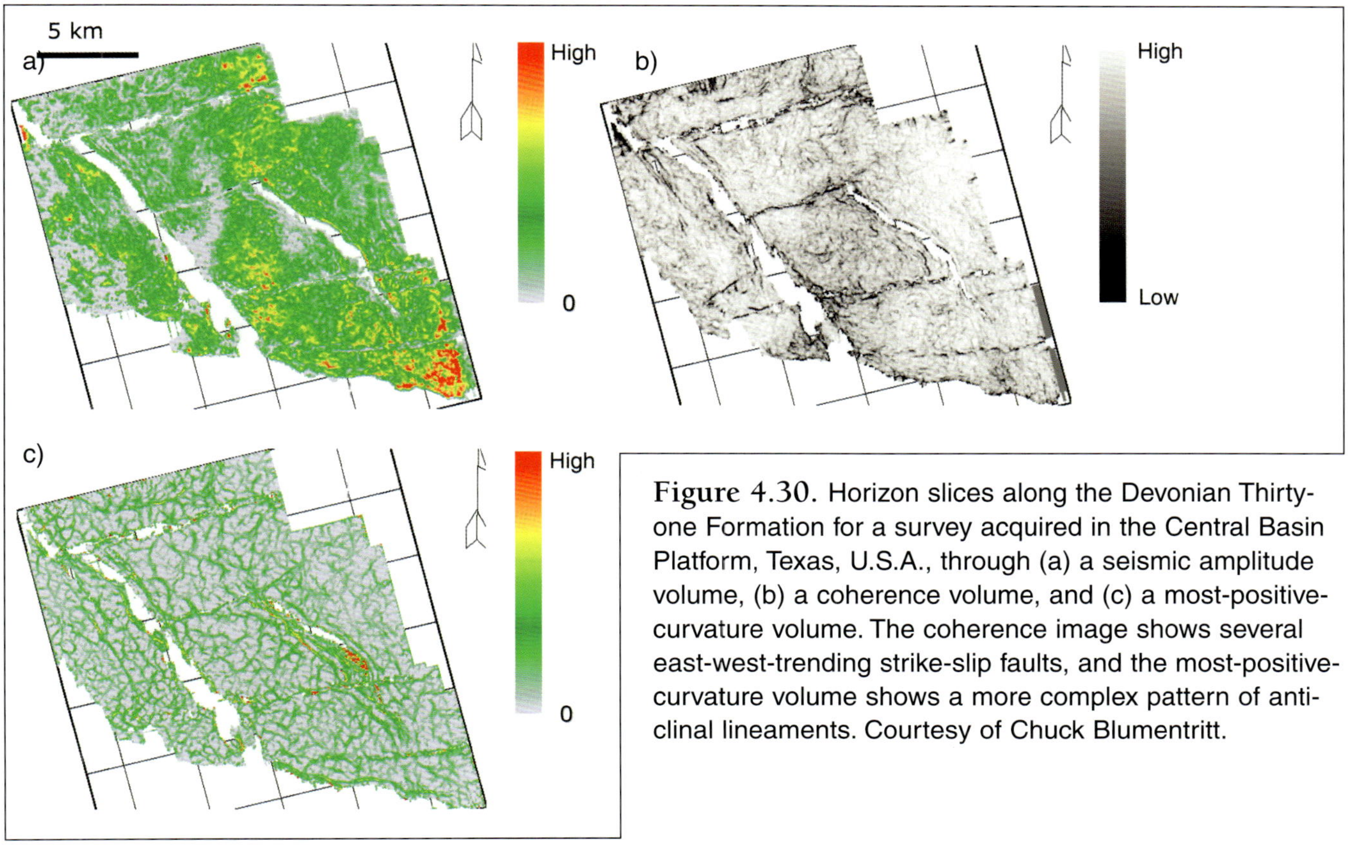

Figure 4.30. Horizon slices along the Devonian Thirty-one Formation for a survey acquired in the Central Basin Platform, Texas, U.S.A., through (a) a seismic amplitude volume, (b) a coherence volume, and (c) a most-positive-curvature volume. The coherence image shows several east-west-trending strike-slip faults, and the most-positive-curvature volume shows a more complex pattern of anticlinal lineaments. Courtesy of Chuck Blumentritt.

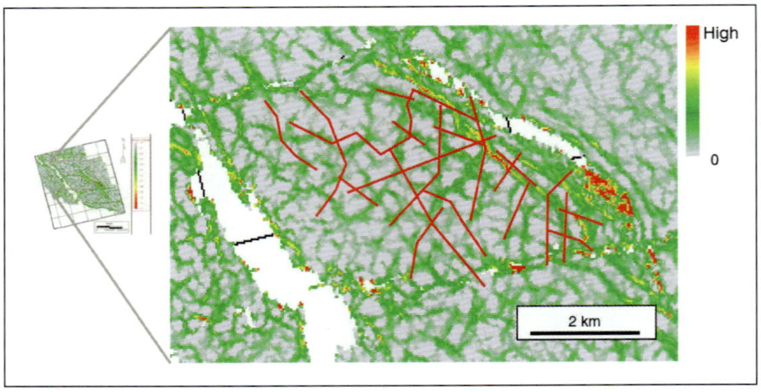

Figure 4.31. Expanded view of a portion of the most-positive-curvature horizon slice shown in Figure 4.30c. To map these lineaments, we pick them and display them as red segments.

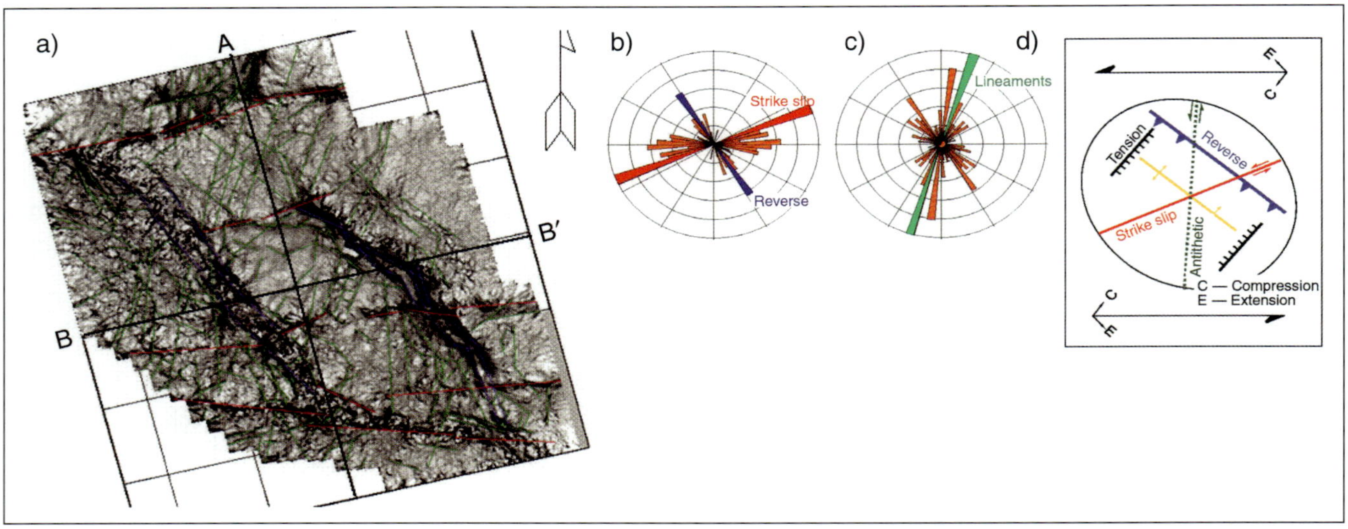

Figure 4.32. Analysis of structural deformation using picked faults and lineaments. (a) An interpreted time slice through most positive curvature near the Devonian horizon in the center of the survey. Blue lineaments represent reverse faults and red lineaments represent strike-slip faults, both of which can be interpreted easily from the original seismic data using traditional techniques. Green lineaments represent trends visible only on curvature attributes. A rose diagram of (b) the picked faults and of (c) the picked curvature lineaments. The most frequently occurring azimuth does not correspond to any azimuth observed for the red and blue trends. The second most frequently occurring azimuth does align with the most frequently occurring azimuth for the blue trend. (d) A left-lateral-strain ellipsoid (modified from Harding, 1974), with the primary trends highlighted. This model suggests that tension fractures are oriented along a strike of N 40–45° E. Courtesy of Chuck Blumentritt.

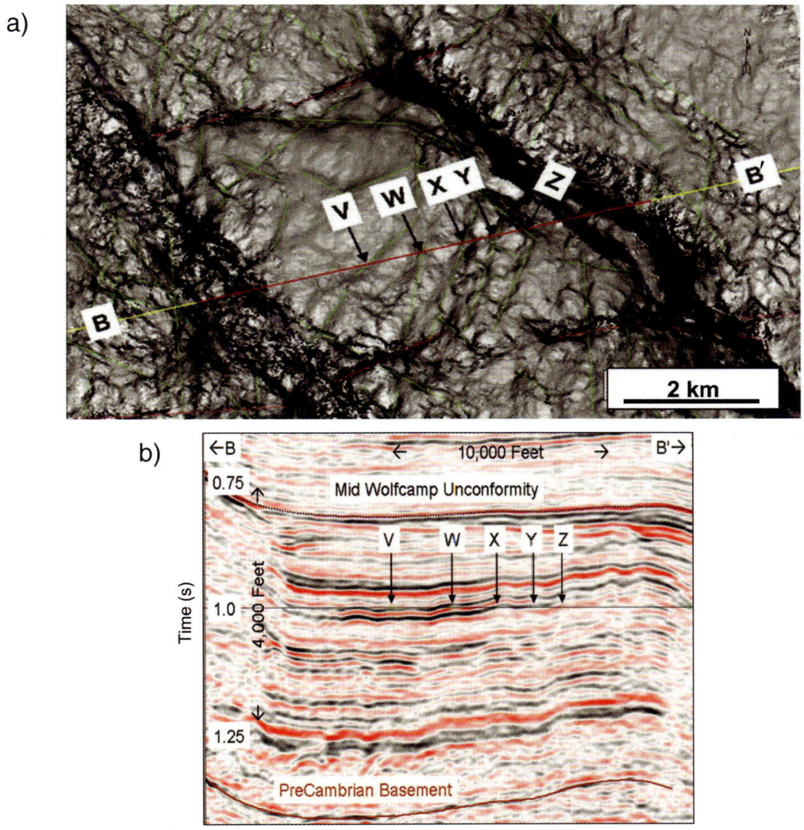

Figure 4.33. Enlarged views of (a) a most-positive-curvature time slice and (b) line BB′, showing locations of five lineaments, indicated by arrows, that are visible on the most-positive-curvature slice. Red indicates high curvature; gray indicates low curvature. Lineaments appear to be associated with undulations on the seismic section but tie to different positions on different undulations. Courtesy of Chuck Blumentritt.

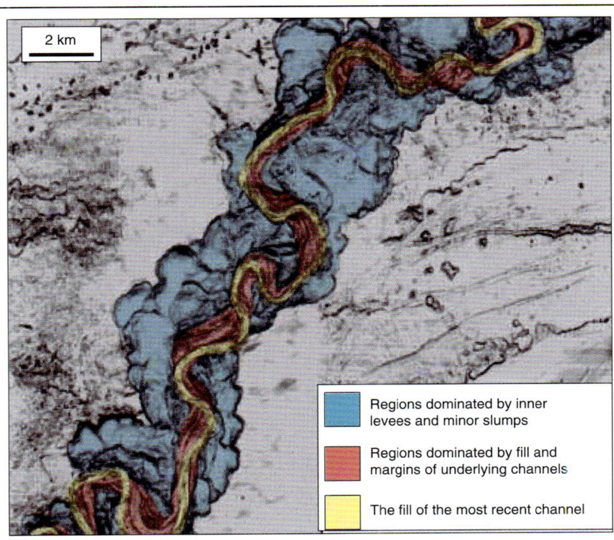

Figure 4.34. A phantom-horizon slice 76 ms below the seafloor through a coherence volume calculated from a survey acquired over the Jeanne d'Arc Basin in eastern Canada. The horizon slice cuts through the inner levee-dominated facies that developed during an earlier phase (in blue) and through the stacked channel forms that occupied the system during a later stage (yellow and red). After Deptuck (2003).

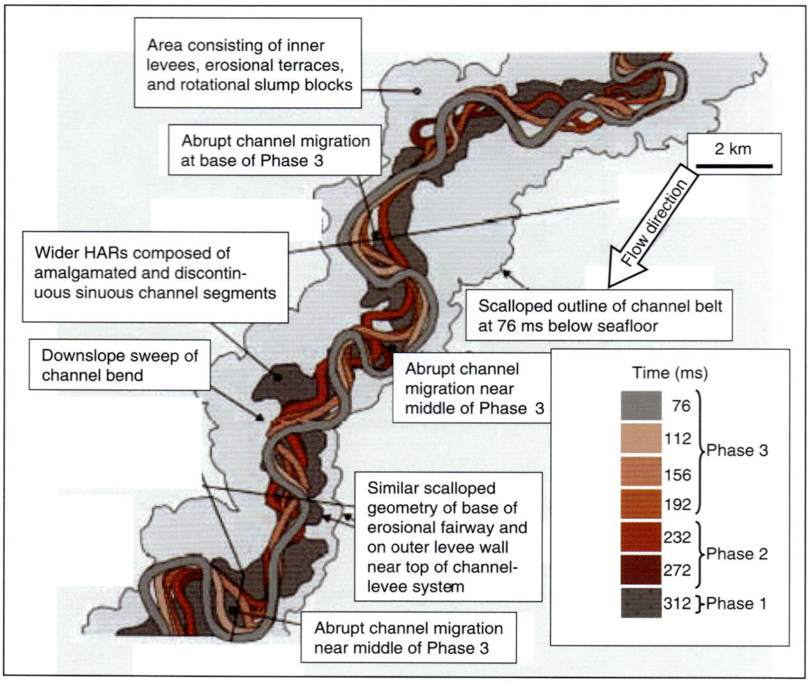

Figure 4.35. A drawing based on data from the Jeanne d'Arc Basin survey in the previous figure, showing here the planform geometry of the stacked channel-axis-deposit high-amplitude reflections (HARs) between 36 and 40 ms below the seafloor. The drawing was derived from a coherence horizon slice. After Deptuck (2003).

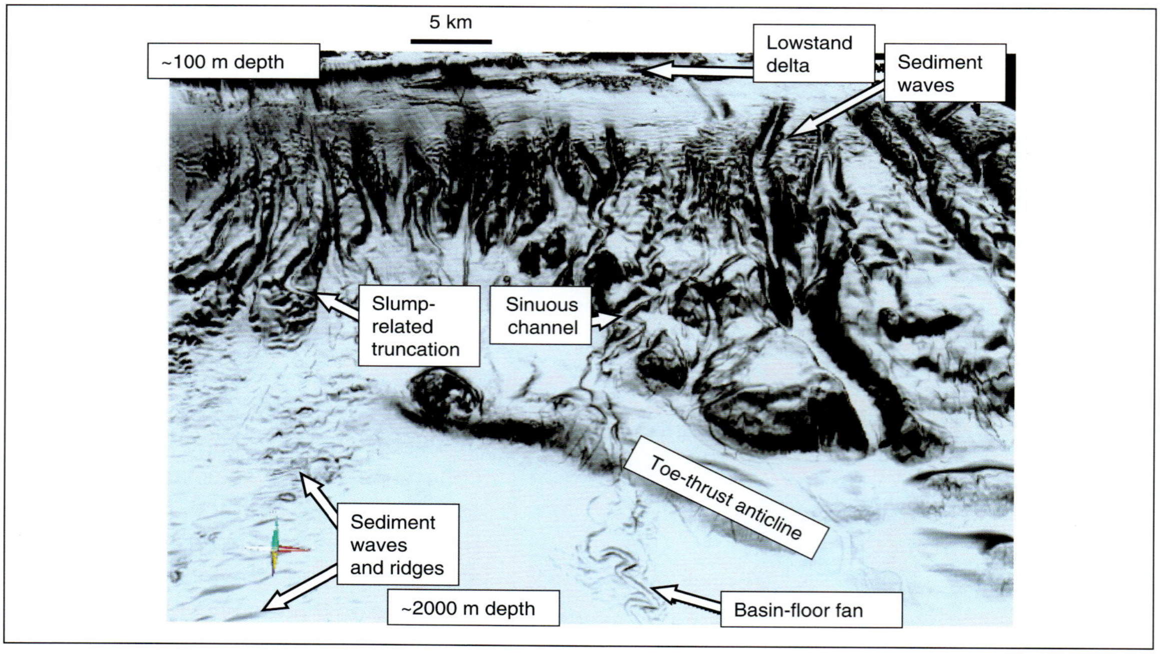

Figure 4.36. Sedimentary features commonly seen on a deepwater shelf margin, imaged by a coherence horizon slice along the water bottom from a survey acquired offshore East Kalimantan, Indonesia. The slope extends from the shelf, which is approximately 100 m deep, to the basin floor, which is approximately 2000 m deep. Most canyons on the slope are straight or gently curving, except one, which has sharp sinuous bends and connects the earlier lowstand delta to a contemporaneous basin-floor fan. Sediment waves are common on the upper slope, upslope from the deep canyons, and on the lower slope. After Saller et al. (2004). AAPG©2004. Reprinted by permission of the AAPG whose permission is required for further use.

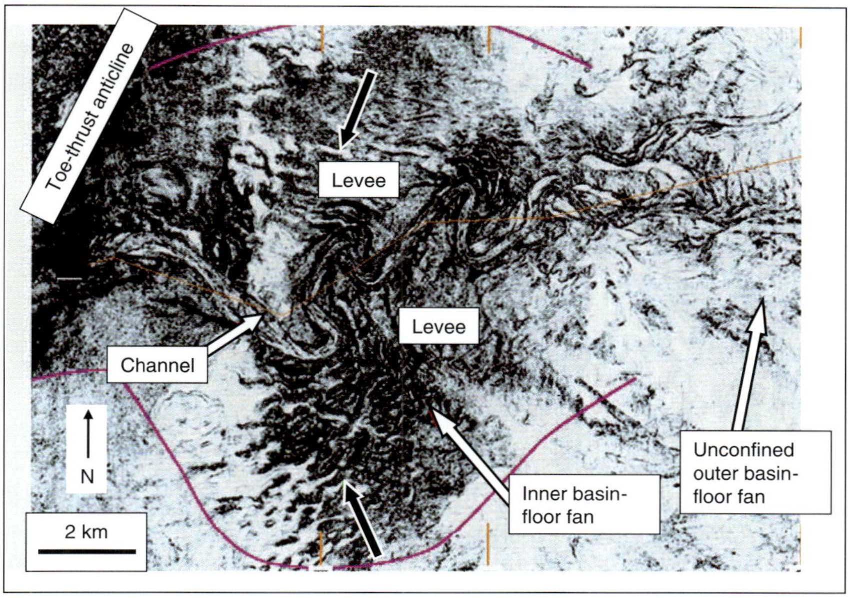

Figure 4.37. A coherence horizon slice that shows a sinuous, meandering channel in an inner basin-floor fan passing into an unconfined outer basin-floor fan. On the outside of the meander loops, the black arrows point to ridges that are sediment waves. After Saller et al. (2004). AAPG©2004. Reprinted by permission of the AAPG whose permission is required for further use.

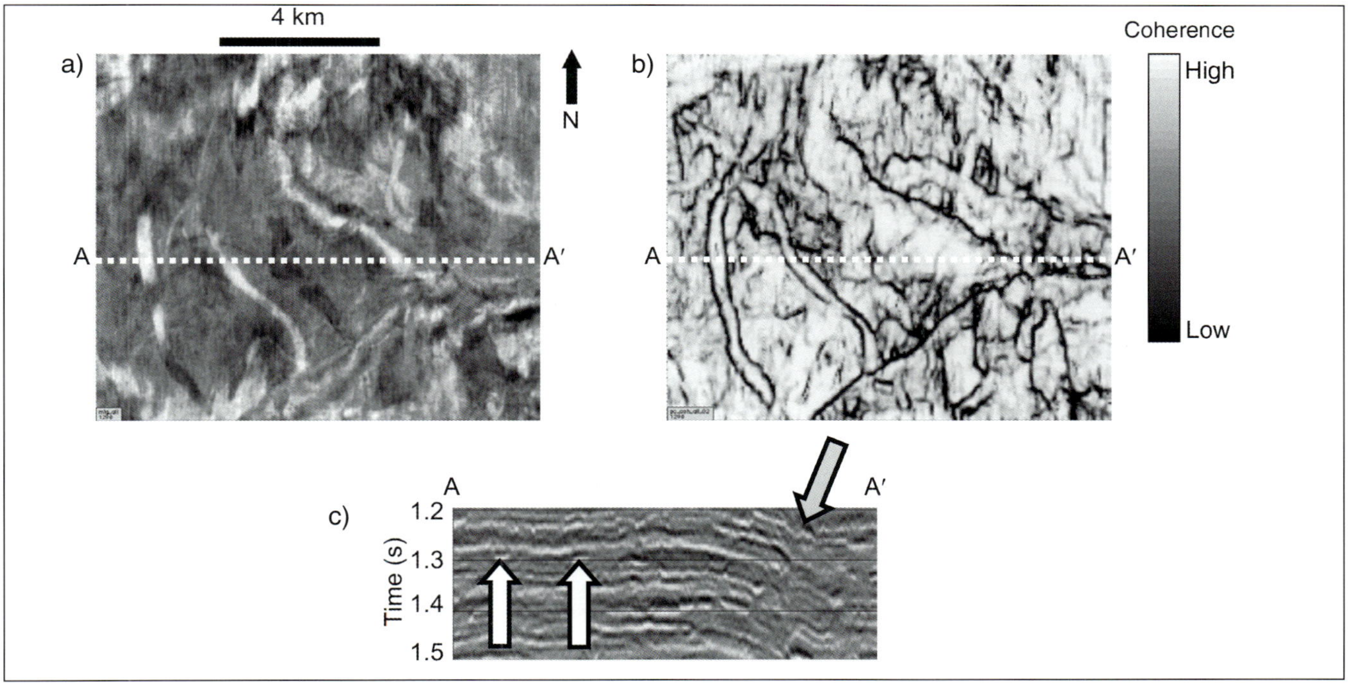

Figure 4.38. (a)-(b). Time slices at 1.300 s showing multiple channels through (a) a seismic data volume and (b) a coherence volume for a survey acquired in the South Marsh Island area, Gulf of Mexico, U.S.A. (c) A vertical section through the seismic data along line AA′. White arrows indicate the two prominent channels seen in the time slices. The gray arrow indicates a fault. Coherence was calculated using a nine-trace, ±10-ms analysis window. Seismic data are courtesy of Fairfield.

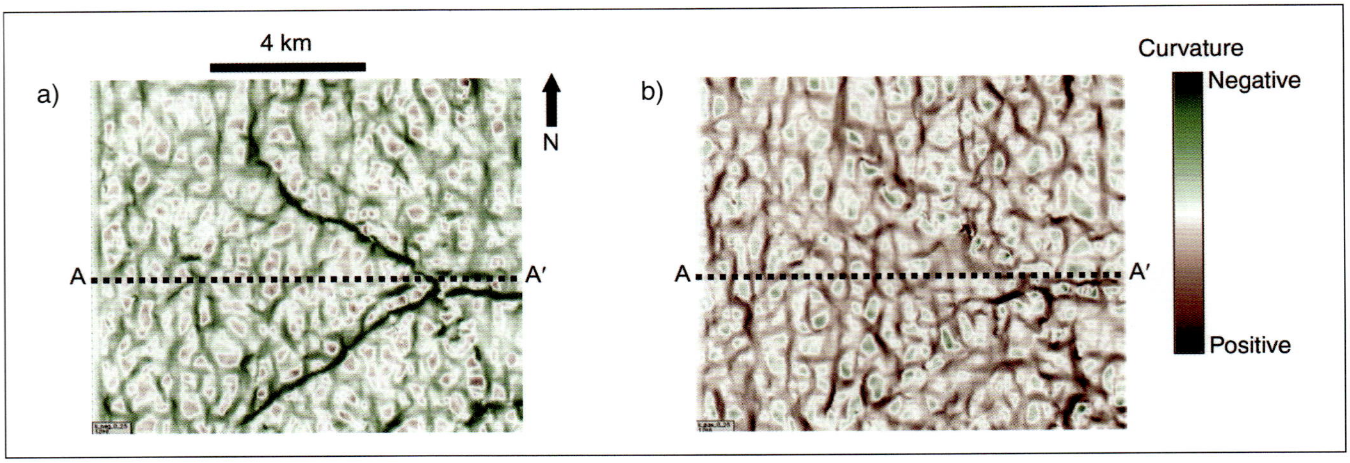

Figure 4.39. Time slices at $t = 1.300$ s, corresponding to those in the previous figure and shown here through (a) a most-positive-curvature volume and (b) a most-negative-structural-curvature volume. Although the two intersecting faults show up well on the most-negative-curvature time slice, that slice bears little indication of the channels that are so obvious on the amplitude and coherence images in the previous figure. Careful inspection of Figure 4.38c reveals that the rather incoherent patterns are real and are correlated perfectly with the hummocky nature of the sediments along seismic line AA′. We suspect that this hummocky pattern is associated with initial deposition or subsequent slumping rather than with differential compaction over the channels.

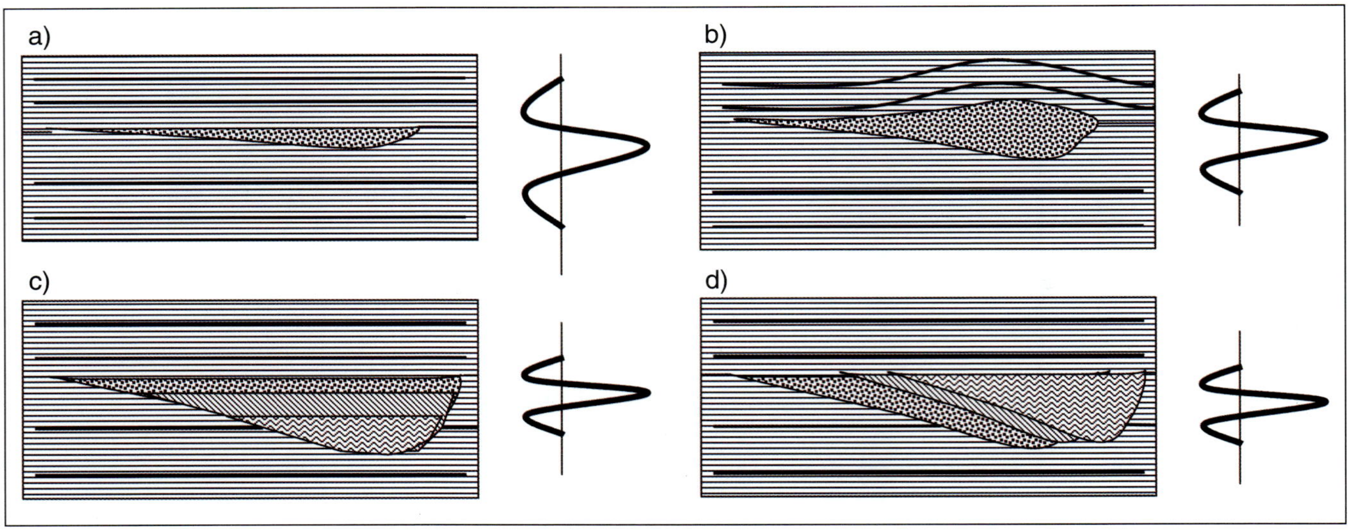

Figure 4.40. Attribute response to channel features preserved in the geologic record. (a) Below thin-bed tuning, with homogeneous fill and no differential compaction. Only the amplitude changes. (b) Below thin-bed tuning, with homogeneous fill and differential compaction. Curvature and amplitude change; waveform does not change. (c) Above thin-bed tuning, with heterogeneous horizontal fill and no differential compaction. Curvature does not change; amplitude and waveform do change. (d) Above thin-bed tuning, with heterogeneous aggradational fill. Curvature, amplitude, and waveform all change.

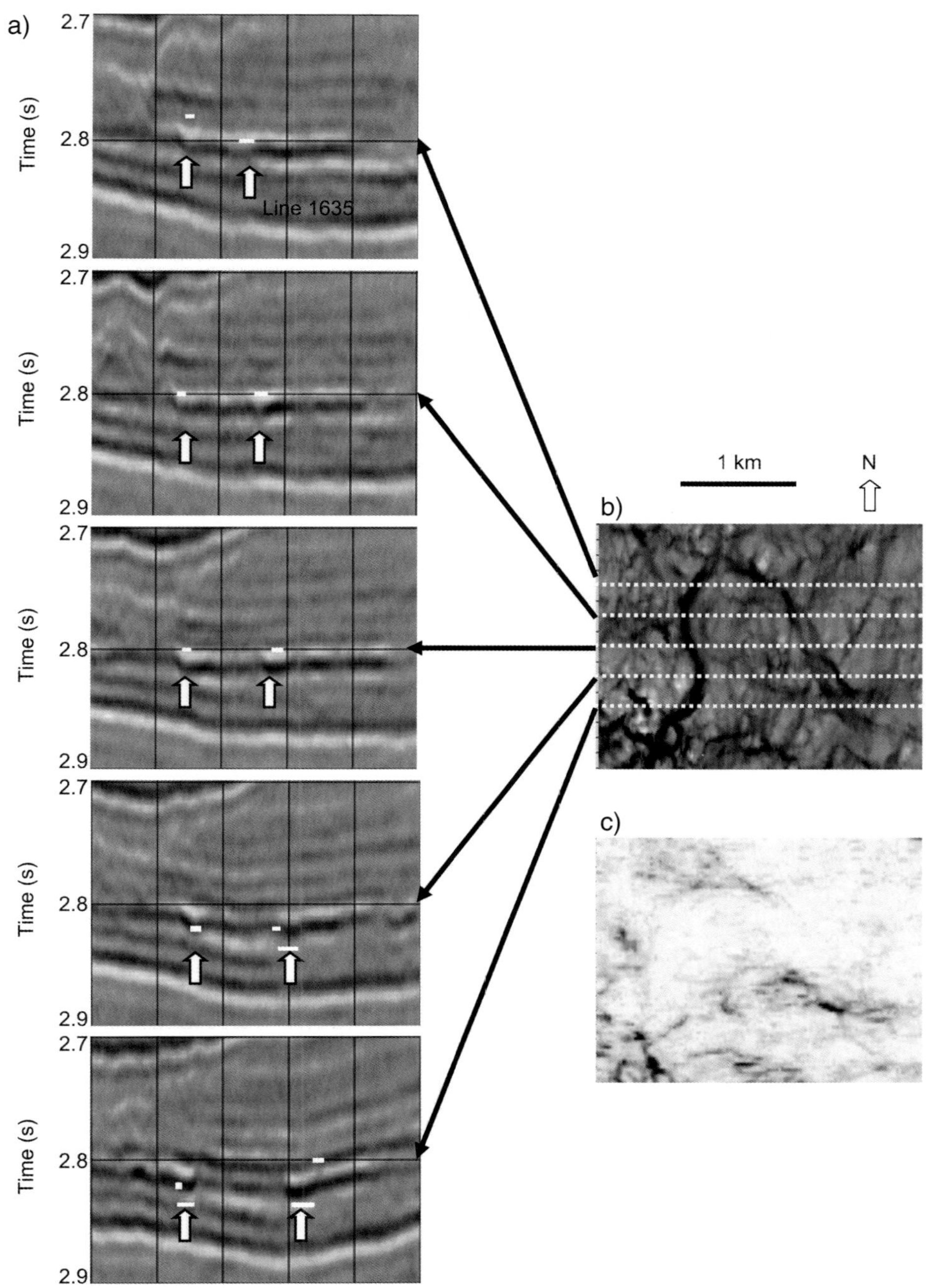

Figure 4.41. (a) Vertical slices through seismic data at 300-m intervals and (b) a time slice at $t = 2.800$ s through a most-negative-curvature volume. The channel, which shows up here because of differential compaction, was painted at 20-ms intervals to show its appearance on the seismic image. (c) A corresponding time slice through the coherence volume.

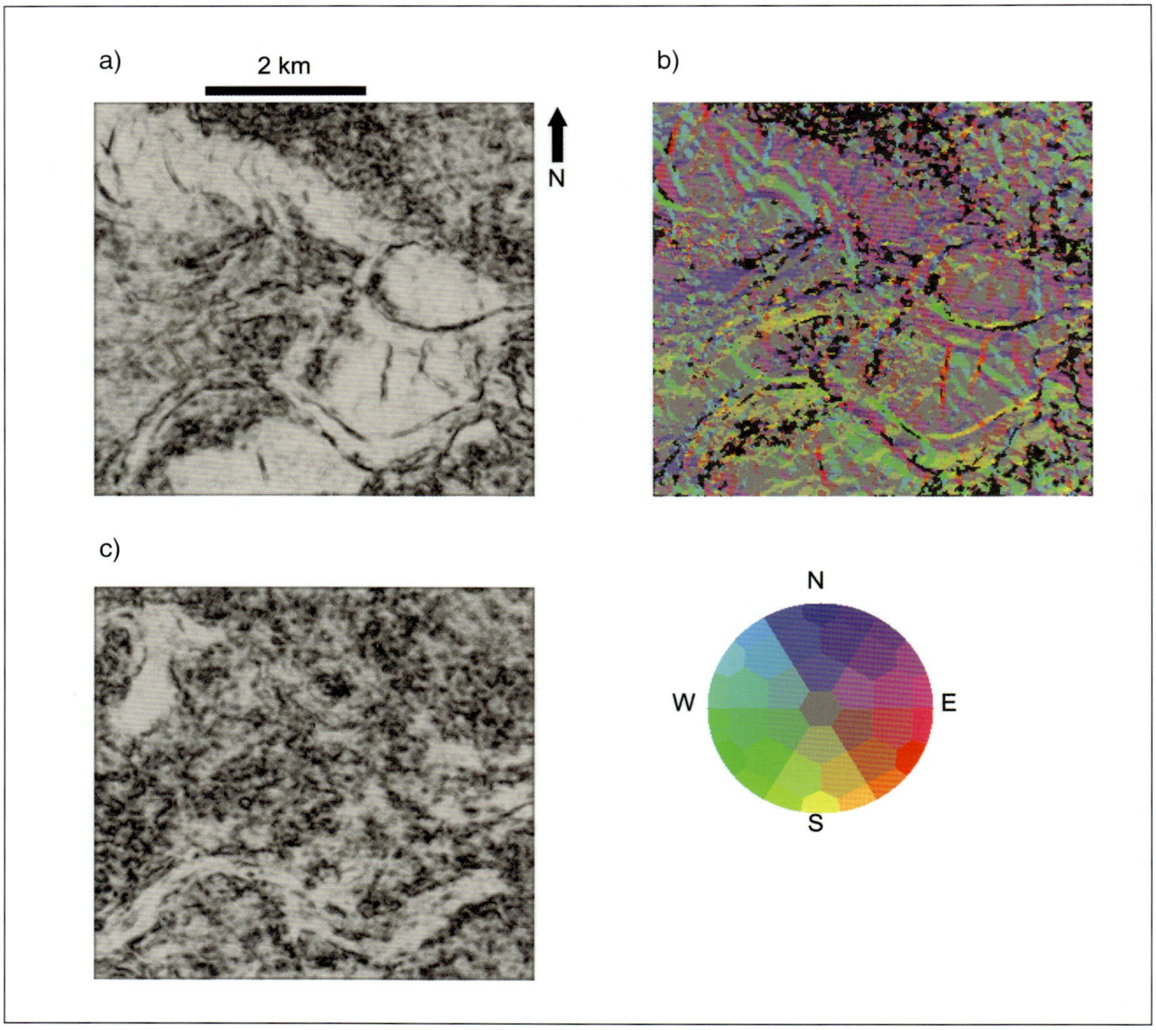

Figure 4.42. The appearance of differential compaction. (a)-(b). Time slices through (a) a coherence volume and (b) a dip and azimuth volume at the suspected channel level. Note that the apparent dip along the channel edges is inward, or toward the north-northeast (blue and magenta) on the south flank, and is toward the south-southwest (yellow and green) on the north flank. (c) A time slice through the coherence volume 220 ms below that shown in (a). We now see that the true channel occurs at this deeper level and that faulting caused by differential compaction produced the congruent image of the channel on the shallow slice. After Hughes et al. (1996). Reprinted by permission of the Geological Society.

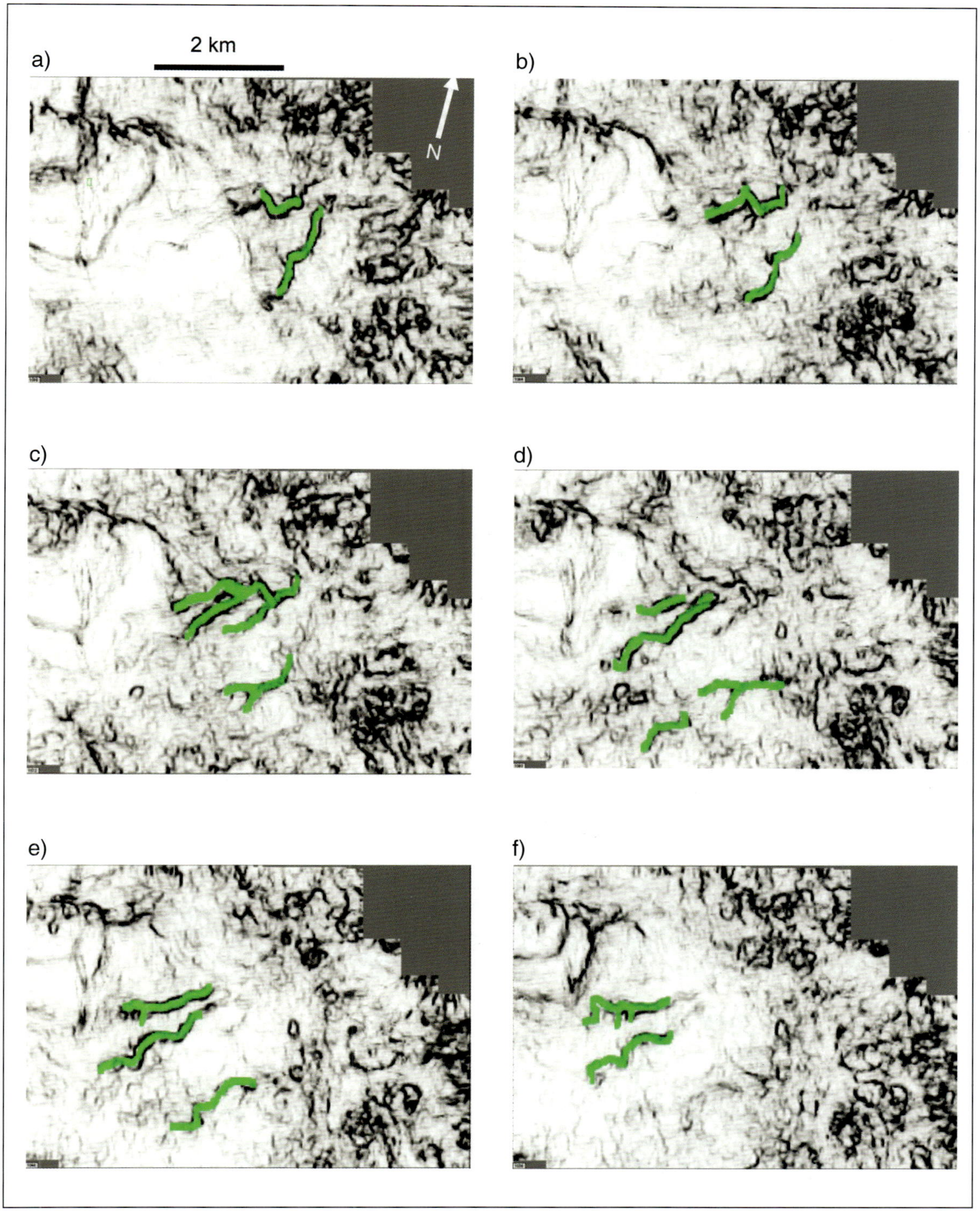

Figure 4.43. A work-flow diagram shows how to quickly obtain a crude map of a channel without first interpreting the horizon in which it is confined. After initial animation through the six time slices, we recognize a channel in a structurally deformed setting. Using commercial interpretation software, we "paint" the channel in green as we see it on each time slice, one slice at a time. Seismic data are courtesy of Burlington Resources.

Figure 4.44. Commercial interpretation software generates a map of the painted picks from each time slice in the previous figure. Thus, we do not need to explicitly pick a horizon above or below the channel to map its location.

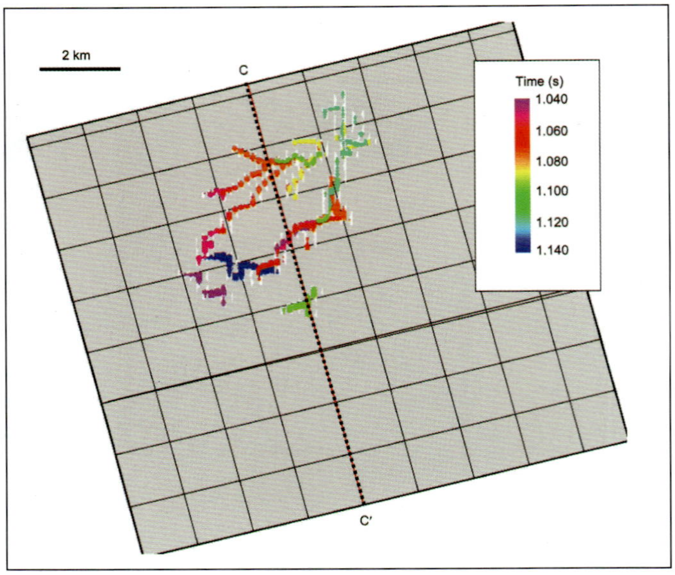

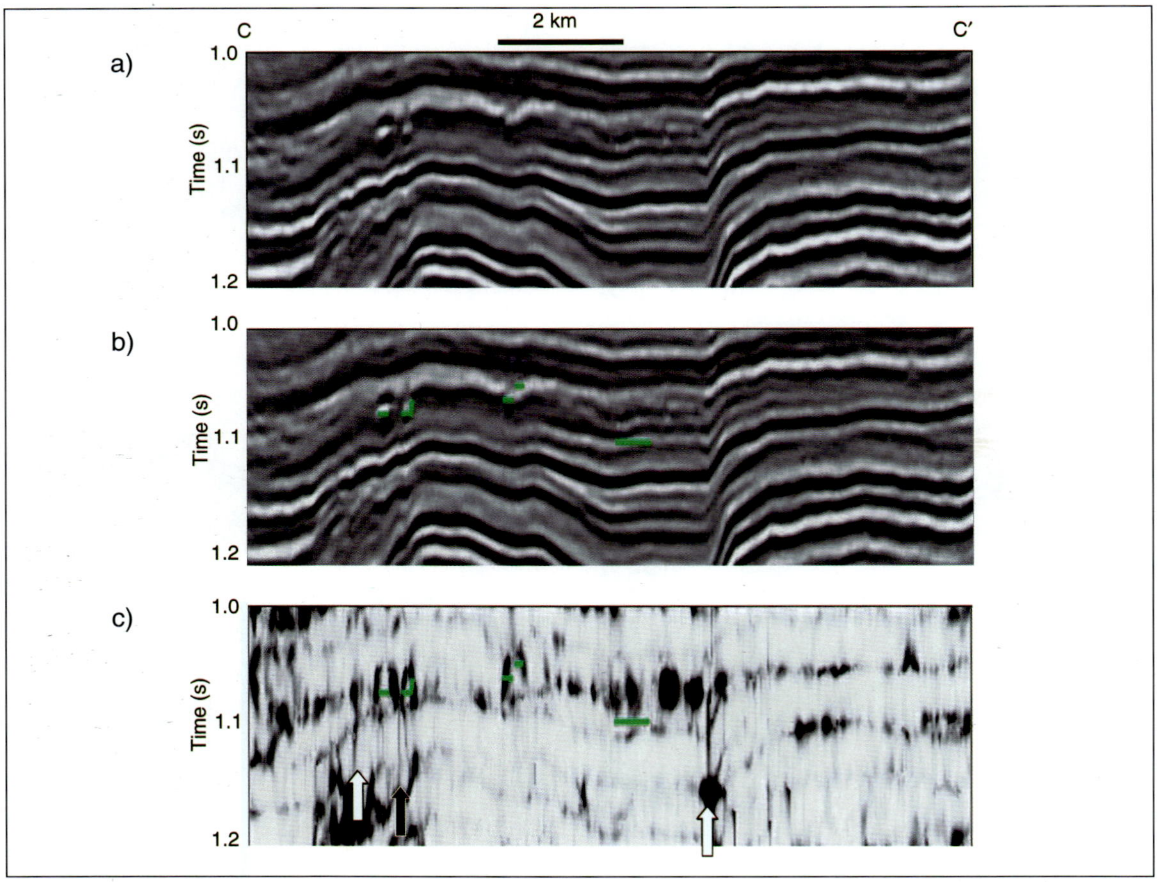

Figure 4.45. Line CC′ from the previous figure, shown here through the seismic data volume (a) without and (b) with the channel picks that were made in Figure 4.43 displayed on the data. (c) A corresponding vertical slice through the coherence volume. The seismic data have been subjected to principal-component, structure-oriented filtering.

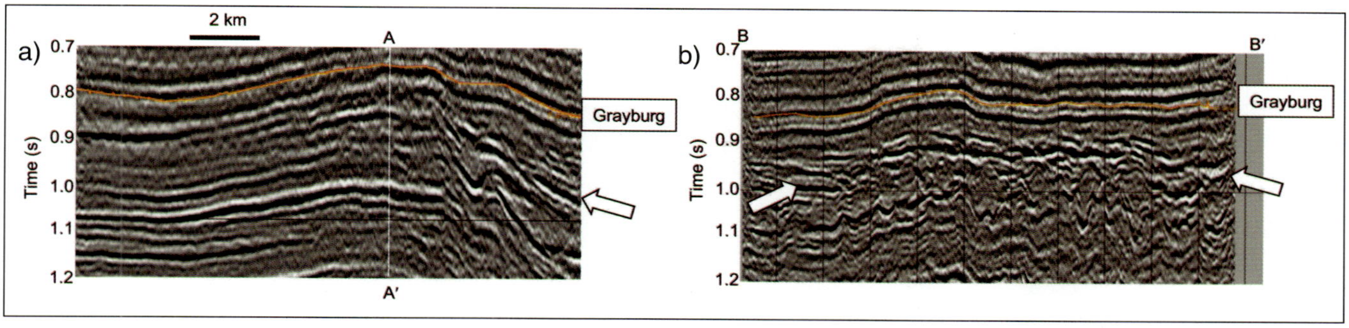

Figure 4.46. Vertical seismic lines (a) perpendicular to and (b) parallel to a platform edge, from a survey acquired over the northern Delaware Basin, New Mexico, U.S.A. A siliciclastic sequence (arrows) is too difficult to pick, so we image it by displaying horizon slices parallel to the Grayburg horizon (shown in the next figure). Data are courtesy of Marathon Oil Company.

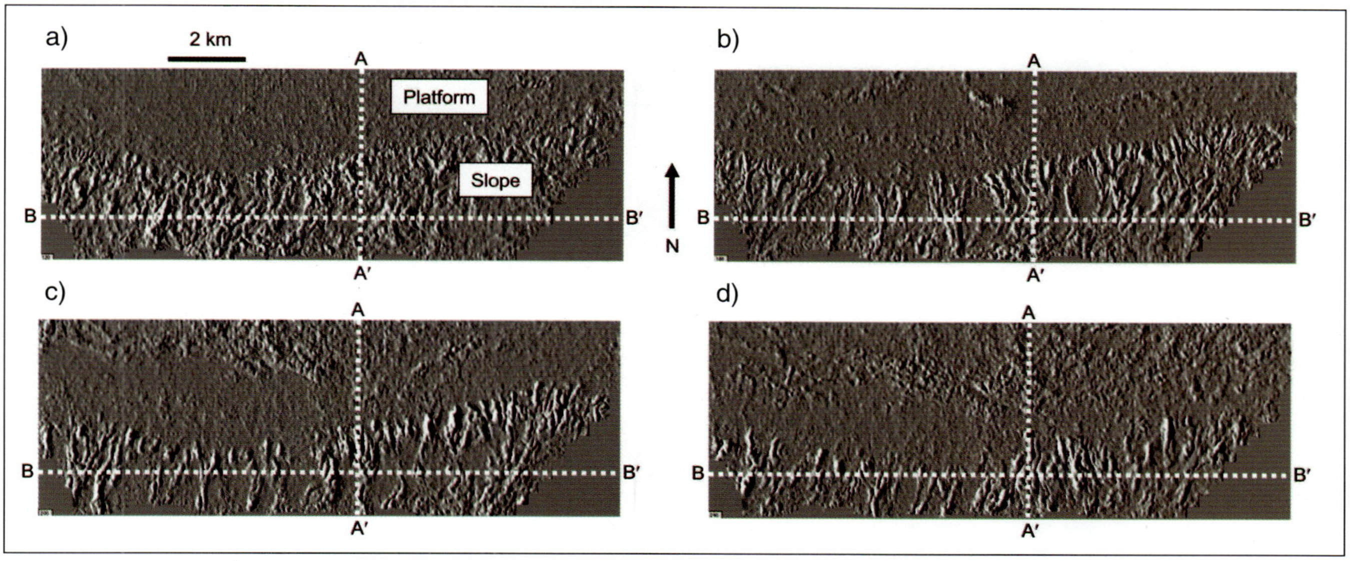

Figure 4.47. Phantom-horizon slices at t = (a) 0.120, (b) 0.160, (c) 0.200, and (d) 0.240 s below the orange Grayburg pick displayed in the previous figure. The slices cut through the east-west component of the energy-weighted coherent-amplitude gradient. The northern part of the slice cuts through the Permian carbonate platform. South of the platform are slumps and channel fill, sourced from eolean dunes, that gave rise to the Brushy Canyon and Cherry Canyon formations of the Delaware Mountain Group.

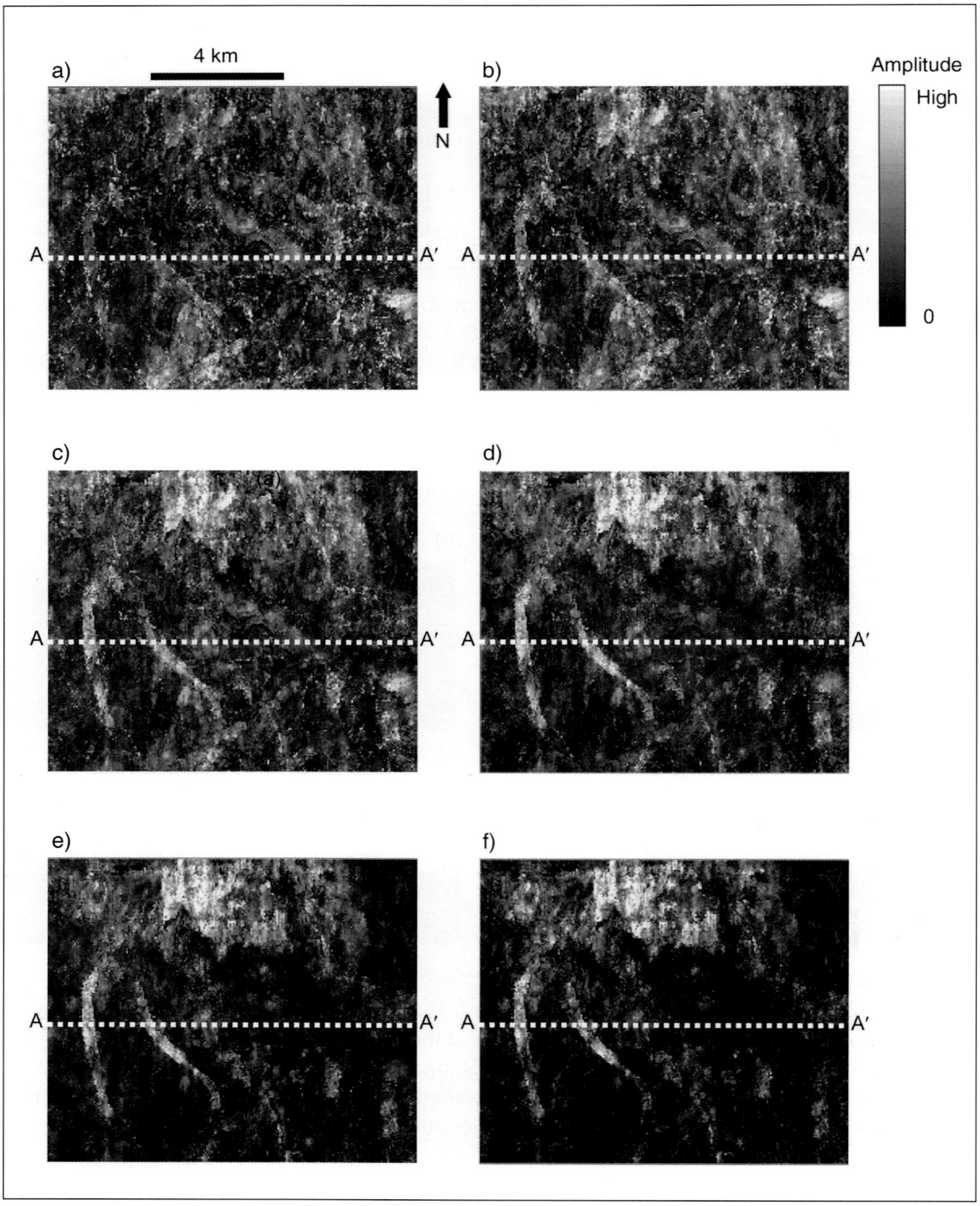

Figure 4.48. Spectral components derived using a matched-pursuit algorithm and corresponding to the time slice at t = 1.296 s, from Figures 4.38 and 4.39, shown here at (a) 30 Hz, (b) 40 Hz, (c) 50 Hz, (d) 60 Hz, (e) 70 Hz, and (f) 80 Hz. Each spectrum is scaled independently. The channel stands out best against the background between 60 and 70 Hz.

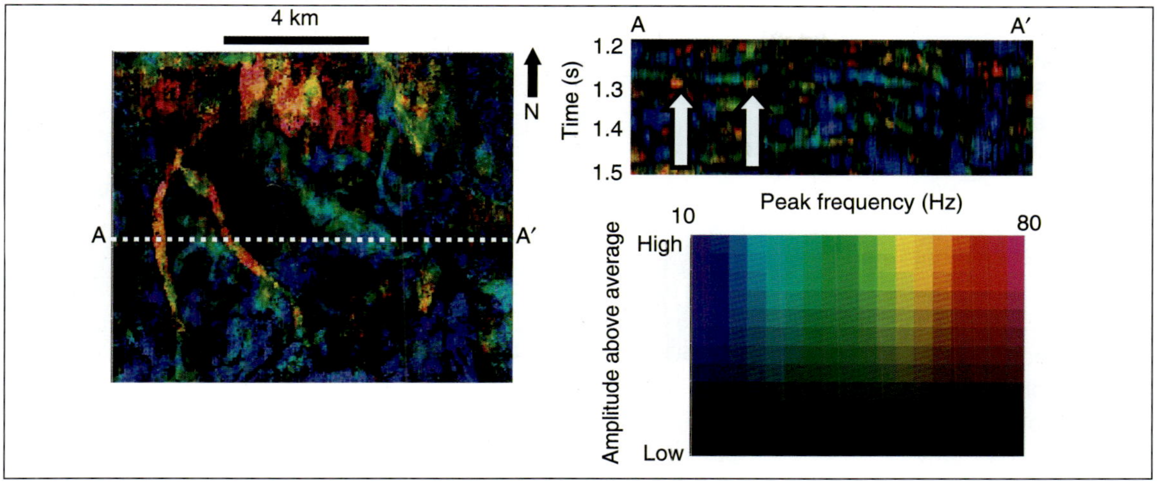

Figure 4.49. A composite image summarizing the information shown in the previous figure, here using a 2D color table and plotting peak frequency against the amplitude of that frequency above the average normalized amplitude for all frequencies at each sample. Note that the channel tunes in as orange, implying a peak frequency of about 65 Hz. Similarly, the bright blue area in the southern part of the image tunes in at 25 Hz. Such composite images can summarize much of the information contained in the individual components.

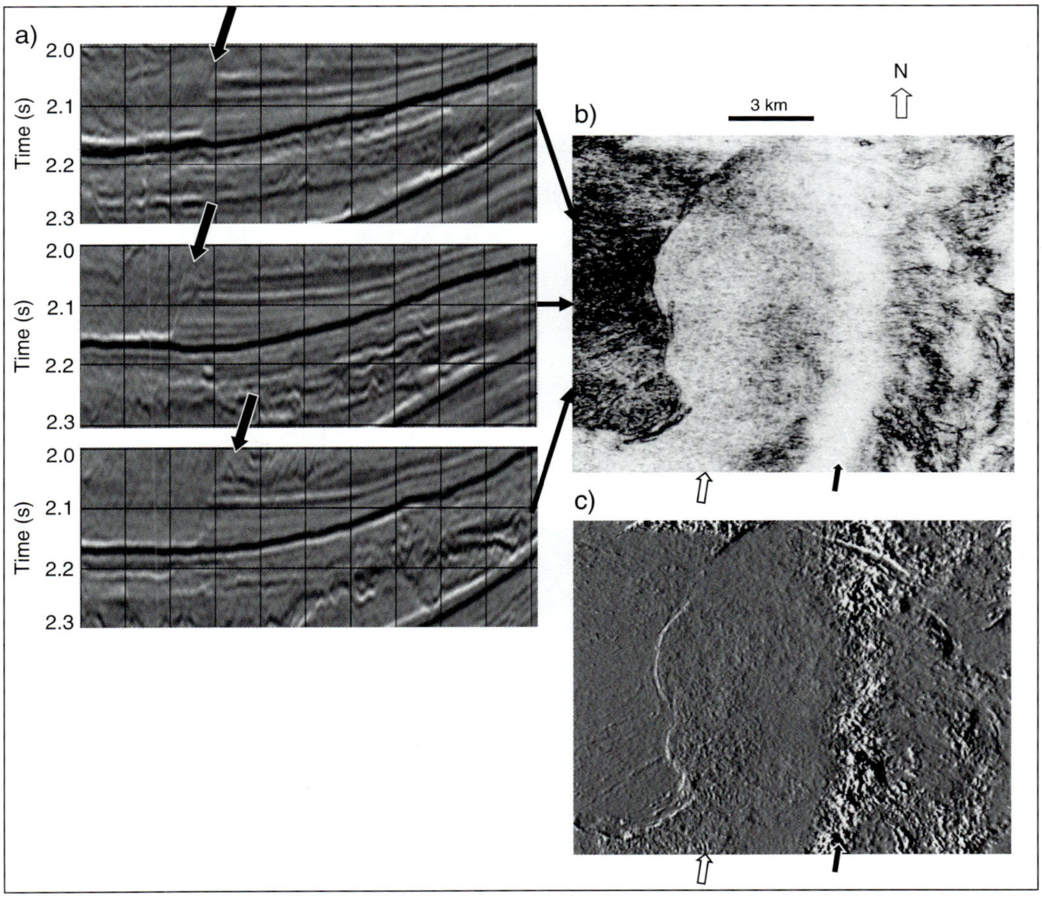

Figure 4.50. The appearance of an erosional scarp on (a) vertical seismic slices and (b)-(c) on time slices at $t = 2.100$ s, through (b) a coherence volume and (c) an east-west energy-weighted coherent-amplitude-gradient volume.

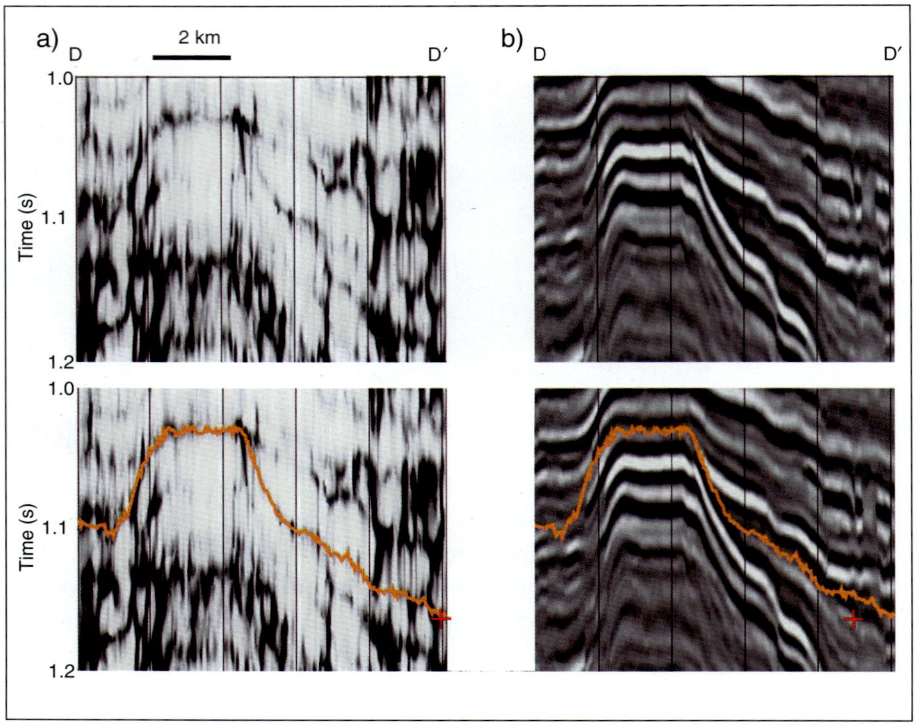

Figure 4.51. Coherence as an aid in picking angular unconformities. Vertical slices through (a) a coherence volume and (b) a seismic data volume, from the Central Basin Platform in west Texas, U.S.A. We recommend displaying the conventional seismic data and coherence data in dual-window mode, with the cursor (red plus sign) linked between the two windows. Although coherence helps us put the unconformity at a precise location, it leaves many choices to be made. Here, we made the choice by tracking where the cursor moved on the seismic data. We then snapped the pick to a nearby low value of coherence. Seismic data are courtesy of Burlington Resources.

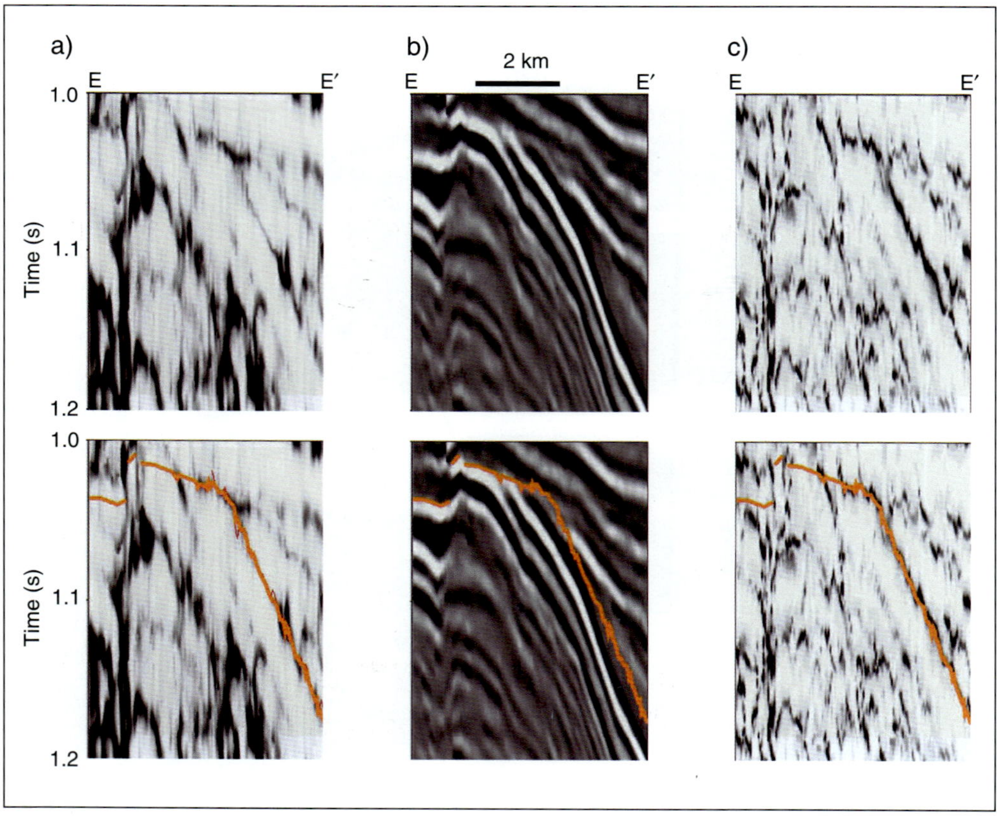

Figure 4.52. Vertical slices through (a) a coherence volume, (b) a seismic data volume, and (c) a measure of angular unconformity computed as the magnitude of the vertical derivative of reflector dip and azimuth. The location of line EE′ is shown in Figure 4.53. The northern end of the platform is cut by a steep fault. Clearly, more than one angular unconformity occurs on the south flank. Here, we have tracked one that we interpret to be the Atoka unconformity.

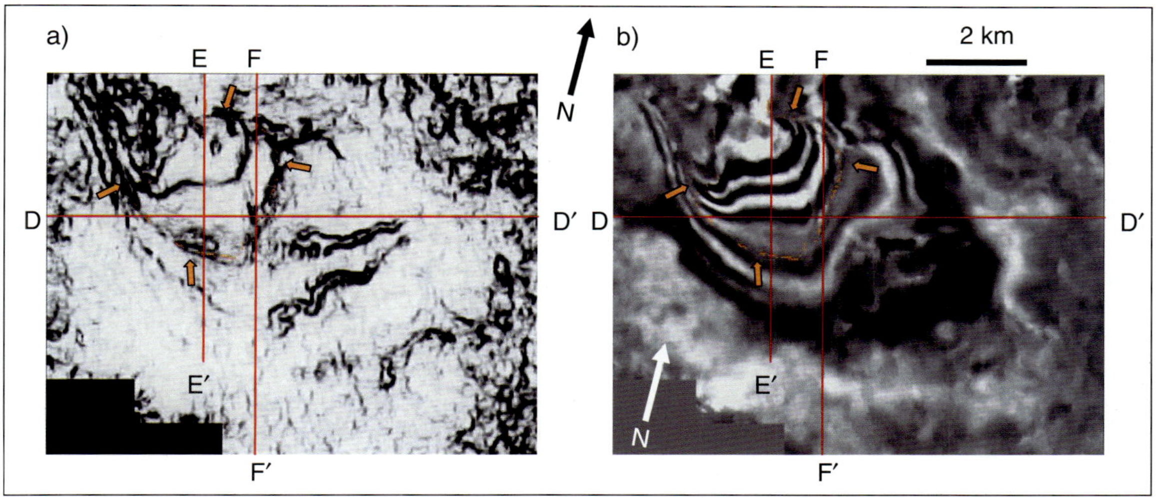

Figure 4.53. Time slices at $t = 1.000$ s through (a) a coherence volume and (b) a seismic data volume. Orange arrows indicate the location of the angular unconformity, which is nearly impossible to see on the seismic time slice. In contrast, the coherence time slice shows the unconformity as a zone of low coherence. Note the channels to the east that drained the unconformity. They were discussed in the captions of Figures 4.43 and 4.44.

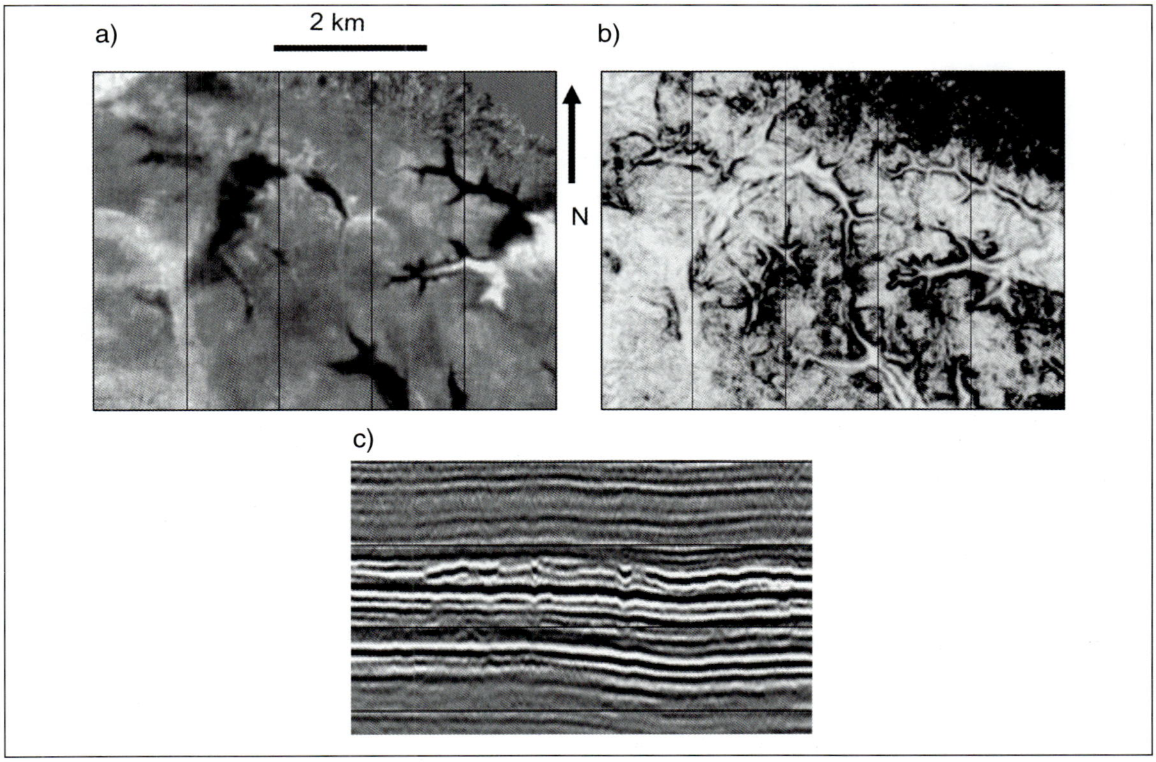

Figure 4.54. (a)-(b).Horizon slices through (a) a seismic data volume and (b) a coherence volume, illuminating a complex series of interrelated drainage channels in an estuary from Alberta. (c) Vertical seismic line AA′. Although the conventional seismic time slice is excellent, it does not show as complete an image as the coherence slice does.

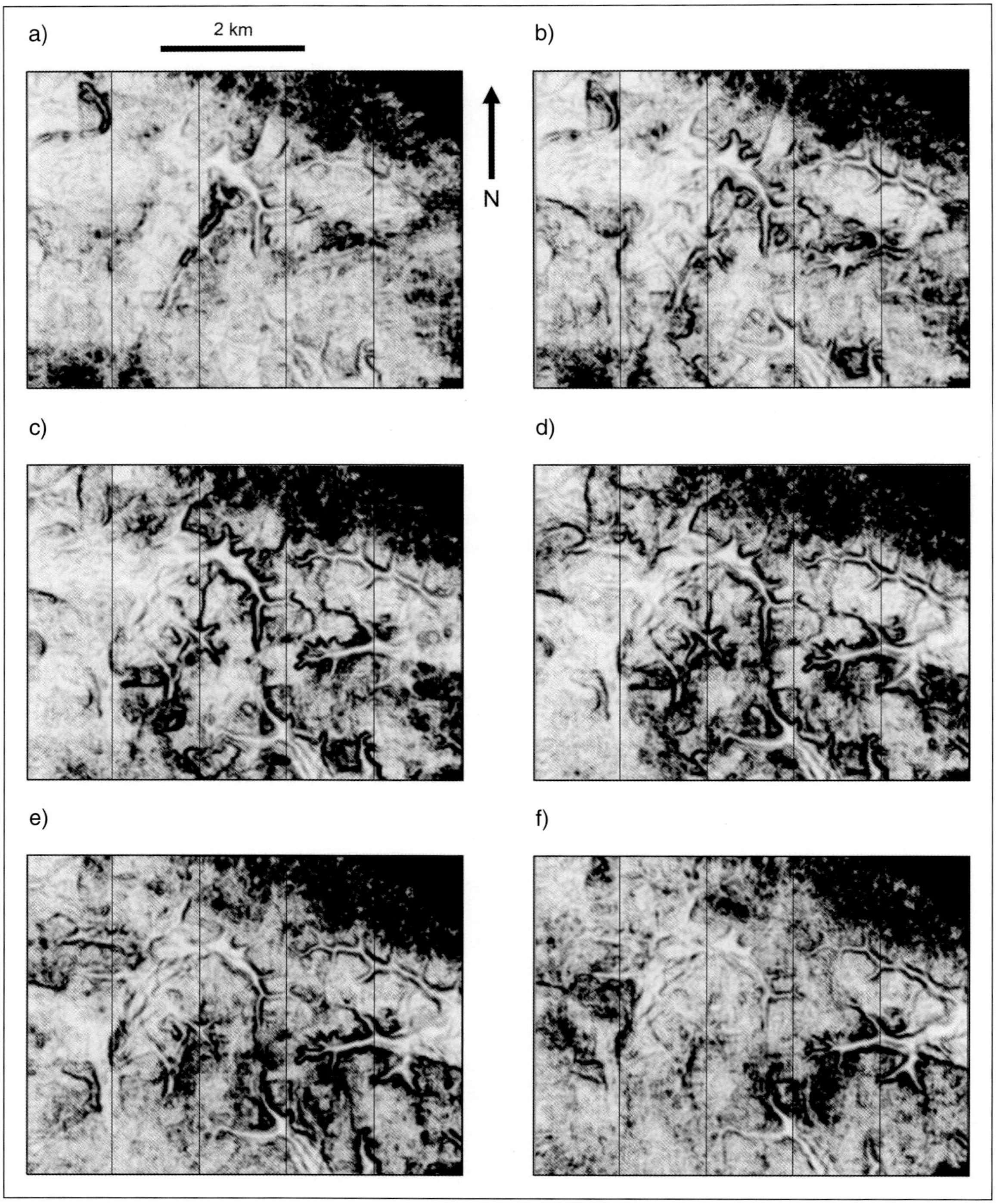

Figure 4.55. A suite of horizon slices, at (a) +12 ms, (b) +8 ms, (c) +4 ms, (d) 0 ms, (e) –4 ms, and (f) –8 ms below the time slice through the coherence volume shown in Figure 4.54b. These images give the areal disposition of the channel system and a qualitative estimate of the depth of channel sediments.

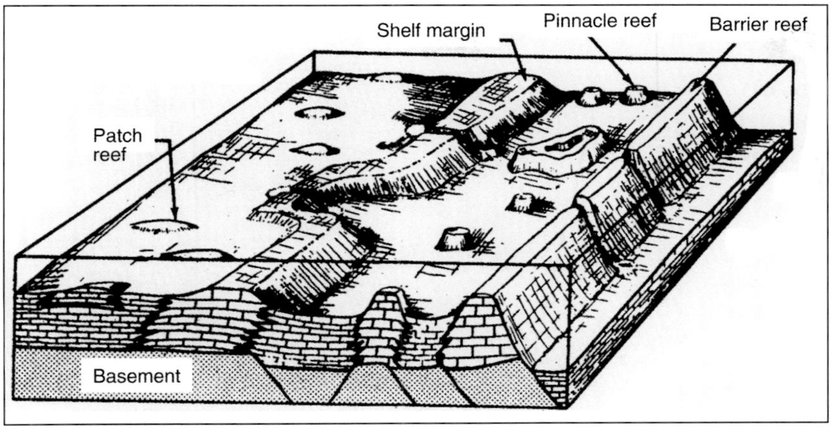

Figure 4.56. The typical environment for reef growth and the types of reefs. After Bubb and Hatledid (1978).

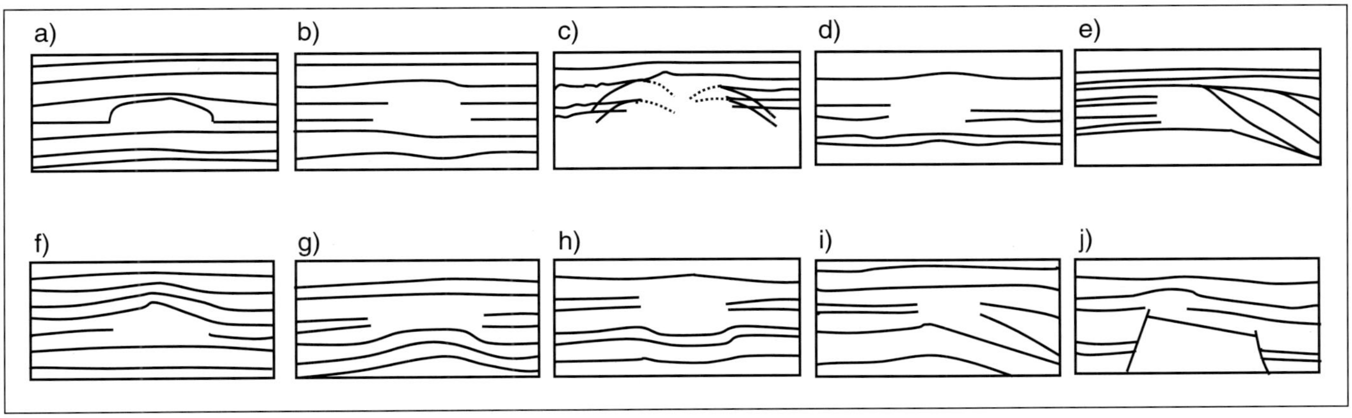

Figure 4.57. Seismic expression of reefs. Unless otherwise noted, all the following figures represent the seismic expression of a reef in 3D migrated data. (a) Reefs often have a mound shape. (b) Porous reefs often display a reflection void, or dimming, caused by a smaller velocity contrast between shale and porous carbonate rock than between shale and tight limestone. (c) Reef edges often create diffractions that can be observed in unmigrated data. (d) On migrated data, reflections often terminate abruptly along the reef's flanks. (e) Reflection patterns differ on different sides of a reef. One side generally is in a basinal environment, whereas the other side is on the shelf. (f) Layers compact differentially, or "drape," over the top of a reef because reefs tend to lithify early in their burial history. (g) Reflectors below a reef generally exhibit a velocity pull-up on a seismic time section so that subreef reflections arrive earlier. This occurs because seismic energy will have passed through high-velocity carbonate rocks rather than through the lower-seismic-velocity rocks that surround the reef. (h) Although it occurs less commonly, porous carbonates may be surrounded by higher-velocity rocks (such as anhydrite), in which case rocks below the reef exhibit a velocity pull-down or delay on the time section. (i) Reefs often initiate on a hinge line or shelf edge. (j) Reefs may grow on a structural high caused by faulting. After Bubb and Hatledid (1978).

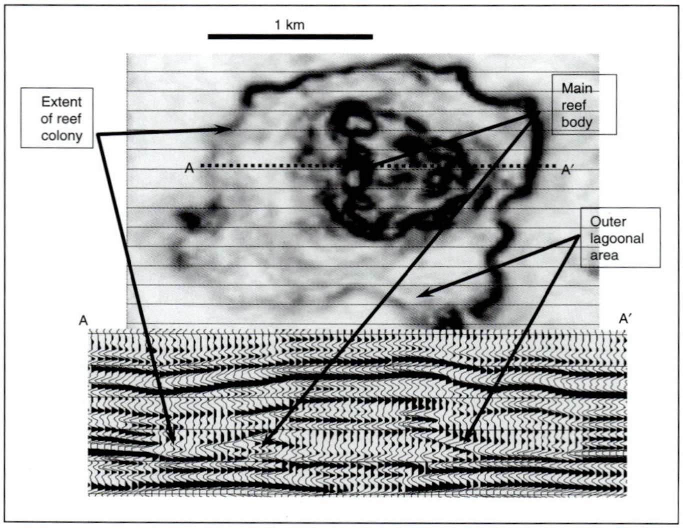

Figure 4.58. (a) A time slice at $t = 1.200$ s through a coherence volume and (b) a vertical slice AA′ through a seismic data volume, from a survey conducted over a pinnacle reef in Alberta, Canada.

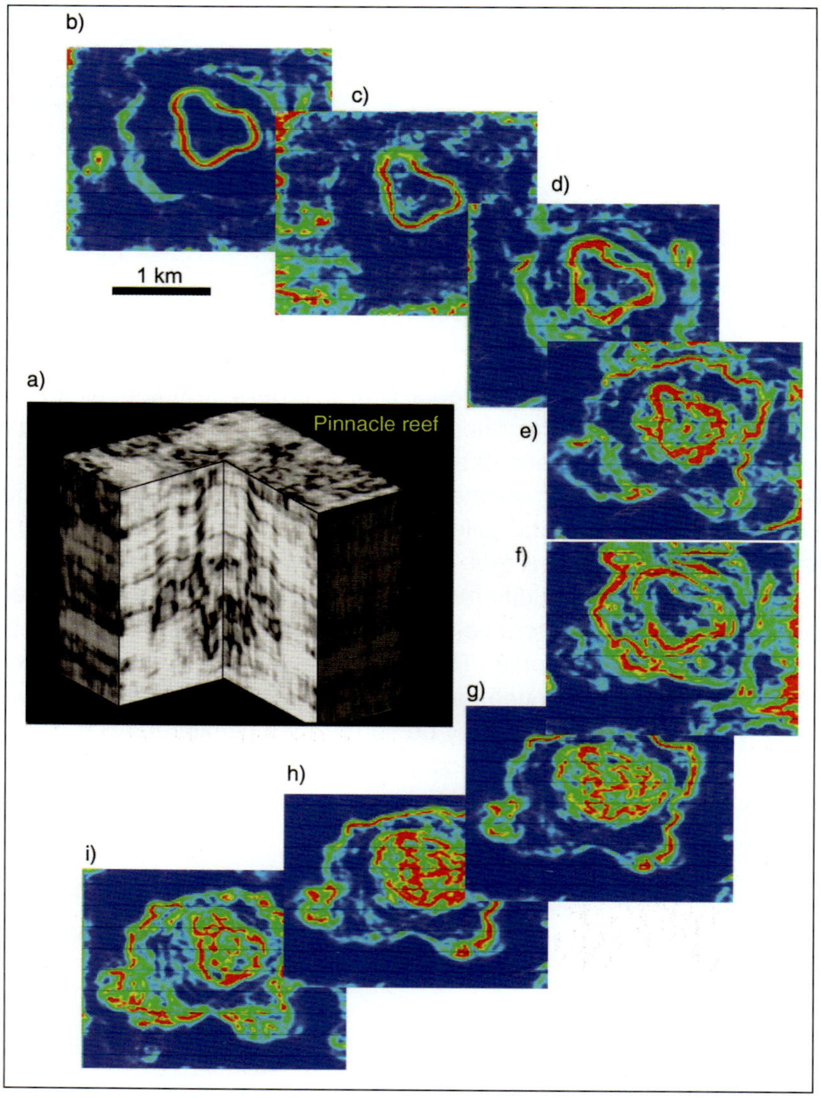

Figure 4.59. (a) A cutaway volume view and (b)-(i) time slices at 0.020-ms increments through a coherence volume generated from a survey over a pinnacle reef in Alberta, Canada. We see a gradual broadening of the reef boundary as we go down in time, which we would expect for a pinnacle reef. At the same time, the reef's inner structural details also become more apparent.

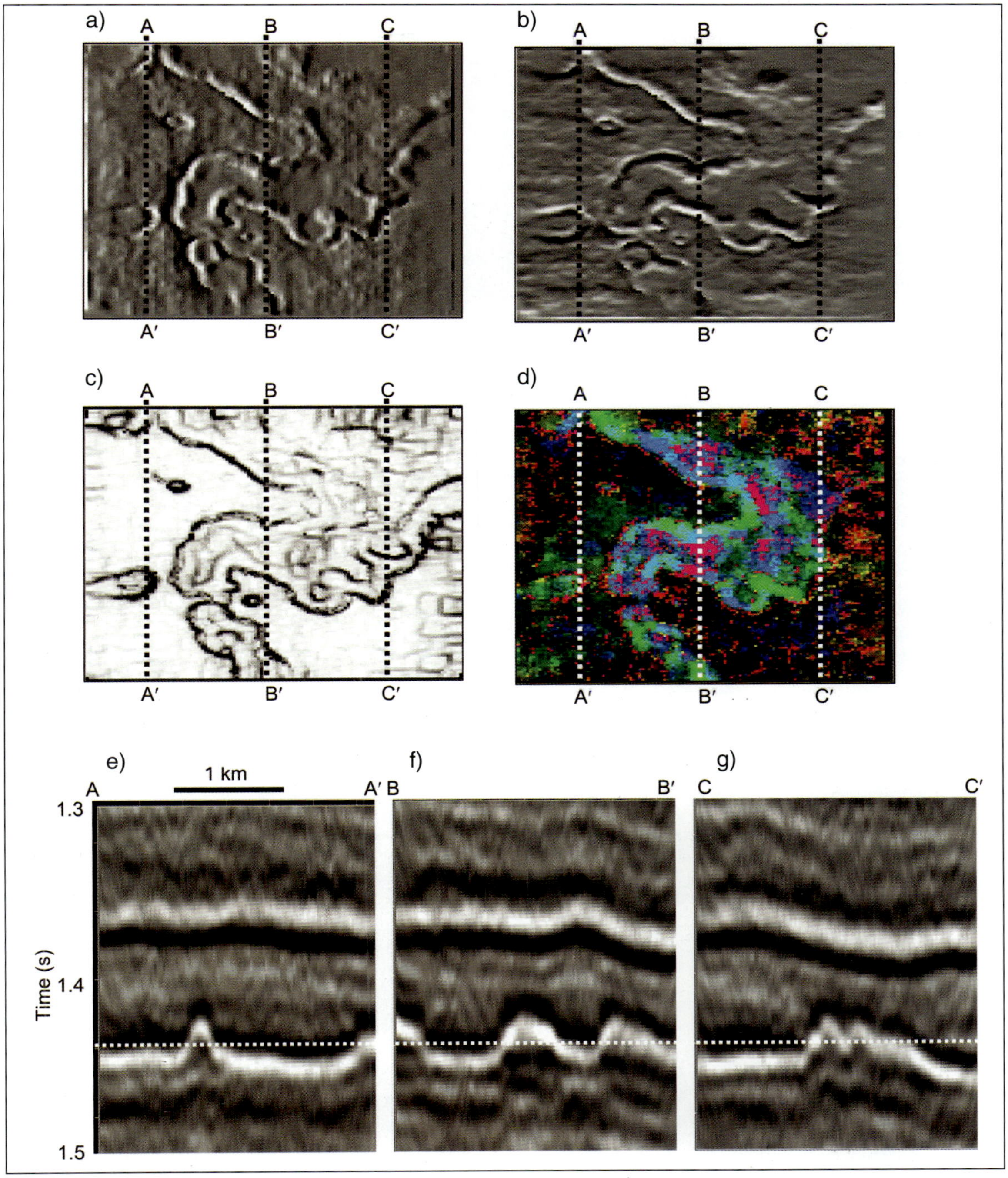

Figure 4.60. (a)-(d). Four time slices at t = 1.438 s through a Winnepegosas reef, from a survey acquired in Saskatchewan, Canada. (a) The east-west component and (b) the north-south component of a energy-weighted coherent-amplitude gradient, (c) a coherence volume, and (d) peak spectral frequency modulated by peak amplitude. (e)-(g). Vertical slices through the seismic data along lines (e) AA′, (f) BB′, and (g) CC′. The 2D color table for peak frequency versus amplitude of peak frequency is identical to the color table in Figure 4.49. Seismic data are courtesy of Talisman.

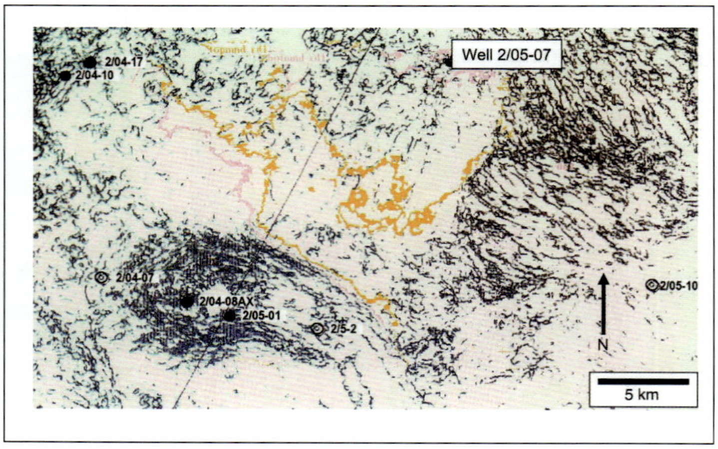

Figure 4.61. A time slice at $t = 3.356$ s through Tor field, Norway. Low-coherence (black) features correspond to massive chalk slumping that formed soon after deposition, while the chalk was still plastic. These "fractures" then are locked into the formation after it lithifies. After Skirius et al. (1999).

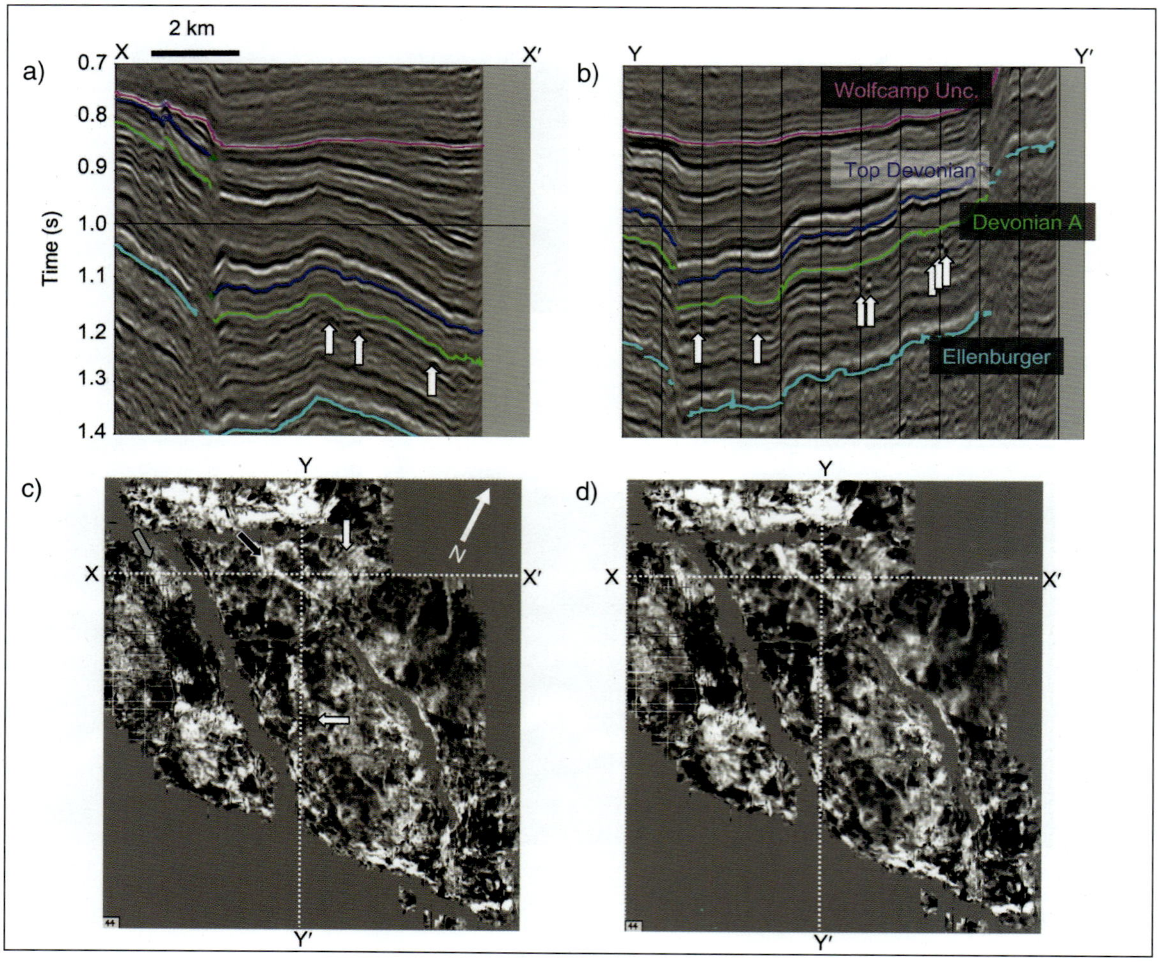

Figure 4.62. (a)-(b). Two representative lines, (a) XX′ and (b) YY′, from a survey acquired over the Central Basin Platform, Texas, U.S.A. The arrows indicate small channel features below the green Devonian A horizon. The data have undergone principal-component, structure-oriented filtering. A horizon slice 44 ms below the green Devonian A horizon (c) without and (d) with structure-oriented filtering.

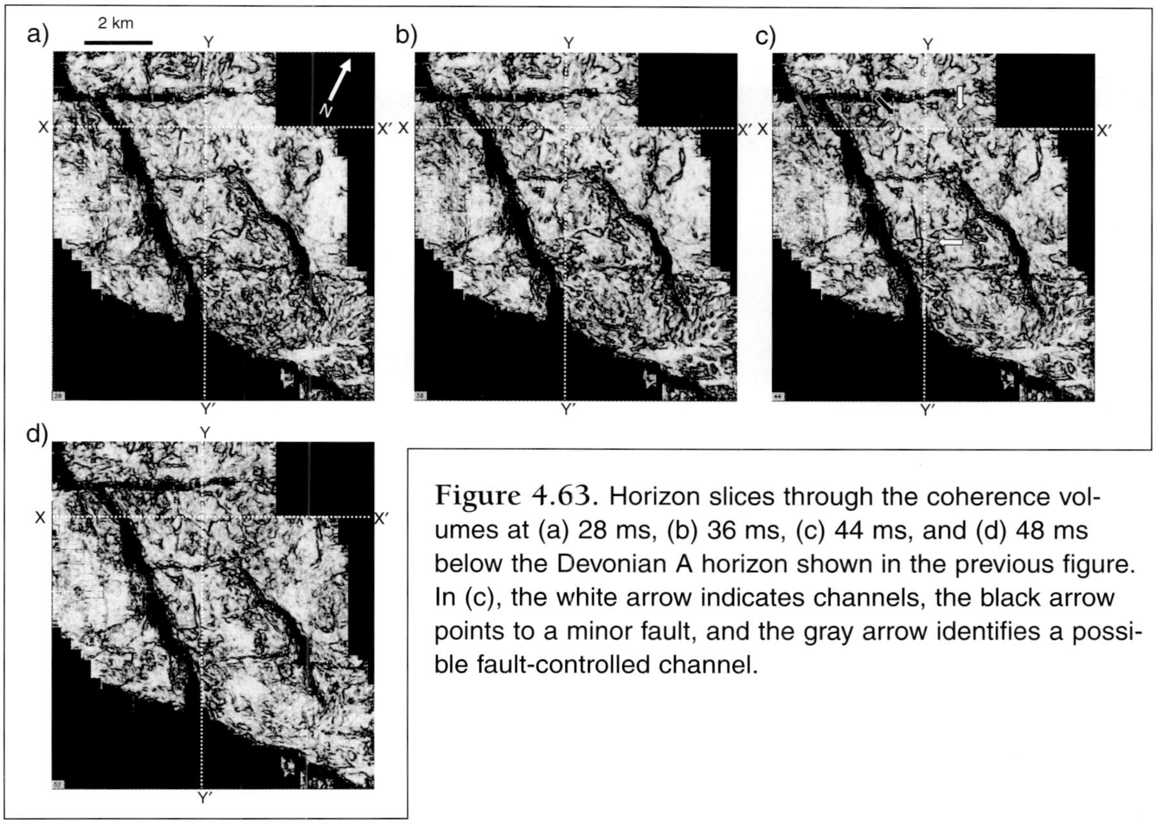

Figure 4.63. Horizon slices through the coherence volumes at (a) 28 ms, (b) 36 ms, (c) 44 ms, and (d) 48 ms below the Devonian A horizon shown in the previous figure. In (c), the white arrow indicates channels, the black arrow points to a minor fault, and the gray arrow identifies a possible fault-controlled channel.

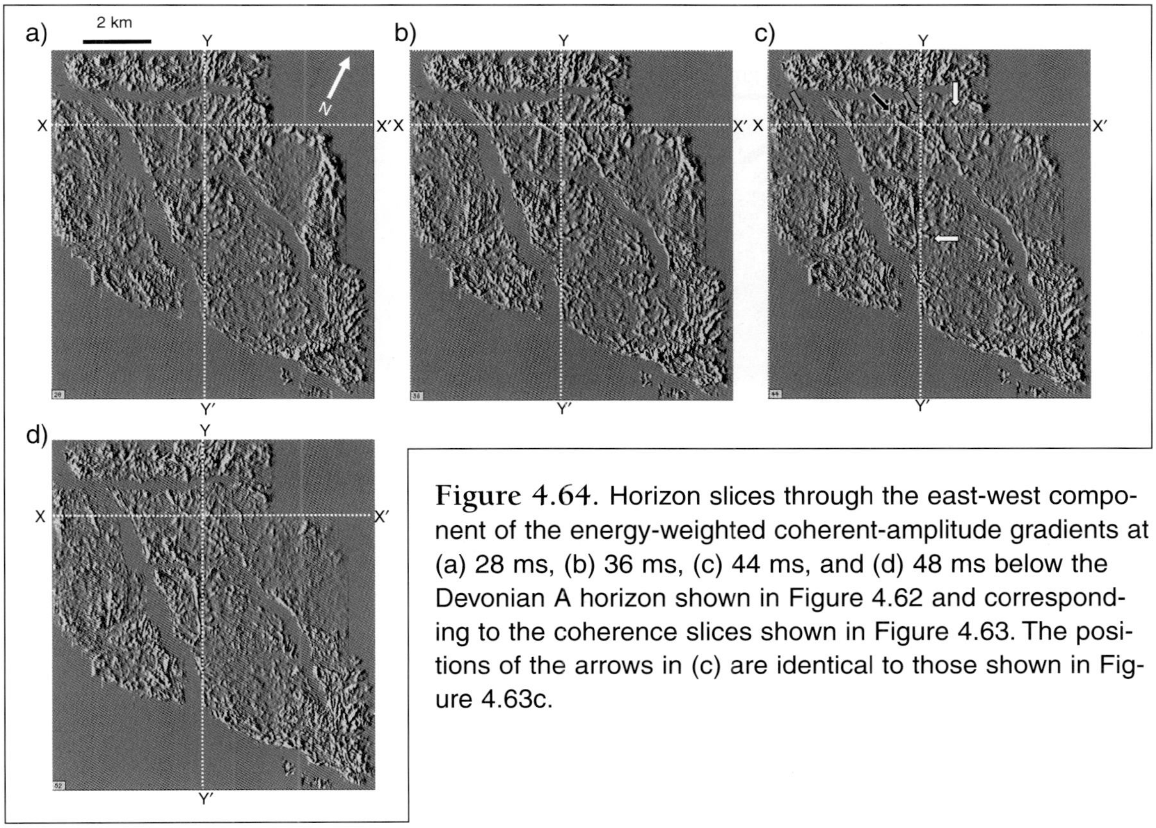

Figure 4.64. Horizon slices through the east-west component of the energy-weighted coherent-amplitude gradients at (a) 28 ms, (b) 36 ms, (c) 44 ms, and (d) 48 ms below the Devonian A horizon shown in Figure 4.62 and corresponding to the coherence slices shown in Figure 4.63. The positions of the arrows in (c) are identical to those shown in Figure 4.63c.

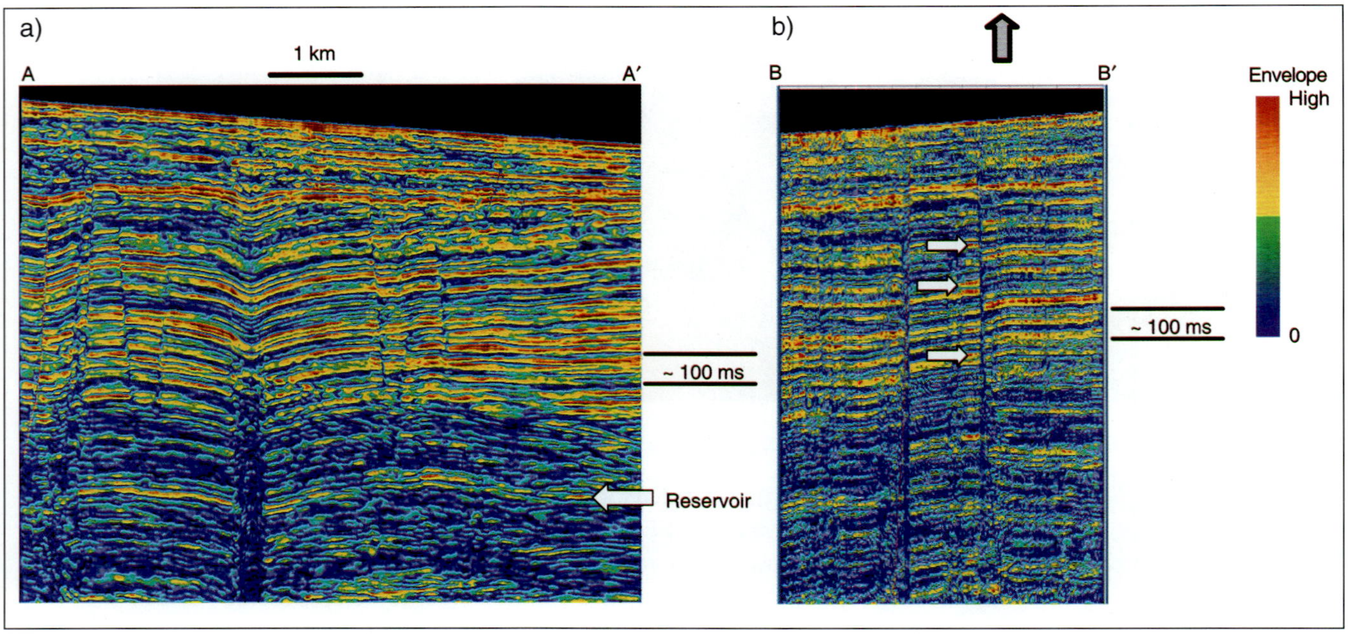

Figure 4.65. A vertical cross section through the envelope of the seismic data corresponding to lines AA′ and BB′ from a survey acquired over Lihua field reservoir offshore China. The white arrow indicates the approximate reservoir level. The gray arrow indicates a gas chimney emanating from a sinkhole. Yellow arrows indicate faults. Note the gas sag resulting from slower velocities and decreased amplitude, probably caused by an incorrect stacking velocity. After Story et al. (1999).

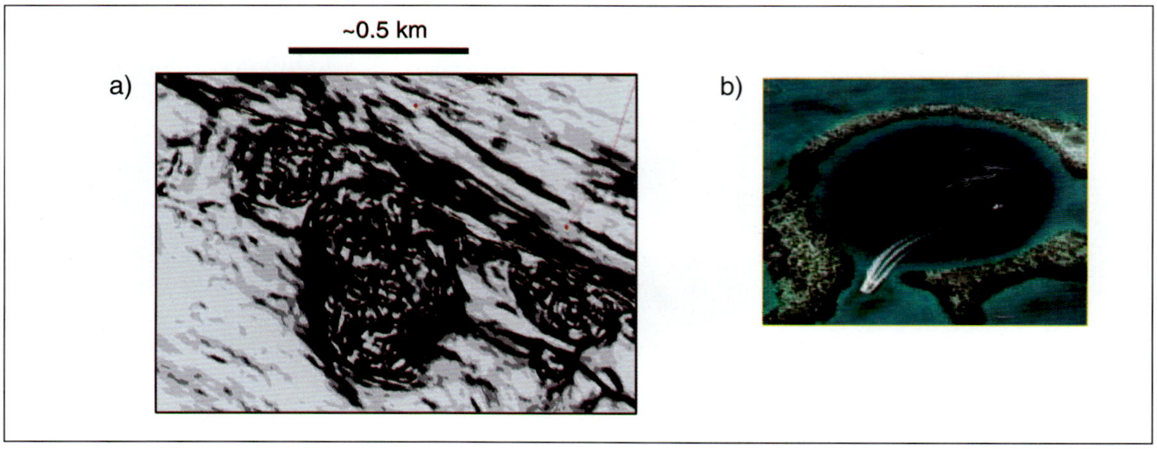

Figure 4.66. A modern analog of (a) a paleosinkhole in the Lihua field offshore China is seen in (b) the 400-m-diameter, 145-m-deep Great Blue Hole, Belize. Note the speedboat for scale. After Story et al. (1999) and http://www.bluedive.com.

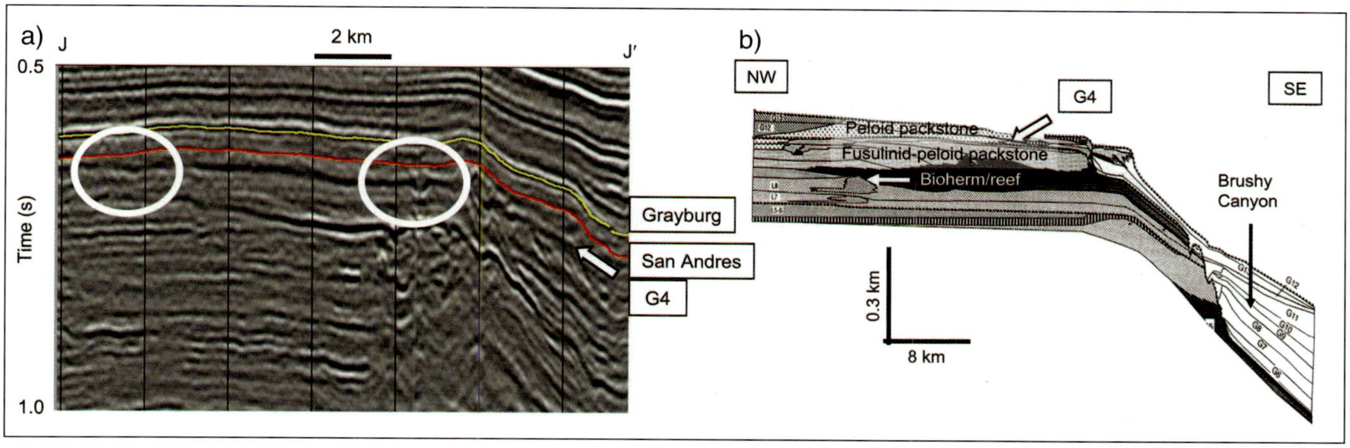

Figure 4.67. (a) Seismic data through a survey acquired over the Central Basin Platform, west Texas, U.S.A., and (b) Kerans et al.'s (1994) sequence stratigraphy for the same data volume, based on an outcrop study of the San Andres Limestone carbonates. The label G4 indicates the top of the Guadalupian 4 sequence, which is the regional bypass surface for the basinal lowstand deposits. The circles in (a) indicate karst features associated with subaerial exposure. The figure is courtesy of Charlotte Sullivan, University of Houston. Data are courtesy of Burlington Resources.

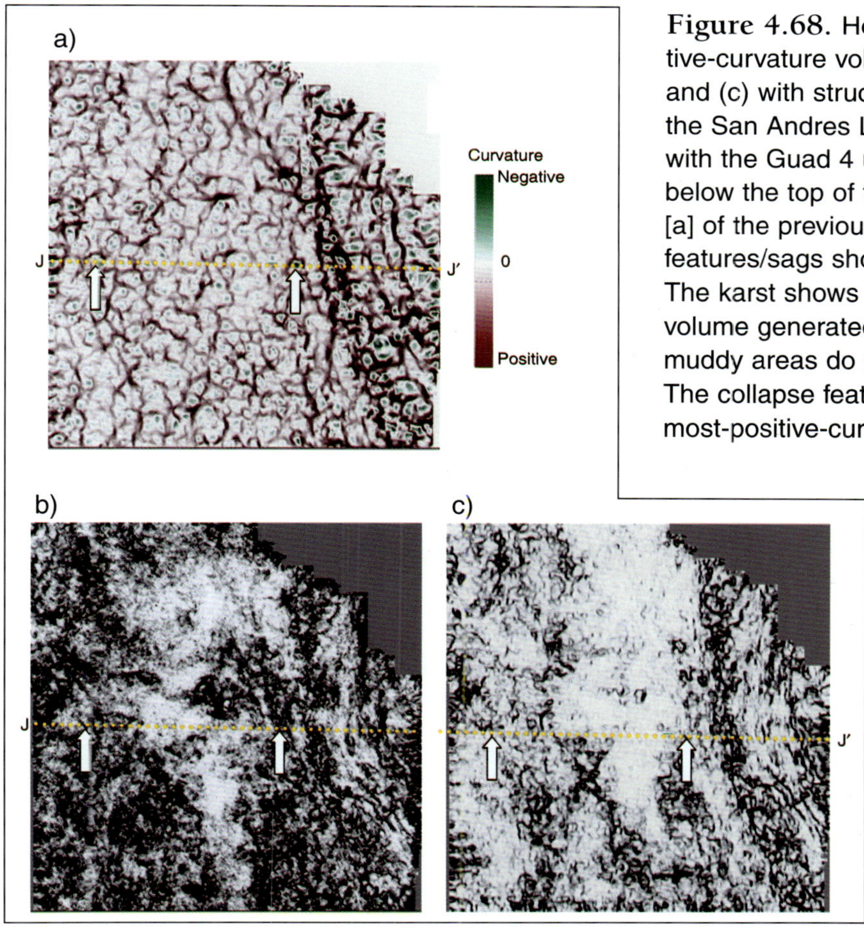

Figure 4.68. Horizon slices through (a) the most-positive-curvature volume, a coherence volume (b) without and (c) with structure-oriented filtering, along the base of the San Andres Limestone reservoir, which coincides with the Guad 4 unconformity (approximately 72 ms below the top of the Grayburg Formation, shown in part [a] of the previous figure). Arrows indicate the two karst features/sags shown in part (a) of the previous figure. The karst shows up as muddy areas on the coherence volume generated without structure-oriented filtering. The muddy areas do not show up well on the filtered version. The collapse features appear as green bowls in the most-positive-curvature volume.

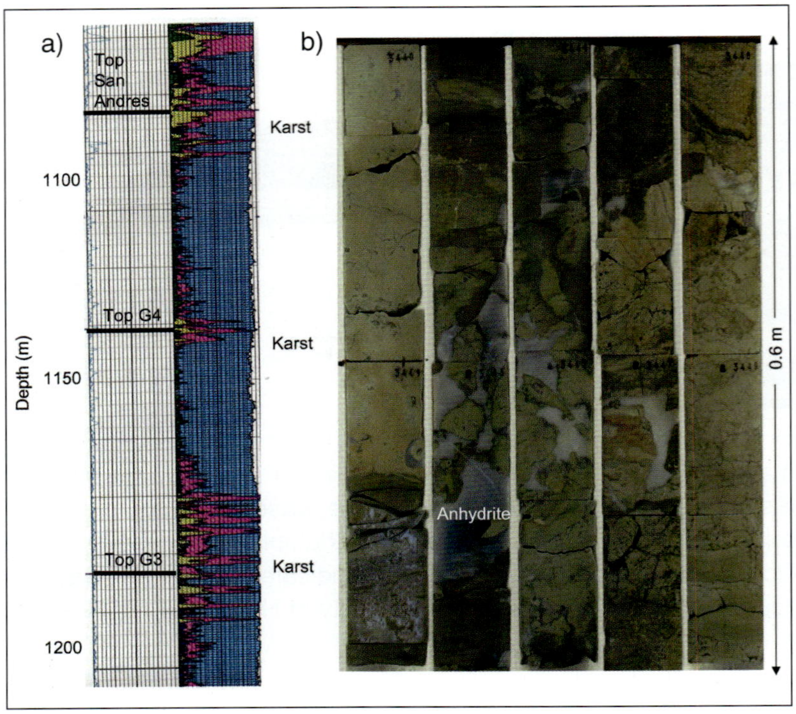

Figure 4.69. (a) Electrofacies through the San Andres limestone carbonates, acquired from a well 20 km downdip from the image shown in Figure 4.68. Karst features are associated with abundance of anhydrite (magenta) at sequence boundaries (G3, G4), and at the top of the San Andres Limestone. (b) Anhydrite-filled karst breccia in core sections from upper San Andres strata. Courtesy of Charlotte Sullivan.

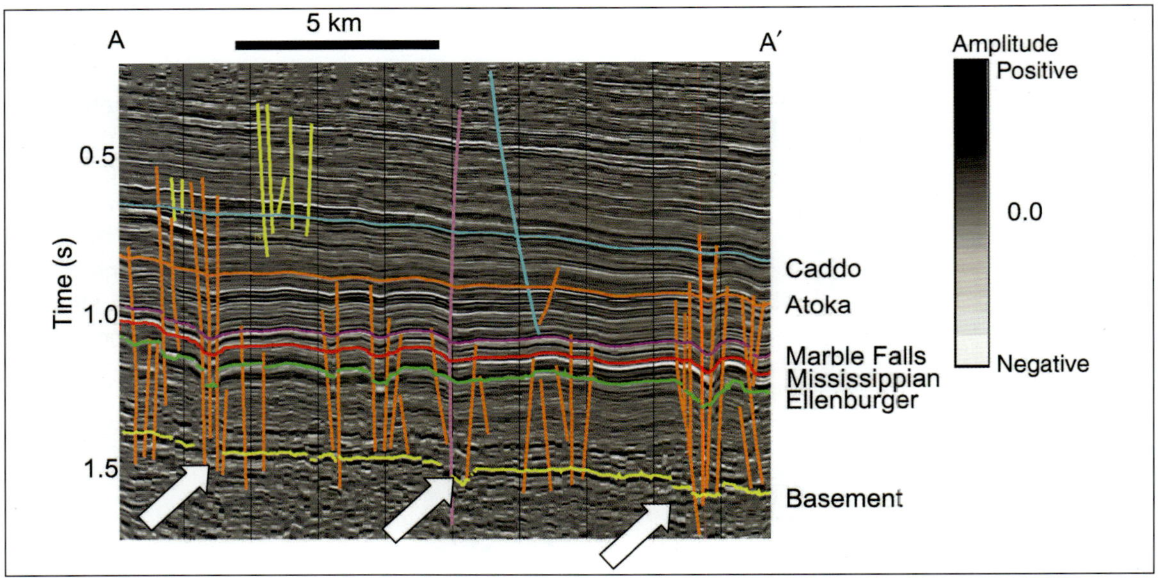

Figure 4.70. A north-south line through a survey acquired over the Fort Worth Basin, Texas, U.S.A. The cyan is the Pennsylvanian Caddo limestone; the green Ellenburger pick is Lower Ordovician. Note the vertical collapse features indicated by arrows. These collapse features are not filled in at later geologic times but can be tracked from the basement to shallower events. Orange faults penetrate the basement; yellow faults are confined to strata above the basement. The pink fault is the major fault crossing the center of the survey. Seismic data are courtesy of Devon Energy. Courtesy of Charlotte Sullivan.

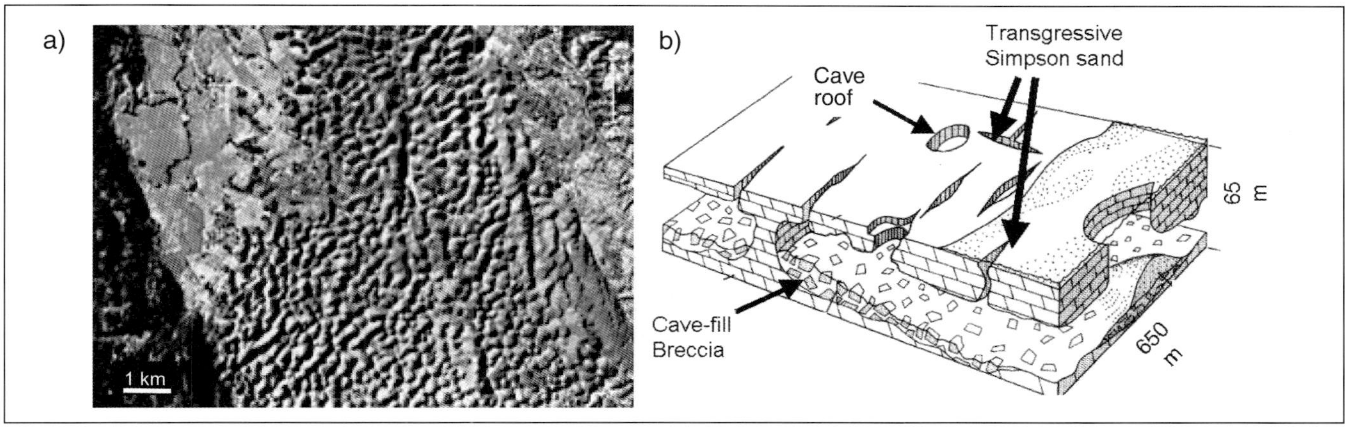

Figure 4.71. Geologic models of features seen in the Fort Worth Basin: (a) Mature "cockpit" karst topography results from a combination of dissolution and preburial collapse of cave systems. In this process, well-cemented, low-porosity intercave areas persist as rounded knolls and hills, as shown in this satellite image of the Cockpit country of Jamaica. After Short and Blair (1986). Reprinted by permission of NASA. (b) A karst model of features associated with subaerial weathering and erosion of west Texas Ellenburger carbonates. Cave systems most commonly fill with collapse breccias and sediments associated with subsequent marine transgressions. After Kerans (1990). Reprinted by permission of the Texas Bureau of Economic Geology.

Figure 4.72. A 3D view of coherence draped over the Ellenburger horizon shown in Figure 4.70. Note the intensity and alignment of the collapse features. The image was generated by Hao Guo, University of Houston.

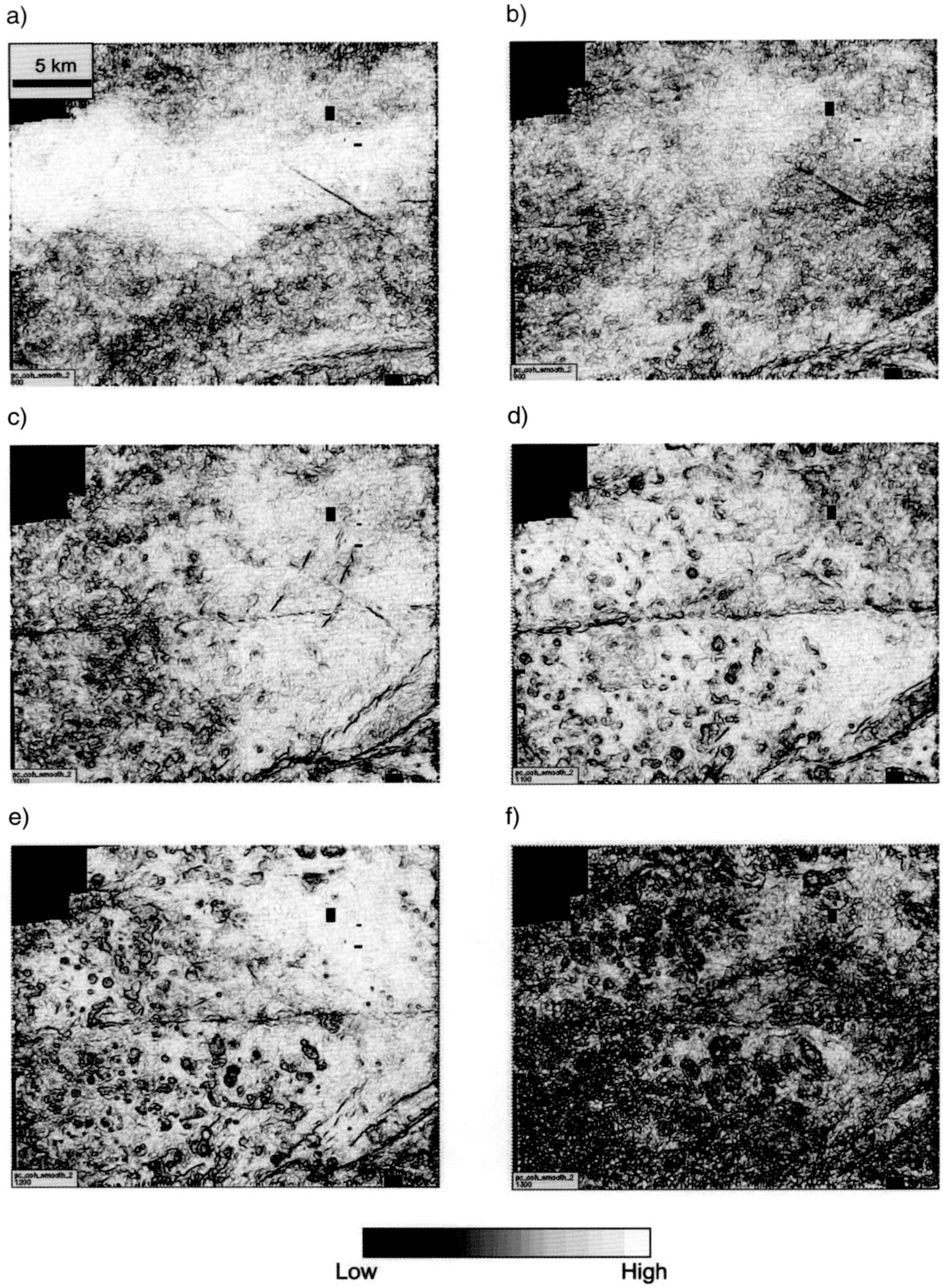

Figure 4.73. Time slices through a coherence volume at t = (a) 0.800, (b) 0.900, (c) 1.000, (d) 1.100, (e) 1.200, and (f) 1.300 s. Black indicates low coherence, and white indicates high coherence. Collapse features appear as circular to elliptical low-coherence rings. Courtesy of Charlotte Sullivan.

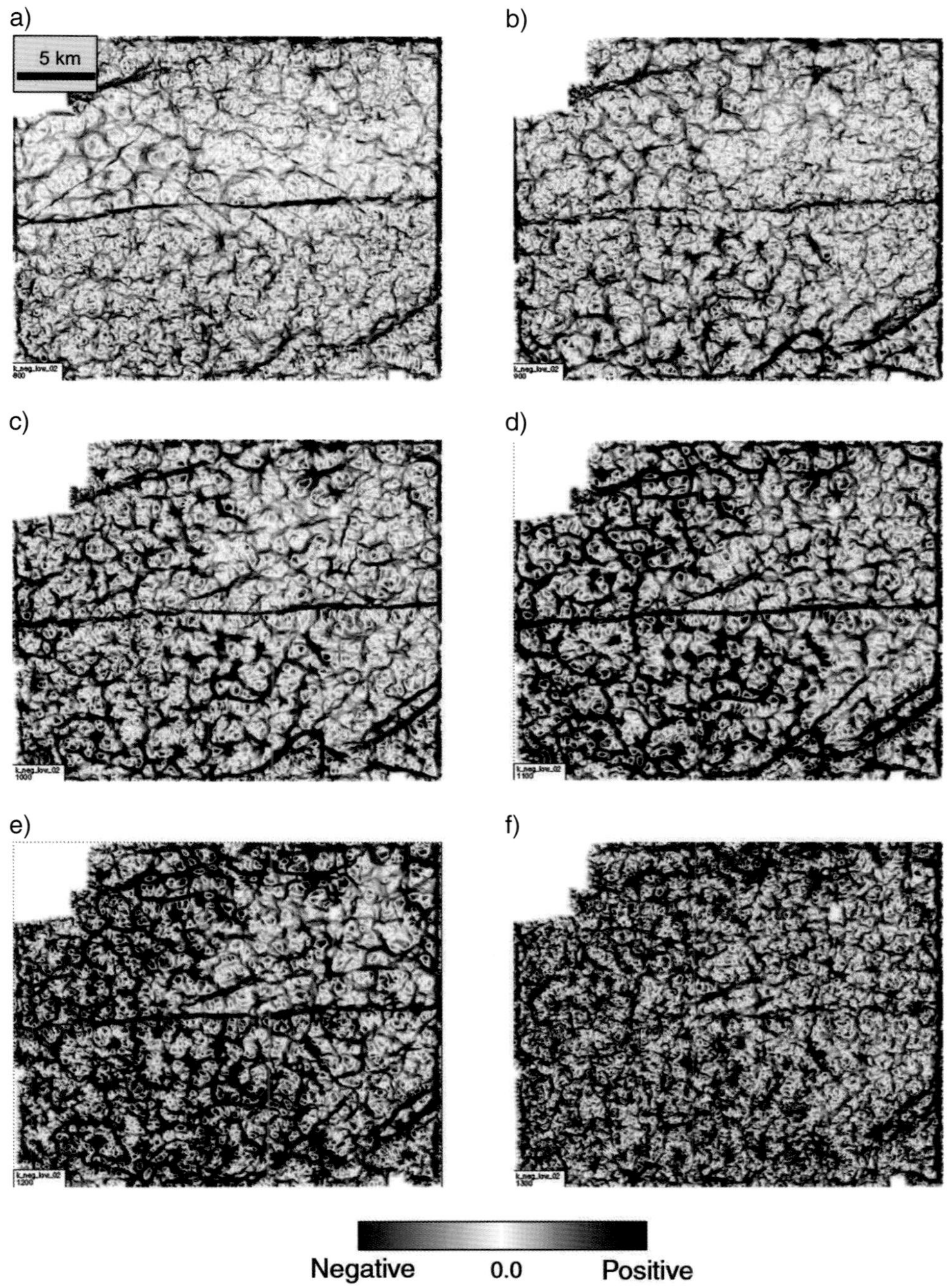

Figure 4.74. Time slices through a most-negatiive-curvature volume at t = (a) 0.800, (b) 0.900, (c) 1.000, (d) 1.100, (e) 1.200, and (f) 1.300 s. We used a long-wavelength calculation with an analysis window of ± 10 ms and 75 traces. Red indicates positive values, white indicates zero values, and green indicates negative values of curvature. Bowls and valleys, including collapse features, show up as green. Domes show up as red. Courtesy of Charlotte Sullivan.

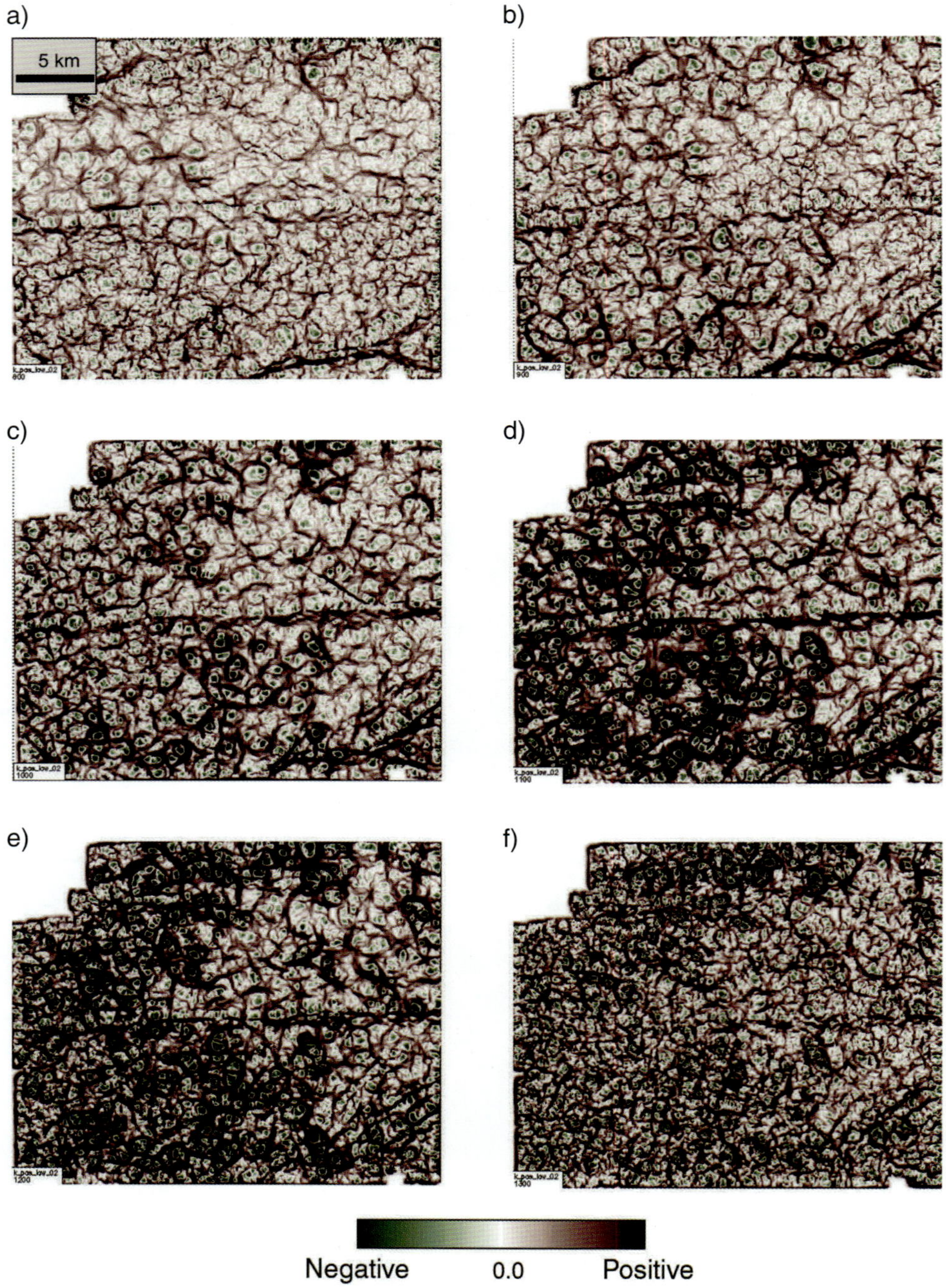

Figure 4.75. Time slices through a most-positive-curvature volume at t = (a) 0.800, (b) 0.900, (c) 1.000, (d) 1.100, (e) 1.200, and (f) 1.300 s. We used a long-wavelength calculation with an analysis window of ± 10 ms and 75 traces. Red indicates positive values, white indicates zero values, and green indicates negative values of curvature. Bowls (collapse features) show up as green. Domes and ridges show up as red. Courtesy of Charlotte Sullivan.

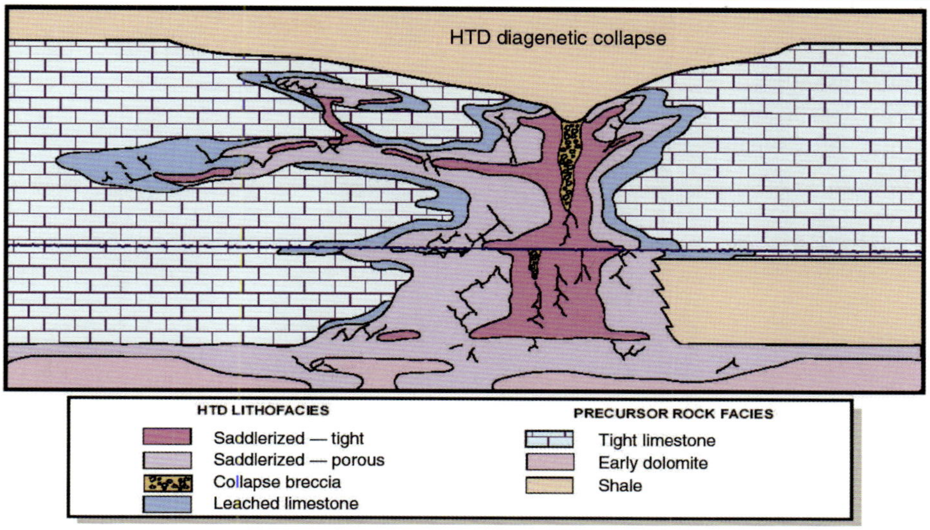

Figure 4.76. A model of hydrothermal alteration of limestone to generate hydrothermal dolomite lithofacies. The seismic and attribute signatures of such a complicated feature are not understood well and require significant modeling and testing. After Reimer and Teare (1991, 1992).

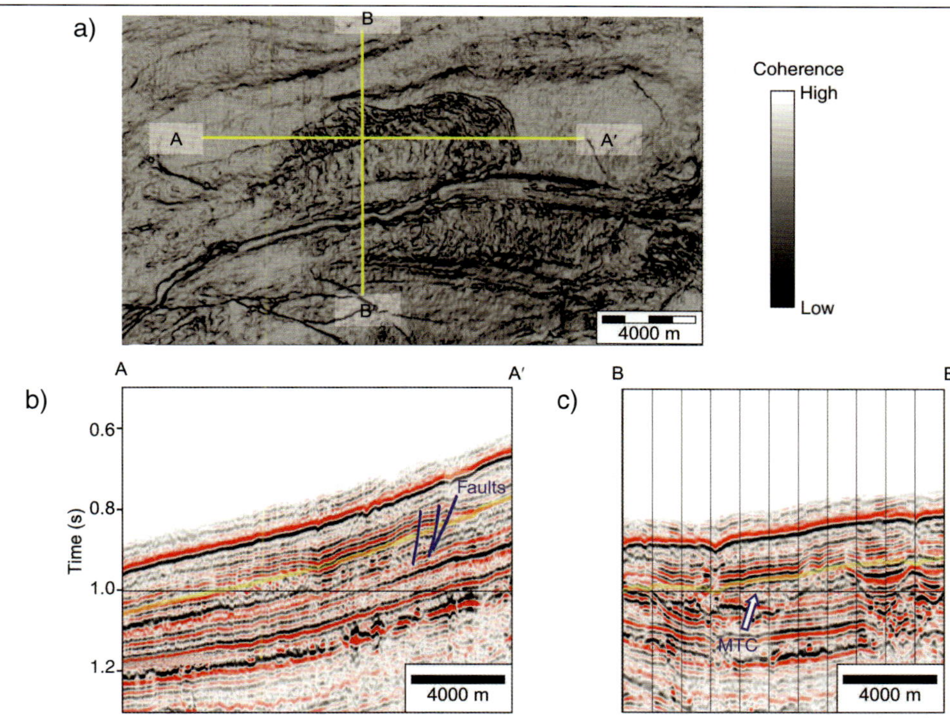

Figure 4.77. (a) A horizon slice through a coherence volume from offshore Nigeria 100 ms below the water bottom, showing a mass-transport complex (MTC). The phantom horizon is indicated in yellow on vertical seismic slices (b) AA′ and (c) BB′. The MTC measures 5 km × 17 km in area. After Haskell et al. (1999).

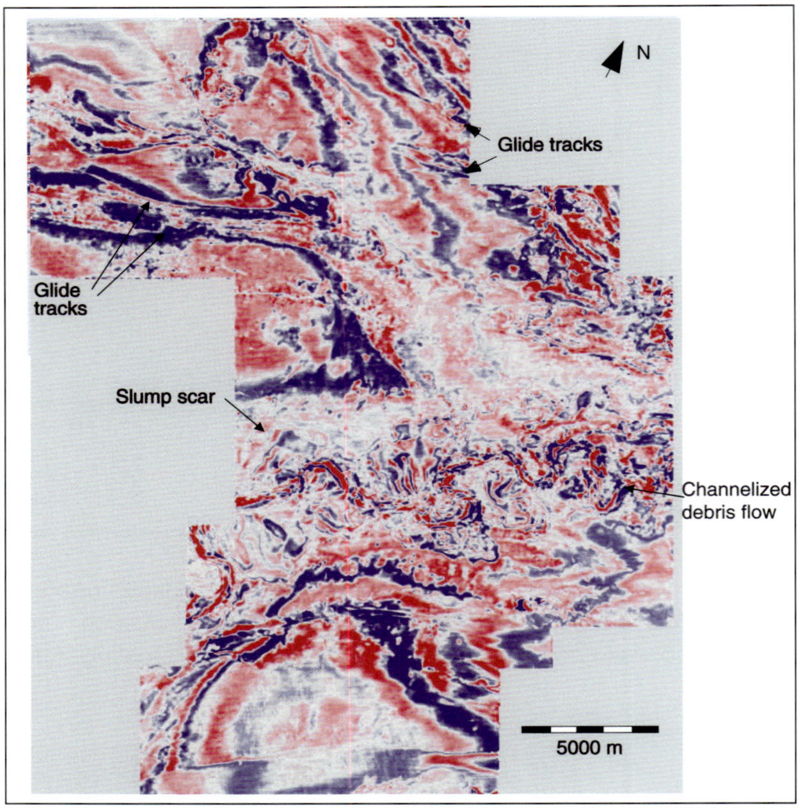

Figure 4.78. A phantom-horizon slice 144 ms below the seafloor, through a seismic data volume offshore Nigeria. After Nissen et al. (1999).

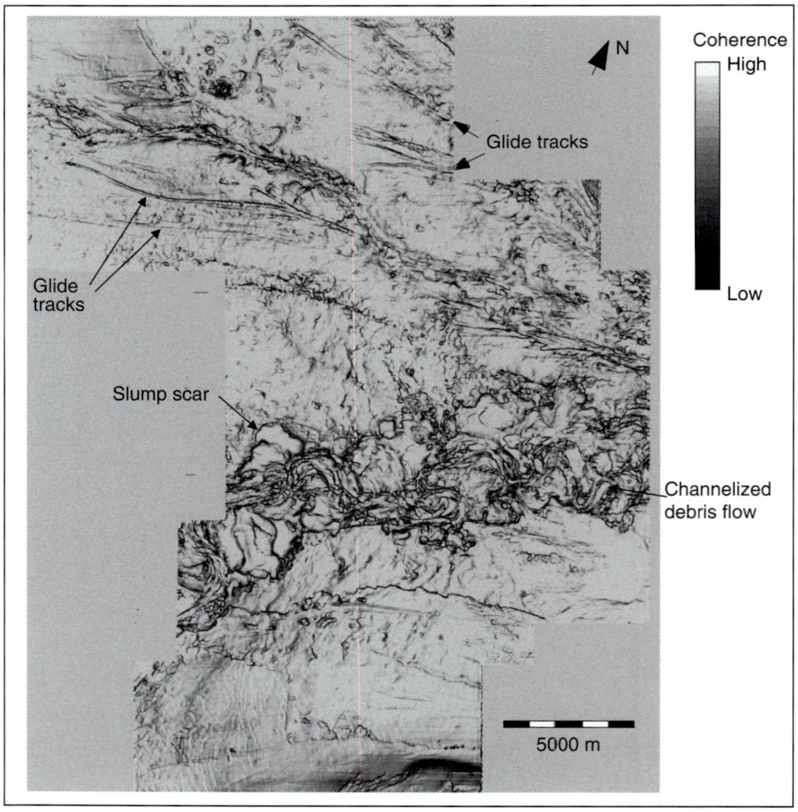

Figure 4.79. A phantom-horizon slice 144 ms below the seafloor, through the coherence volume corresponding to the previous figure. In addition to faults, we see a slump scar and channelized debris flow. We will discuss the glide tracks in the next figure. After Nissen et al. (1999).

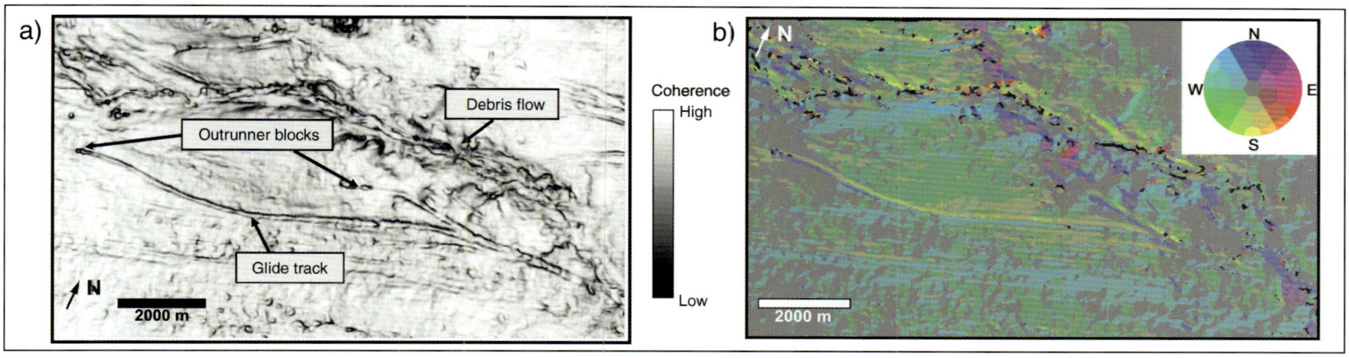

Figure 4.80. A phantom-horizon slice 144 ms below the seafloor through (a) a coherence volume and (b) a dip and azimuth volume, showing outrunner blocks, glide tracks, and other debris-flow features. Note how the northern flanks of the east-west-trending glide tracks dip to the south, whereas the southern flanks dip to the north. After Nissen et al. (1999).

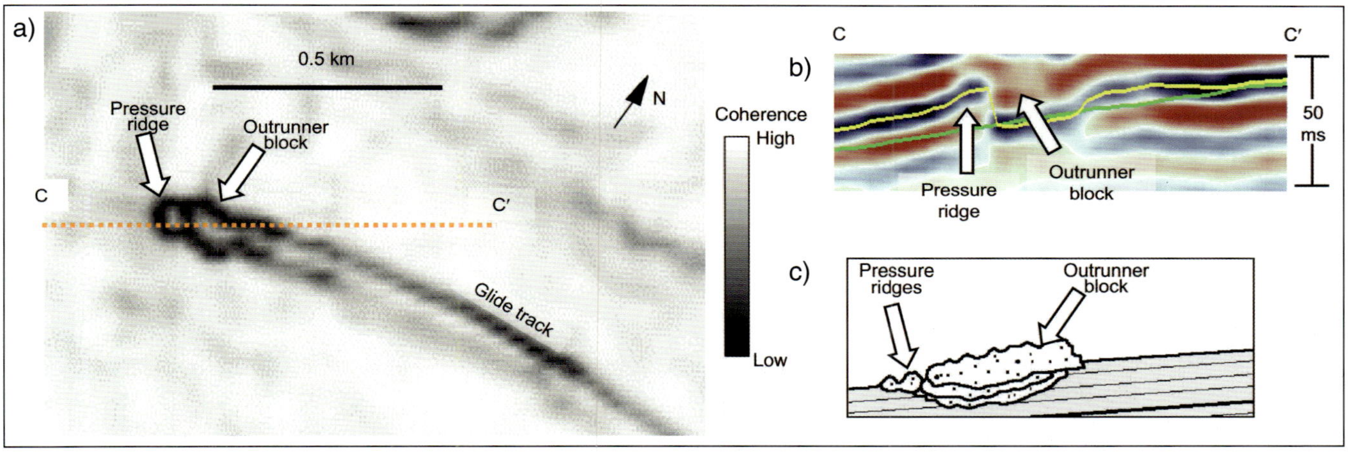

Figure 4.81. (a) An enlargement of one of the outrunner blocks shown in Figure 4.79. (b) A vertical seismic slice through the outrunner block. Yellow indicates a stratigraphic horizon, and green indicates a phantom horizon 144 ms below the seafloor. (c) A schematic diagram of an outrunner block in a modern mass-transport complex (MTC) from Kitimat Fjord, British Columbia, Canada, courtesy of David Prior, Dalhousie University. After Nissen et al. (1999).

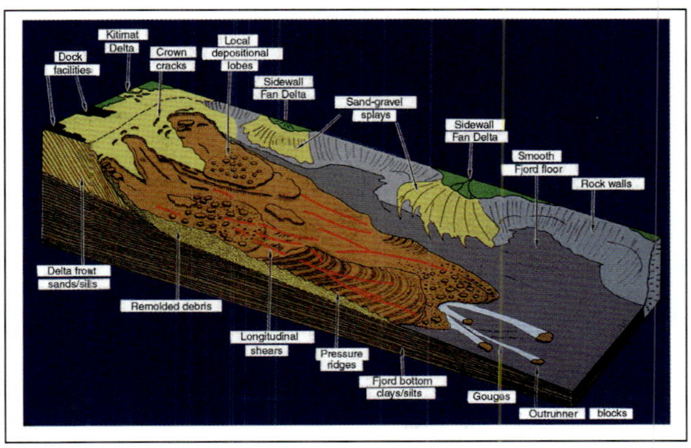

Figure 4.82. A modern analog of a mass-transport complex, from Kitimat Fjord, western British Columbia, Canada, showing the relationship among crown cracks, debris-flow lobes, pressure ridges, glide tracks (gouges), and outrunner blocks. A schematic section through an example of an outrunner block from Kitimat Fjord was shown in part (c) of the previous figure. This image was drawn from sidescan sonar data. After Prior (1984). University of Chicago Press©1984 whose permission is required for further use.

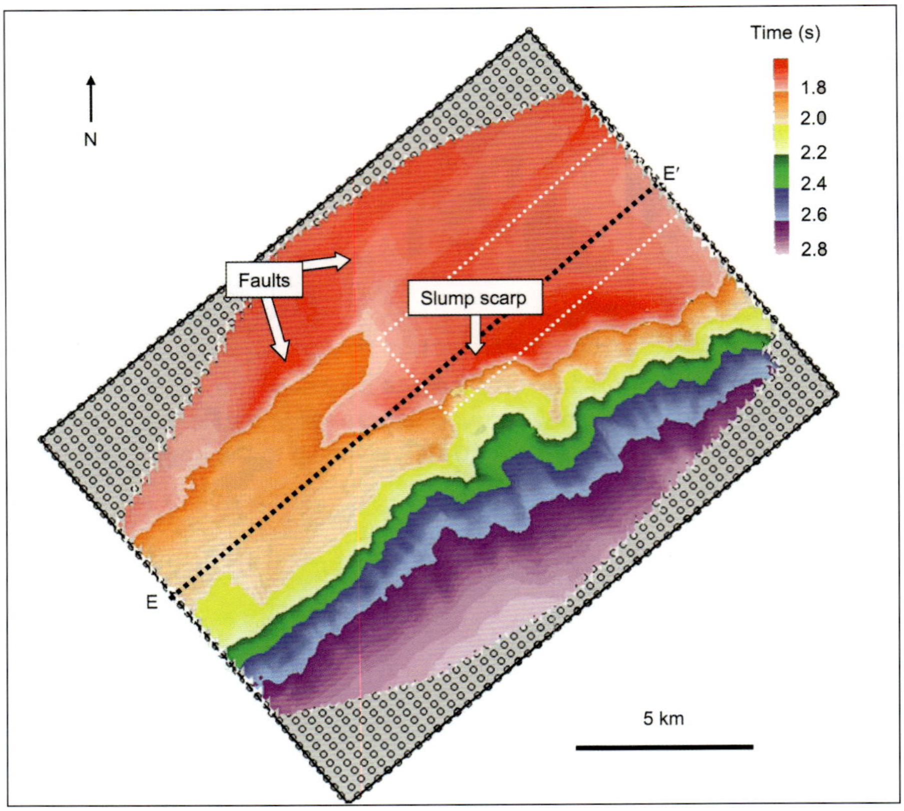

Figure 4.83. A time-structure map of the water bottom, generated from a 3D hazard survey acquired over the Sigsbee Escarpment, Green Canyon, Gulf of Mexico, U.S.A. The escarpment parallels the yellow contour. After Oyedele (2005).

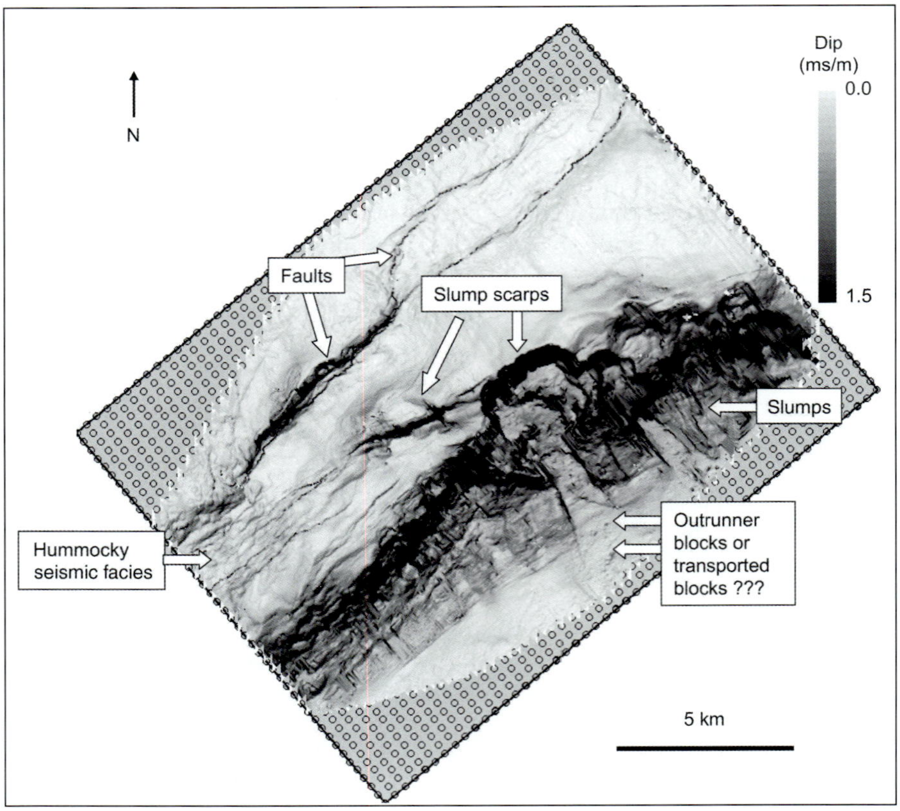

Figure 4.84. A conventional dip-magnitude map derived from a time-structure map of the water bottom shown in the previous figure. In addition to the Sigsbee Escarpment, we note several faults in the northwest part of the figure on the continental shelf, several slump scarps near the escarpment, and slumps at the base of the escarpment. After Oyedele (2005).

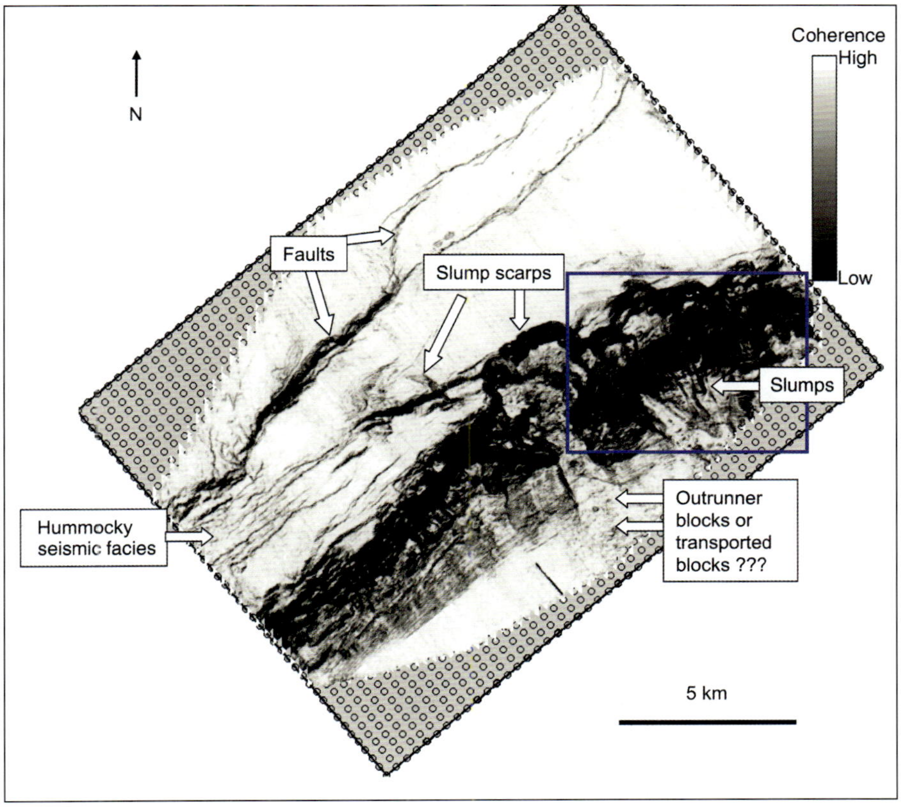

Figure 4.85. A horizon slice along the water bottom through a coherence volume. Because coherence is calculated on the seismic volume before interpretation, it exhibits fewer picking errors than does the dip-magnitude map shown in the previous figure. Although the dip-magnitude and coherence images are similar, they measure different features of the seismic data, with coherence measuring lateral changes in waveform. After Oyedele (2005).

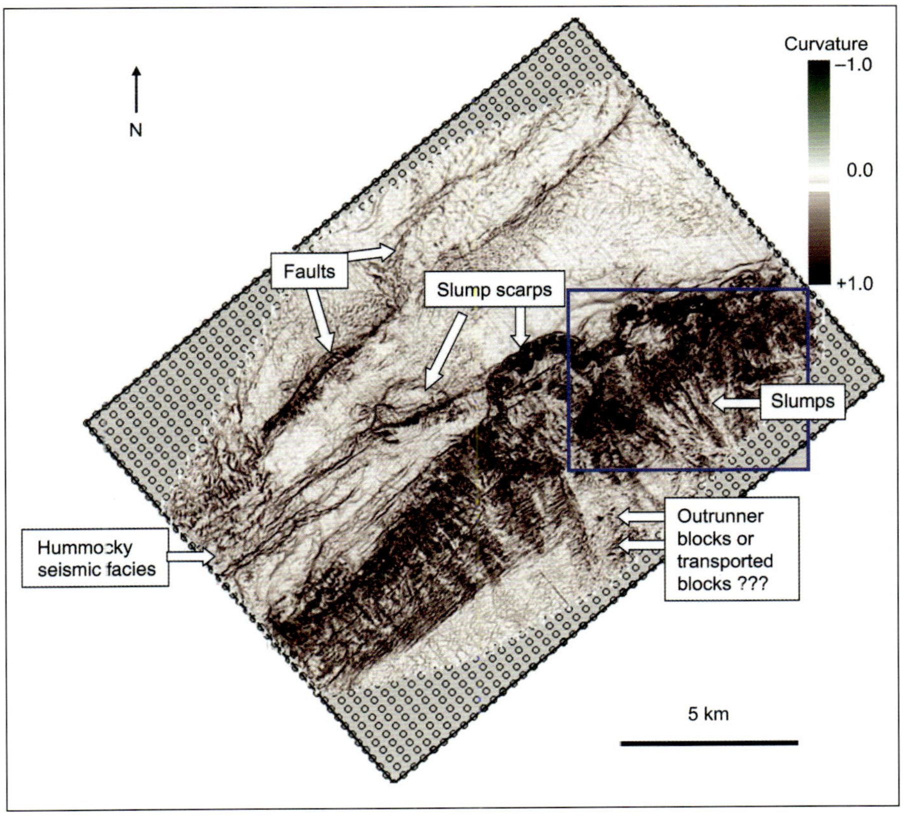

Figure 4.86. A horizon slice along the water bottom, through a most-positive-curvature volume. This image highlights in red the features that are anticlinal and provides significantly greater detail of the slump scarps and slumps at the base of the escarpment than did the previous three figures. After Oyedele (2005).

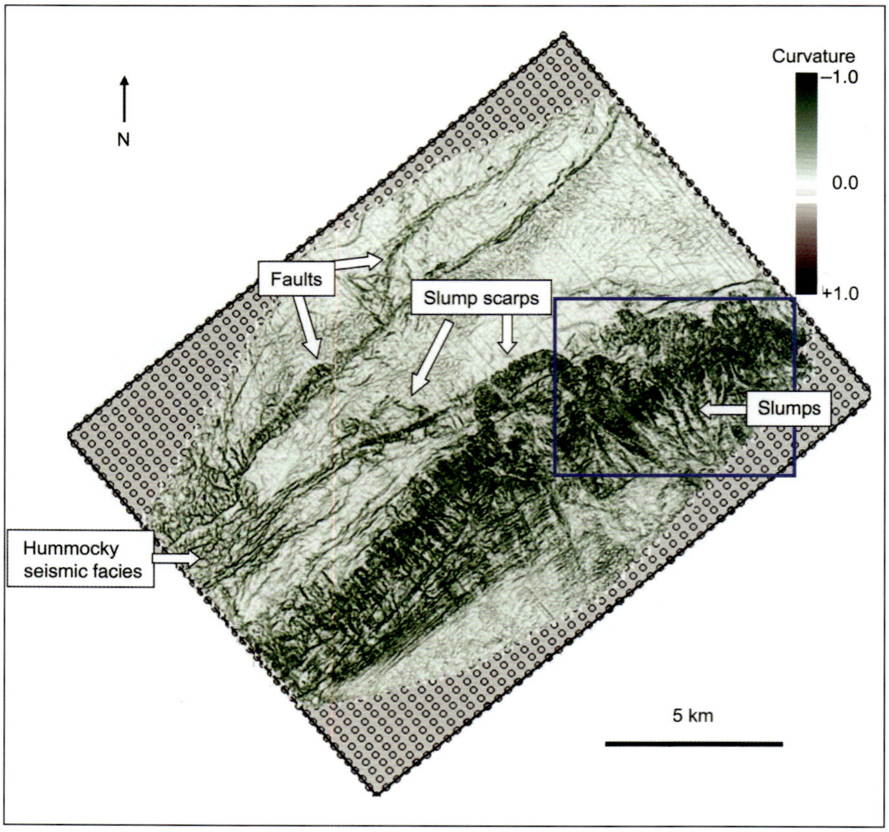

Figure 4.87. A horizon slice along the water bottom, through a most-negative-curvature volume. This image highlights in green the features that are synclinal and, like the most-positive curvature in the previous figure, provides significantly greater detail of the slump scarps and slumps at the base of the escarpment than do dip-magnitude or coherence horizon slices. After Oyedele (2005).

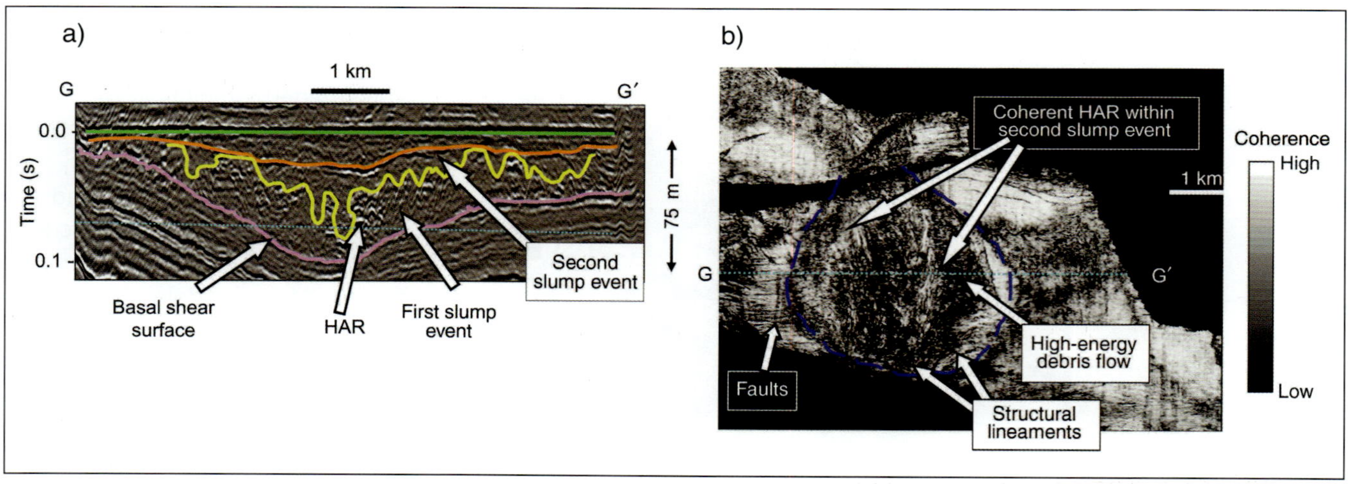

Figure 4.88. (a) A southwest-northeast seismic section showing an MTC. (b) A phantom-horizon slice 70 ms below the green horizon and through a coherence volume. In general, the lobate chaotic slump shows up as a region of low coherence. However, high-amplitude reflections (HARs) show up as coherent striations. After Oyedele (2005).

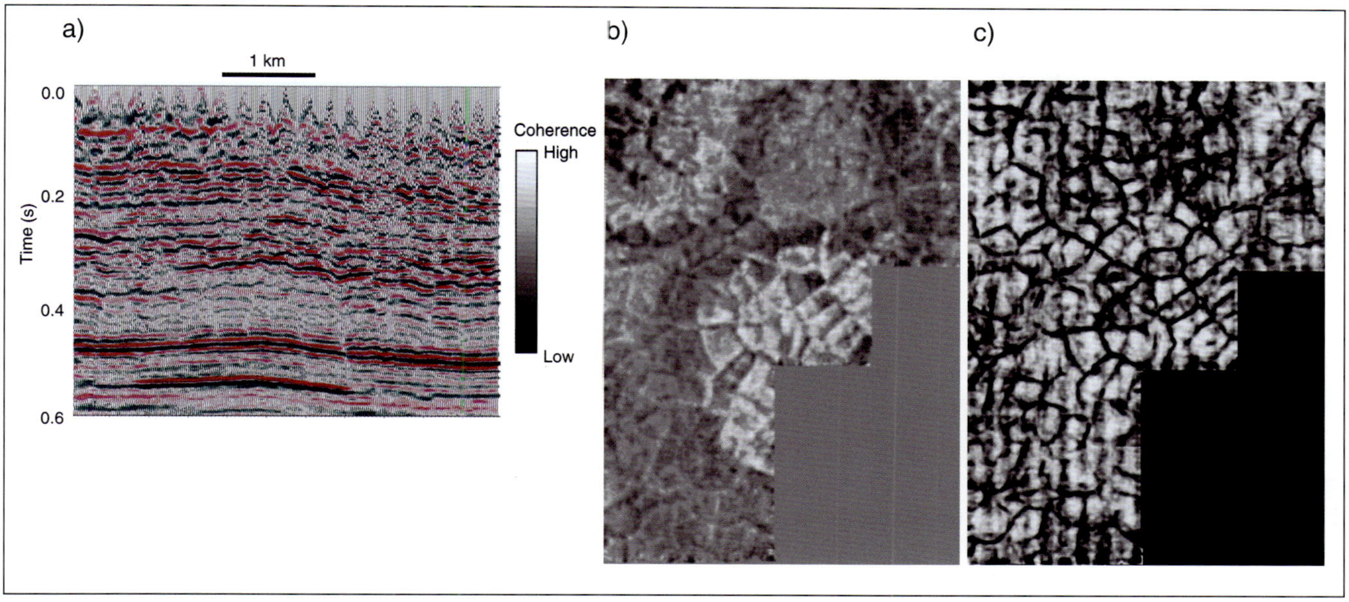

Figure 4.89. (a) A vertical slice through seismic data and (b)-(c) time slices at $t = 0.420$ s through (b) a seismic data volume and (c) a coherence volume, from a survey in Alberta, Canada. Polygonal fault patterns, delineating cells 1–5 km in diameter, are seen clearly on the coherence slice. Such faults are believed to be caused by dewatering in overpressured shales. Their geometries cannot be recognized on vertical seismic sections and are extremely difficult to map on conventional 3D time slices. Because surfaces exhibiting polygonal fault patterns indicate areas with a potential for overpressure, recognition of these fault patterns on coherence slices can be important in planning wells.

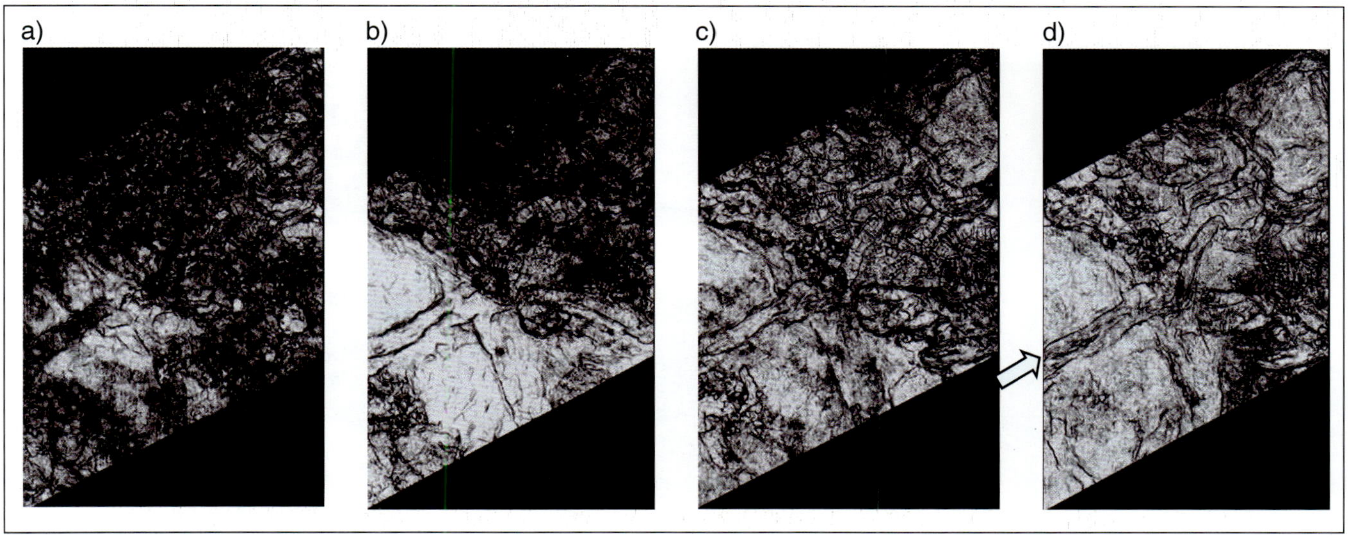

Figure 4.90. Rapid evolution of shallow faults in the North Sea, as seen via time slices through coherence volumes obtained at (a) 0.320 s, (b) 0.124 s, (c) 0.164 s, and (d) 0.200 s below the water bottom. The larger faults apparently were caused by differential compaction across deeper channels (arrow), whereas the very complex polygonal pattern seen in (c) appears to be polygonal faulting associated with shale dewatering.

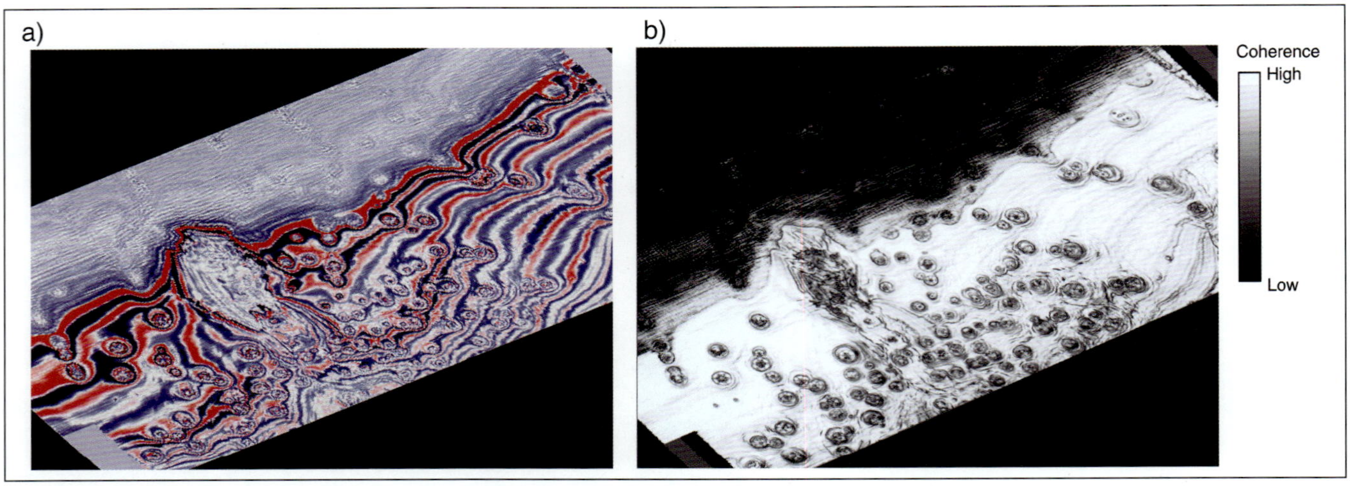

Figure 4.91. A time slice cutting the seafloor at $t = 0.300$ s through (a) a seismic data volume and (b) a coherence volume, from a survey acquired offshore West Africa. Note the circular pockmarks associated with shallow flow of water and gas.

Figure 4.92. A chair diagram through the survey shown in the previous figure. Note the velocity pushdowns below the pockmarks cut by the vertical slice, which indicate that the pockmarks are gas charged. Note the shale diapir indicated by the arrow.

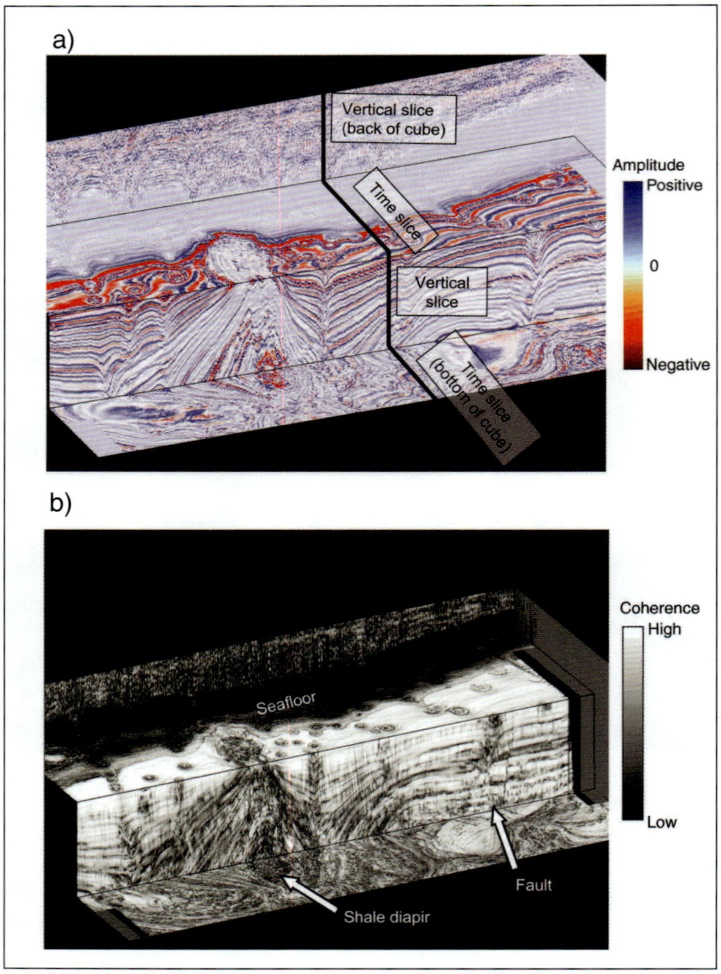

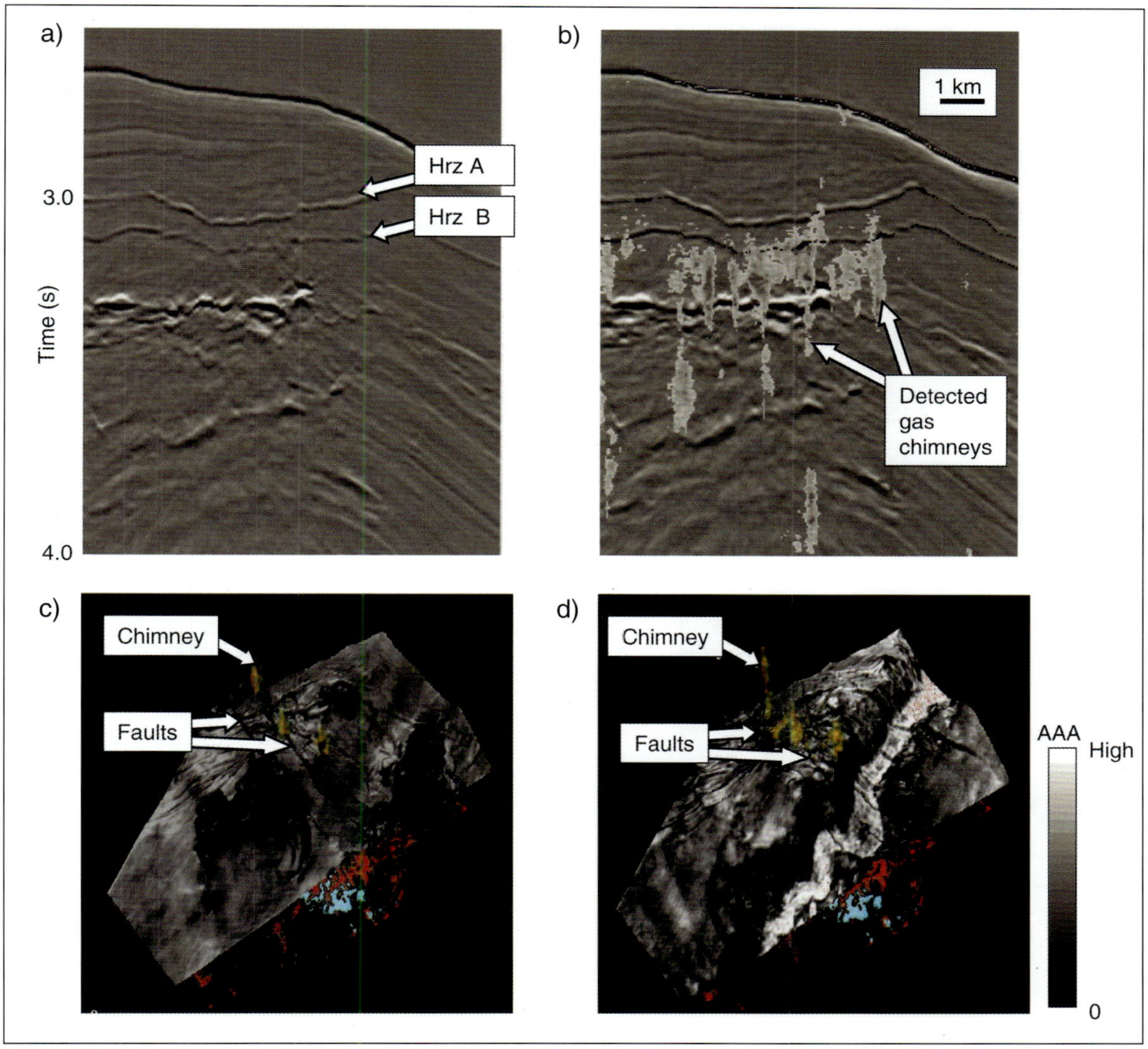

Figure 4.93. An example showing the use of a meta-attribute to map chimneys. First a suite of texture attributes is generated from the seismic data. Next an interpreter identifies zones with and without gas chimneys. This interpretation is used to train a neural network that can be used after validation to identify chimneys throughout the uninterpreted volume. (a) A vertical seismic slice showing horizons A and B. (b) The same slice with predicted gas chimneys. (c)-(d). A 40-ms average absolute amplitude (AAA) map (c) along horizon A and (d) along horizon B. Chimneys are plotted in yellow and line up with the faults, which indicates that the faults are leaking gas. After Heggland (2004). AAPG©2004. Reprinted by permission of the AAPG whose permission is required for further use.

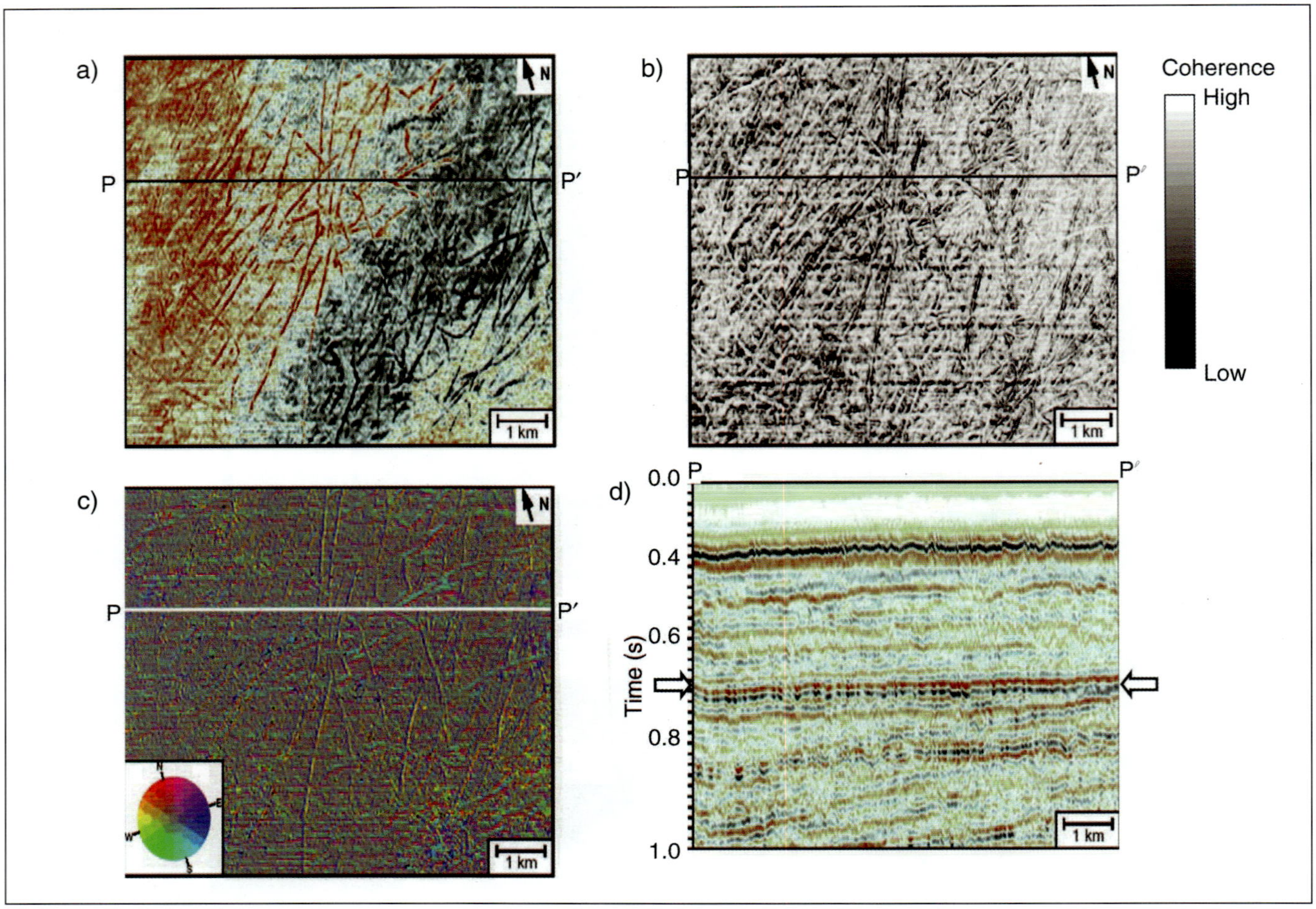

Figure 4.94. Glacial plow marks seen (a) on an amplitude time slice, (b) a coherence time slice, (c) a dip and azimuth time slice, and (d) a vertical seismic section. At first, the washed-out streaks and ripples on the vertical seismic section might be thought to be an acquisition footprint, but the organization on the time slice shows the origin of the streaks and ripples to be geologic. The slight dip of the horizon causes us to see troughs on the west and peaks on the east side of the amplitude time slice. The dip and azimuth image shows clear depressions along each lineament, confirming that they are plow marks. Although the coherence and dip and azimuth images are interesting, the amplitude data provide all the information we might need to make this interpretation and avoid such hazards. Seismic data are provided by Statoil. After Gallagher and Heggland (1995).

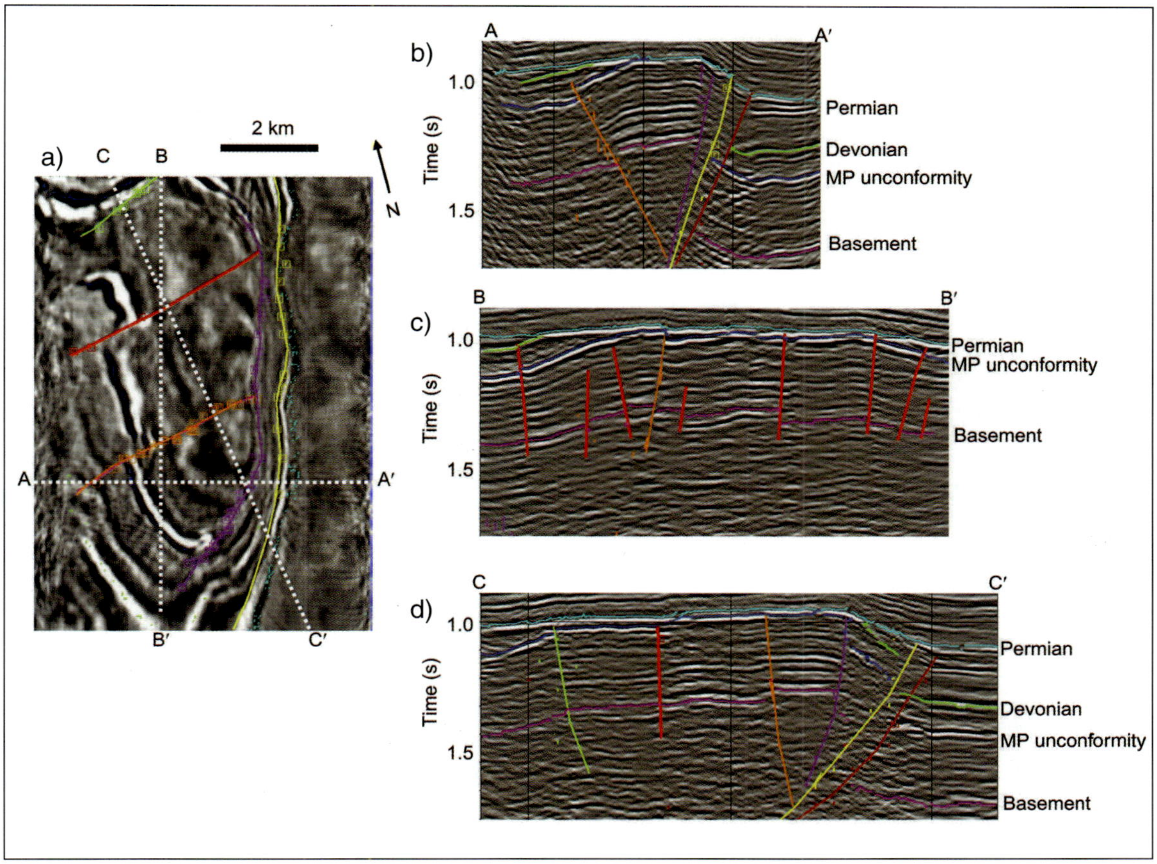

Figure 4.95. (a) A time slice at $t = 1.060$ s, showing the locations of vertical seismic traverses (b) AA′, (c) BB′, and (d) CC′ at Dollarhide field. Reverse faults that bound the field to the east are shown in purple, yellow, and maroon on vertical sections AA′ and CC′. High-angle, northeast-striking reverse faults with right-lateral displacement are shown in orange, red, and green. The conjugate northwest-striking faults have not been interpreted on the upthrown block. The right-lateral strike-slip faults seen on the seismic time slice mask the presence of these lineaments. On the vertical seismic lines, the horizons are indicated by the following colors: Cyan = the base of an Early Permian unconformity, blue = a Mississippian to Pennsylvanian (MP) unconformity, light green = the top of the Devonian Thirtyone Formation, and magenta = the top of Precambrian granite basement. After Serrano et al. (2003).

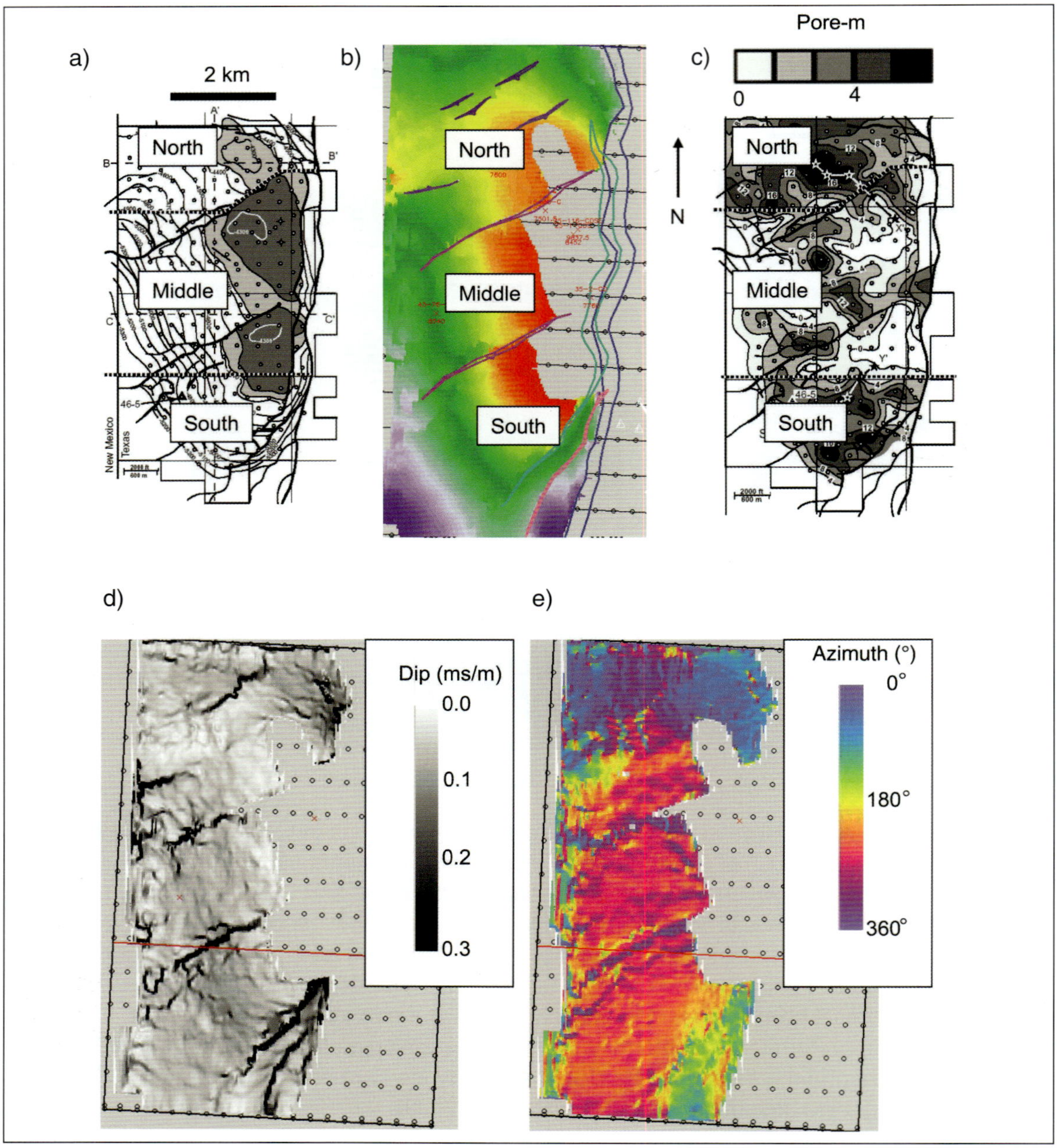

Figure 4.96. (a)-(b). Time-structure maps of the top of the Devonian in Dollarhide reservoir, Central Basin Platform, Texas, U.S.A., (a) after Saller et al. (2001) and (b) after Serrano et al. (2003). Dark-gray shading in (a) indicates the zone of truncation of the upper reservoir and part of the lower reservoir. Light-gray shading indicates the zone where the upper reservoir is partially truncated. (c) Porosity of the lower zone, obtained from well control (after Saller et al., 2001). (d) A conventional dip-magnitude map and (e) a dip-azimuth map, both derived from the time-structure map in (b) and neither of which indicates the lateral heterogeneity seen in (c). Image (a): AAPG©2001. Reprinted by permission of the AAPG whose permission is required for further use.

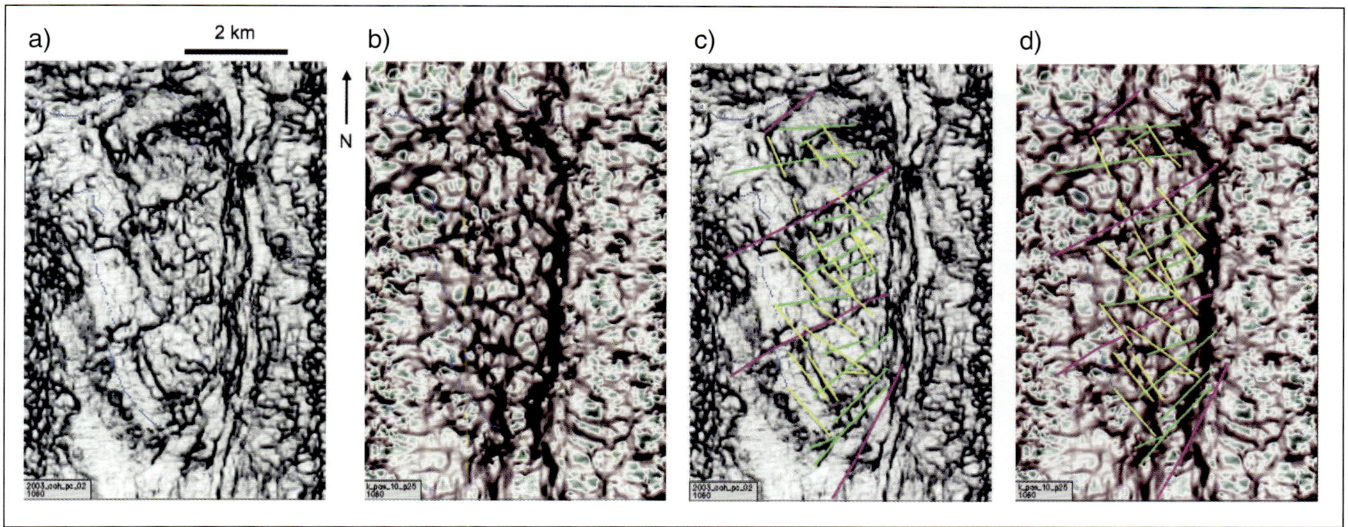

Figure 4.97. (a)-(b). Time slices at $t = 1.060$ s through (a) a coherence volume and (b) a most-positive-curvature volume for the Dollarhide field. Figures (c) and (d) are the same images with interpretation. The northeast-southwest-trending green lineaments, interpreted to be smaller strike-slip faults, correlate positively with fluid flow in reservoir production. The magenta faults were picked directly from the migrated seismic image. The northwest-southeast yellow faults are antithetic to the strike-slip faults. Most of the east-west lineaments are the result of an acquisition footprint associated with the early 1990s 3D-swath seismic-acquisition program. The trends on these lineaments correlate well with porosity trends. After Serrano et al. (2003).

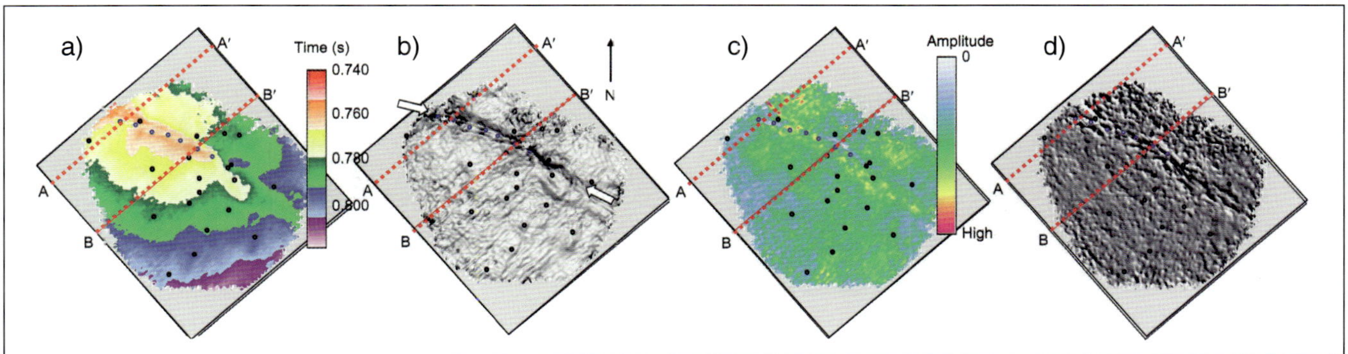

Figure 4.98. (a) A time-structure map of the top Trenton Limestone horizon from a survey acquired in Appalachian Basin, northeastern Ohio, U.S.A, and reported on by Sagan and Hart (2004). (b) The corresponding dip-magnitude map and (c) the corresponding horizon slice through the seismic data. (d) A northeast-southwest energy-weighted coherent-amplitude gradient. An acquisition footprint contaminates the dip-magnitude map and the northwest-southeast energy-weighted coherent-amplitude gradient map (not shown). Arrows on (b) indicate a fault. Seismic lines AA′ and BB′, discussed by Sagan and Hart (2004), are displayed in the next figure. Data are courtesy of CGAS.

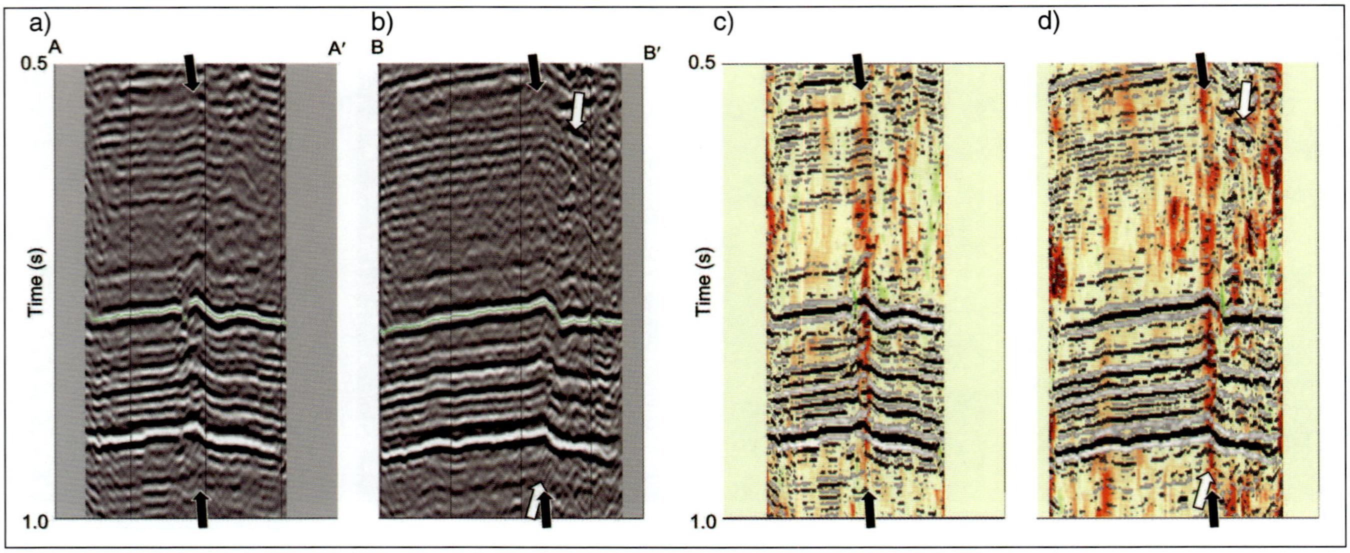

Figure 4.99. Seismic lines (a) AA′ and (b) BB′ (which correspond to lines 33 and 93, respectively, as discussed by Sagan and Hart, 2004), whose locations are shown in the previous figure. Images (c)-(d) are the corresponding composite images of the seismic lines in (a) and (b), respectively, overlaid with most positive curvature. Positive values are red, zero values are pale yellow (to distinguish them from the white troughs of the seismic), and negative values are green. Note that the fault indicated by white arrows in (b) lies in front of an anticlinal nose, which is indicated by black arrows in (d). AAPG©2004. Reprinted by permission of the AAPG whose permission is required for further use.

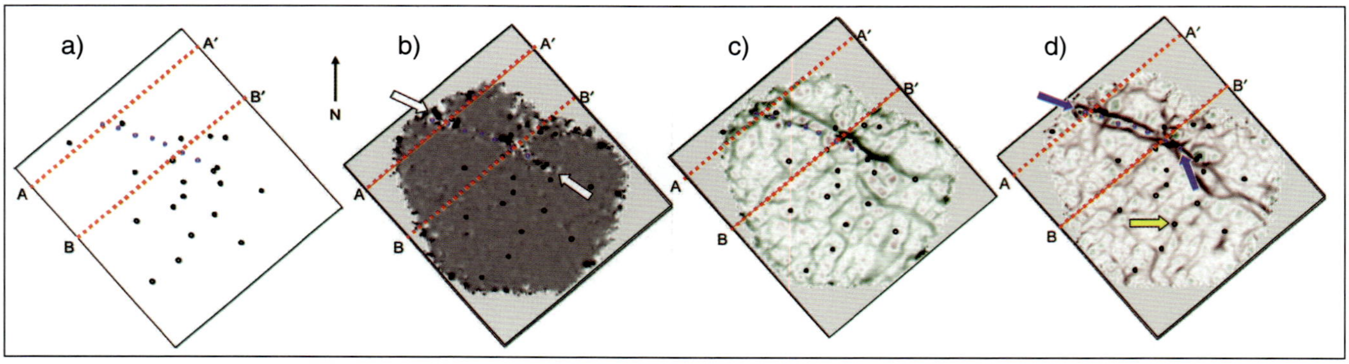

Figure 4.100. Horizon slices corresponding to the top of the Trenton Limestone shown in Figure 4.98a, through (a) a coherence volume, (b) a reflector-rotation map, (c) a most-negative-curvature volume, and (d) a most-positive-curvature volume. The coherence slice does not provide significant information beyond that seen in the dip-magnitude map shown in Figure 4.98b. The rotation map shows that the fault has rotational movement (white arrows), which allows hydrothermal fluids to rise and diagenetically alter the limestone into porous dolomite. Note how the producer wells, indicated by blue circles between the two arrows, correlate to negative values of the most-positive-curvature (green areas in [d]), which correspond to collapse features. The dry holes shown in (d) indicate that the operator drilled subtle anticlinal features such as that indicated by the yellow arrow but failed to find porous dolomite. Curvature was calculated using $\alpha = 0.25$.

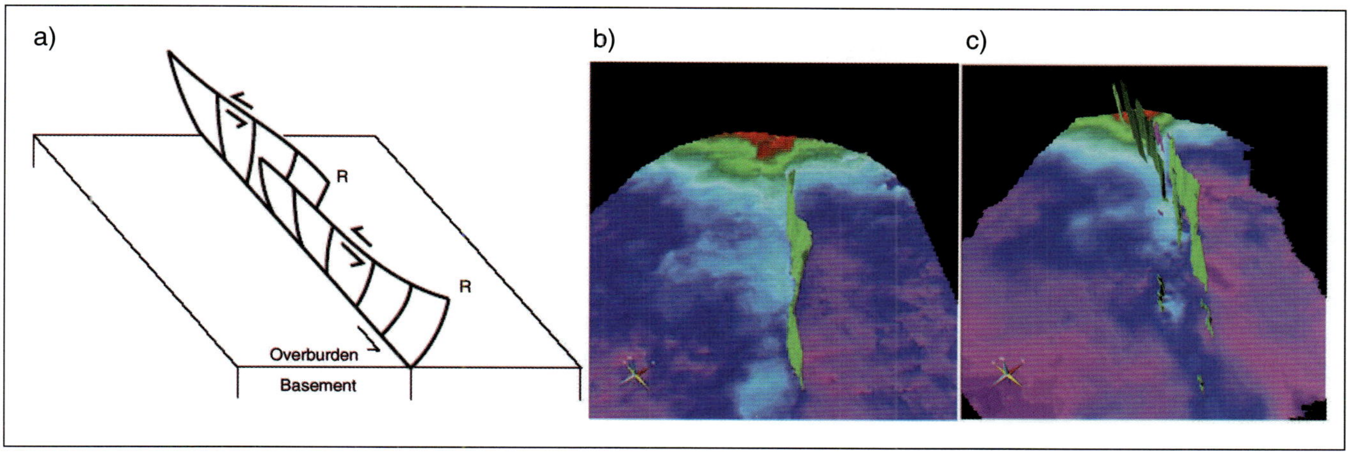

Figure 4.101. (a) Mandl's (1988) block diagram of helicoidal deformation above a single basement fault. (b) The main Riedel shear, exhibiting a helicoidal nature, shown here with the top-of-basement structure map. (c) The main synthetic Riedel shear (light green) and less developed Riedel shears (in darker greens), shown here with the shallower Trempealeau map. The en echelon nature of these faults is apparent from this angle. After Sagan and Hart (2004). AAPG©2004. Reprinted by permission of the AAPG whose permission is required for further use.

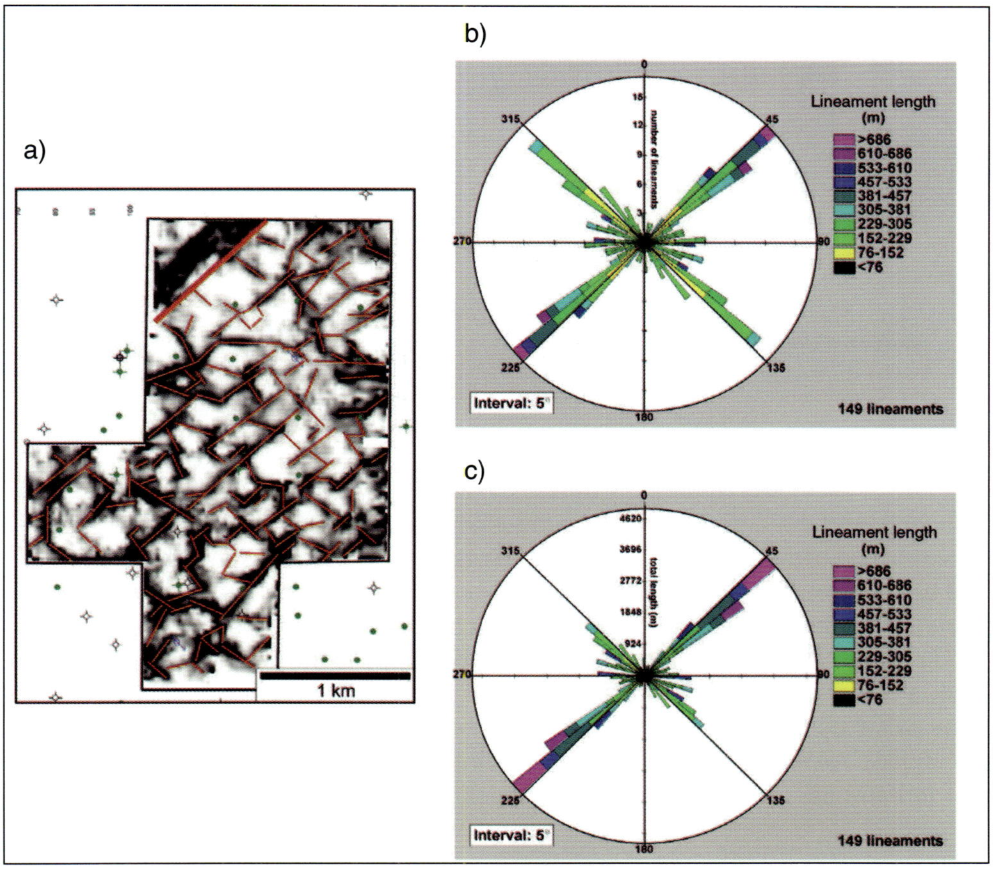

Figure 4.102. (a) An interpretation of lineaments seen on a most-negative-curvature volume along the base of the Mississippian aquifer. (b) A rose diagram of frequency (number of lineaments) versus azimuth. (c) A rose diagram of length (of all lineaments) versus azimuth. There are two preferred lineament directions, northeast-southwest and northwest-southeast. After Nissen et al. (2005). AAPG©2005. Reprinted by permission of the AAPG whose permission is required for further use.

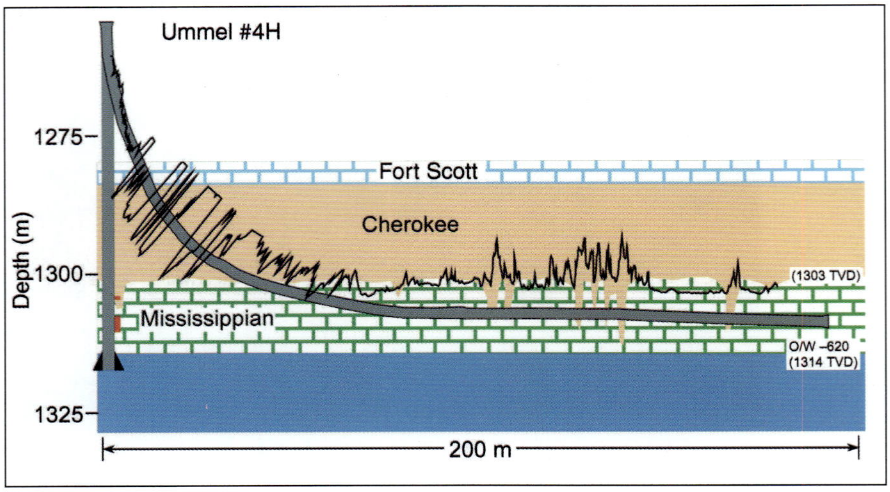

Figure 4.103. A cross section showing shale-filled fractures intersected by a horizontal well through a Mississippian reservoir, Ness County, Kansas, U.S.A. Data are courtesy of Mull Drilling Co. After Carr et al. (2000).

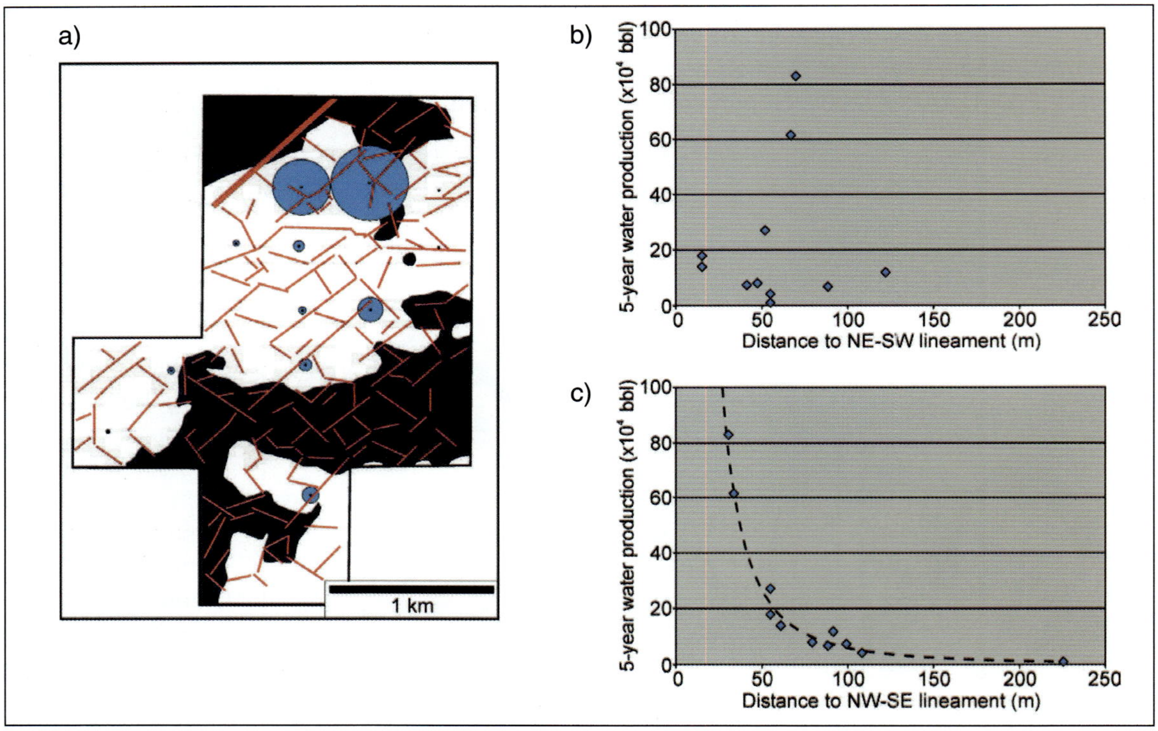

Figure 4.104. (a) Lineaments and the amount of water produced during the first five years of production (water production is indicated by blue circles, with the largest circle corresponding to approximately 830,500 bbl). (b)-(c). The correlation between water produced and the distance from (b) the nearest northeast-southwest lineament and (c) the nearest northwest-southeast lineament. Good correlation exists between water production and proximity to the northwest-southeast lineaments, with a power-law function (dashed line in [c]) providing a good fit to this relationship. No correlation exists between water production and the northeast-southwest lineaments. After Nissen et al. (2005). AAPG©2005. Reprinted by permission of the AAPG whose permission is required for further use.

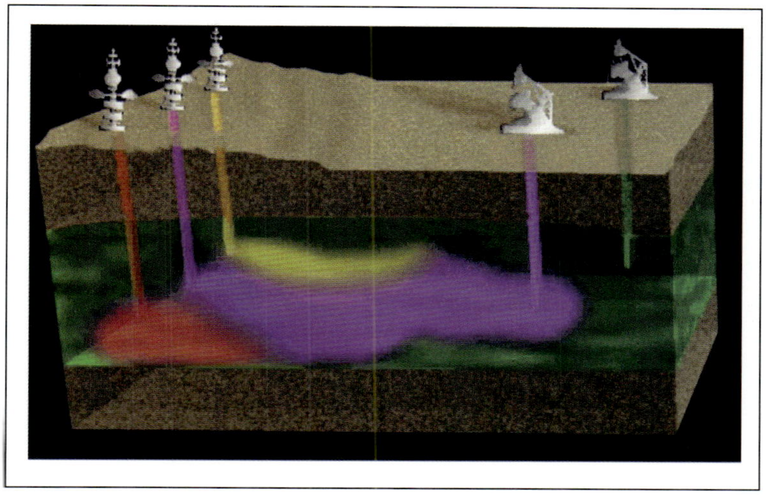

Figure 4.105. Chemical tracers track fluid movement. In this example, the purple tracer injected in the middle well (on the left) reaches the producing well to the right, whereas the yellow and orange injected tracers do not. Green indicates fluid untouched by tracers. After Chopra and McConnell (2004). Reprinted by permission of *Oil & Gas Journal.*

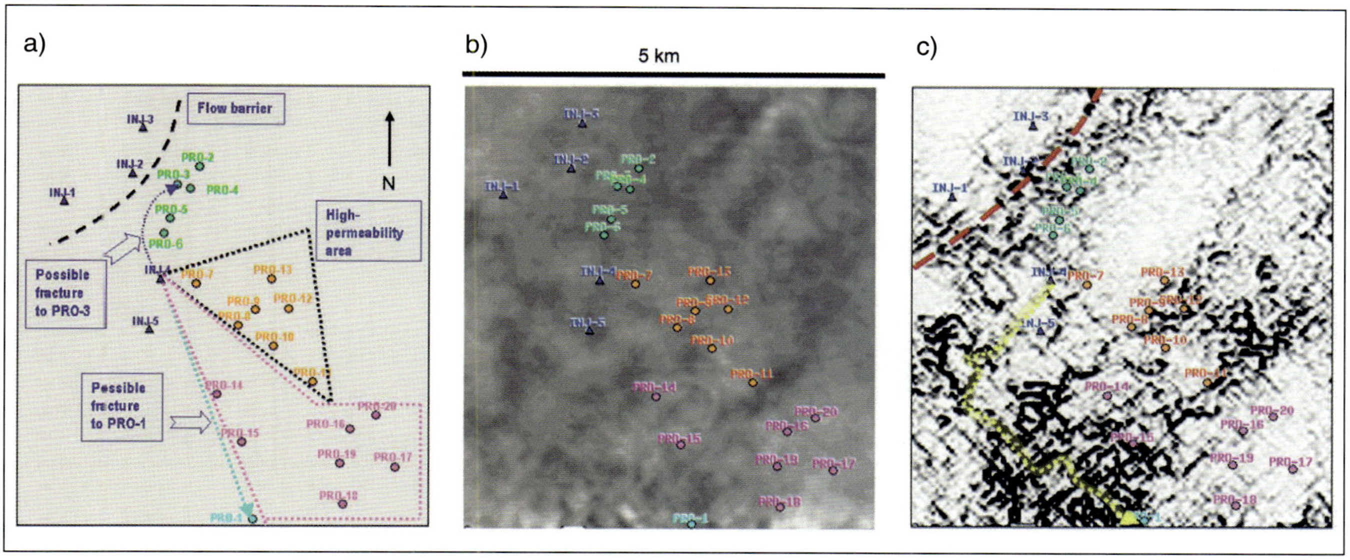

Figure 4.106. (a) A summary map of chemical-tracer responses, showing injectors in blue and delineating four zones of the reservoir that had different behaviors, indicated here by green, yellow, magenta, and cyan producer wells. (b)-(c). Horizon slices below the reservoir through (b) a seismic data volume and (c) a sharpened coherence volume. We note a northeast-southwest low-coherence trend in the upper left corner, encompassing injectors INJ-1, INJ-2, and INJ–3, which do not communicate with any of the producers. We interpret this zone of low coherence to be a flow barrier (graphically shown as a dotted line) that prevents fluids (tracers) from flowing away from the wells toward the right. The quick breakthrough to PRO-1 (teal arrow) in the south is explained by darker lines that depict the potential channel pathway for the tracer to arrive from INJ-4 in such a short time (the potential channel is highlighted by the yellow line). After Chopra and McConnell (2004). Reprinted by permission of *Oil & Gas Journal.*

Unit 5 Attributes and the Seismic Processor

Overview

Seismic attributes are only as good as the seismic data from which they are derived. Errors in geometry, statics, wavelet shaping, and velocities all reduce the spectral bandwidth, signal-to-noise ratio, and lateral resolution of the resulting seismic image. Coherent noise — including multiples, backscattered head waves, and ground roll — produces events that cross the true reflection events of interest and that may exceed the true reflection events in amplitude. Dip and azimuth calculations may track those higher-amplitude crossing events erroneously. Coherence may treat as discontinuities the events that cross the true reflector, and in so doing coherence may produce erroneous time-slice images that might be misinterpreted as faults or stratigraphic boundaries.

We begin by evaluating the sensitivity of geometric-attribute images to the most basic processing parameters — in particular, to statics and velocities. Then we examine the cause and appearance of an acquisition footprint, as well as the partial solutions for a footprint. Next we review the limits to lateral resolution and demonstrate the value of prestack migration. A discussion follows, examining the improvement of vertical resolution using impedance inversion (which uses well control) and of waveform estimation and deconvolution using vertical seismic profiles.

Often, interpreters have neither the option nor the time to reprocess their seismic data. Thus, we discuss two means of improving attribute volumes computed from seismic data that have already been migrated: structure-oriented filtering of the migrated seismic data volume, and 3D image processing of the attribute volumes. We conclude by showing the value of applying attributes to offset- and azimuth-limited seismic data volumes, and of applying attributes to converted waves in conjunction with conventional compressional wave volumes.

Summary

Spectral decomposition and geometric attributes can be important tools to help minimize data-quality issues. We recommend examining time slices through attribute volumes that are sensitive to the exploration objective, at each stage of the seismic-processing flow, to control the quality of the results and to drive the choices of processing parameters and appropriate noise-rejection algorithms.

Acquisition footprint plagues almost all land data volumes as well as marine volumes in which obstacles or currents exist that prevent uniform coverage. An acquisition footprint degrades seismic attributes and autotrackers, clustering algorithms, geostatistics, and neural nets. A footprint can be particularly troublesome in any attempt to map fractures and other lineaments. In general, however, an acquisition footprint diminishes with depth. Therefore, we advise interpreters always to begin by examining attribute time slices in the shallowest part of the section and then to form a mental image of what not to interpret deeper in the section.

Seismic migration collapses diffraction hyperbolae and increases lateral resolution. Given a good velocity model, prestack migration algorithms provide higher lateral-resolution images than do poststack migration algorithms. Poorly focused seismic data, such as those that occur below salt, provide nearly useless attribute images.

Impedance inversion and other processes that use well control (such as VSPs) to improve the frequency content of the resulting image yield improved attribute volumes. Higher-resolution data produce less mixing of overlying and underlying lithologic units and thereby provide attribute images that have improved vertical and lateral resolution.

Structure-oriented filters (SOFs) internally use a measure of the presence of discontinuities to avoid smoothing across faults and other features of interest. Unlike conventional dip filters that pass or reject a band (or fan) of dips at any analysis point, structure-oriented filters accept only one value of dip and azimuth — that of the estimated reflector. For that reason, structure-oriented filters do not suffer from the aliasing that plagues conventional dip filters. Structure-oriented filtering improves the behavior of 3D autotrackers and of attribute calculations. Seismic data that have been subjected to structure-oriented filtering generally have a higher signal-to-noise ratio, sharper discontinuities, and enhanced images of faults, fractures, and stratigraphic features such as channels. Interpreters should be forewarned, however, that not all features of geologic interest appear as discrete boundaries. The appearance of gas chimneys, karsts, diagenetically altered dolomites, and other diffuse features may be suppressed by structure-oriented filtering.

All the attributes described in this book operate within a small, localized analysis window. In contrast, a human interpreter can recognize discontinuities within the context of the entire seismic section and thus can connect seismic discontinuities through zones that appear to be quite continuous locally, while avoiding the inclusion of discontinuities that run subparallel to the geologic strata. Recent advances in 3D image processing mimic this work flow by applying a sequence of dip-sensitive filters to 3D coherence volumes. Computer-assisted extraction and labeling of fault planes is available in advanced interpretation software. However, computer-assisted autotracking of morphologically more complex features, such as channels and unconformities, is still an area of active research.

Using an AVO (amplitude variation with offset) analysis framework, we can apply spectral decomposition, coherence, and coherent-amplitude gradients to offset-limited seismic data volumes to better delineate the edges of hydrocarbon reservoirs and lithologic boundaries. Although the variability between different offset images generally is only moderate, we observe detailed stratigraphic features that may help us to map reservoir heterogeneity better and thereby improve reservoir management.

Seismic reflections also are sensitive to fractures and regional stress, both of which can contribute to P-wave and S-wave anisotropy. Attributes applied to azimuth-limited volumes can provide superior lateral resolution of fault systems in orientations that differ from those of the same attributes applied to all-azimuth volumes.

A volumetric estimate of velocity anisotropy is an attribute that provides information not measured by the dip and azimuth, curvature, coherence, texture, and spectral-decomposition attributes that are the primary focus of this course. In particular, volumetric estimates of anisotropy can be used to estimate both fracture density and orientation. At present, we believe anisotropic effects resulting from fractures are seismically indistinguishable from anisotropic effects caused by regional stress.

In addition to offset and azimuth, we can acquire seismic data that have different modes of propagation. For the same surface acquisition, P-P waves and P-S waves illuminate different regions of the earth and have different reflectivities, so they generate independent and complementary attribute images of the subsurface. Each of these volumes results in complementary attribute maps that can help to delineate faults, fractures, and stratigraphy.

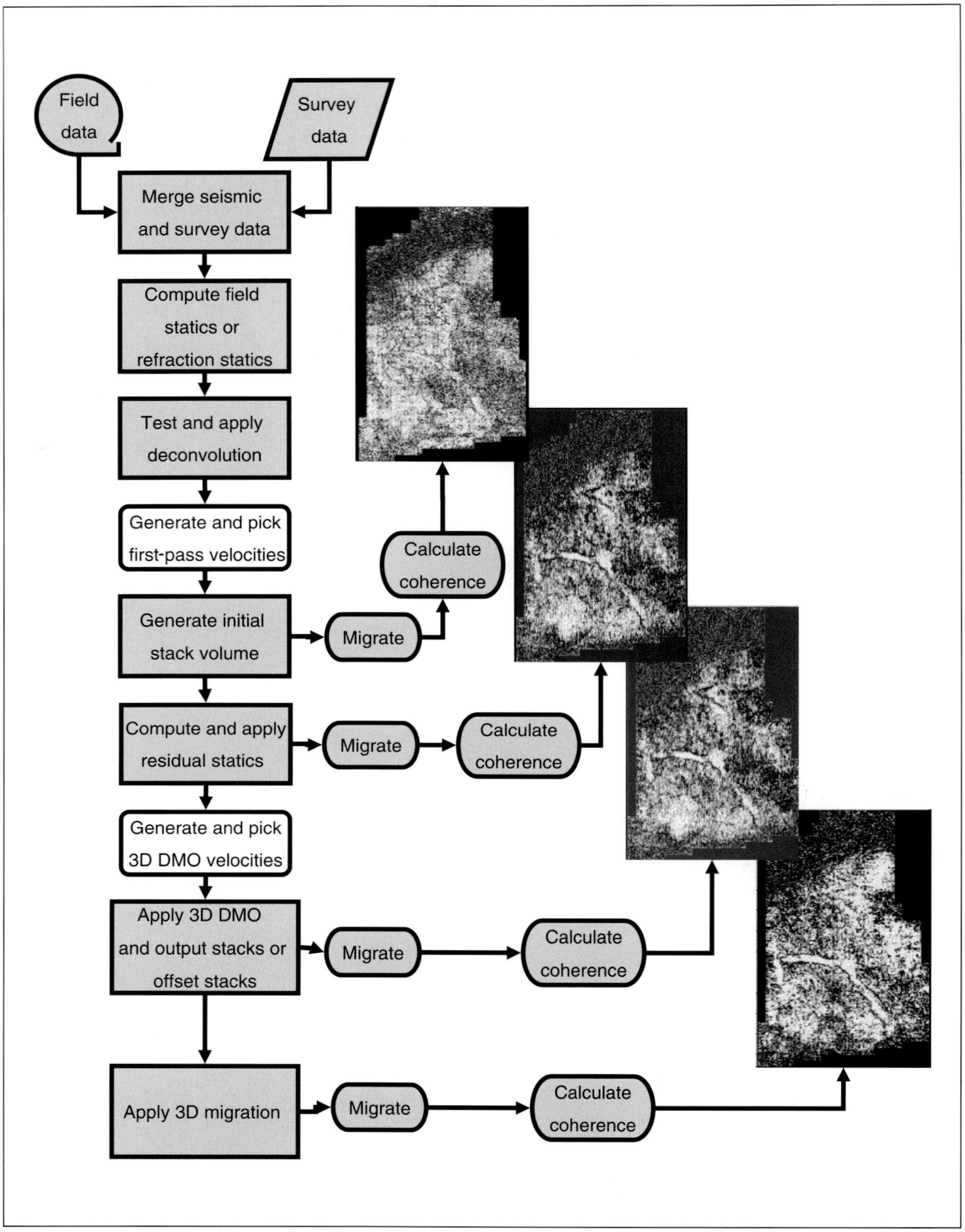

Figure 5.1. A typical processing sequence, using coherence as a processing quality-control tool. Poorly imaged volumes are less focused and therefore are more coherent. Properly imaged volumes are more focused and therefore are less coherent.

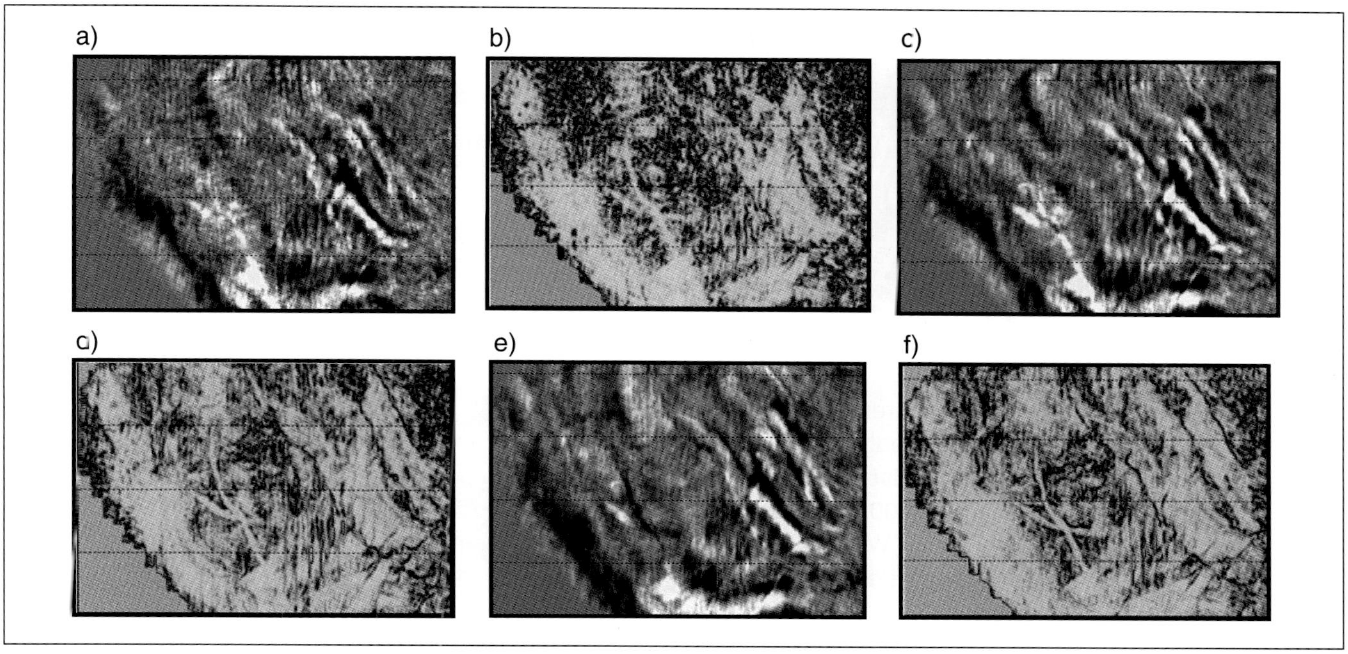

Figure 5.2. Time slices through seismic data generated at different stages in processing: (a) a brute stack followed by migration, (b) residual statics followed by stack and migration, (c) an image after final migration with an improved velocity model, and (d)-(e) slices through the three corresponding coherence volumes.

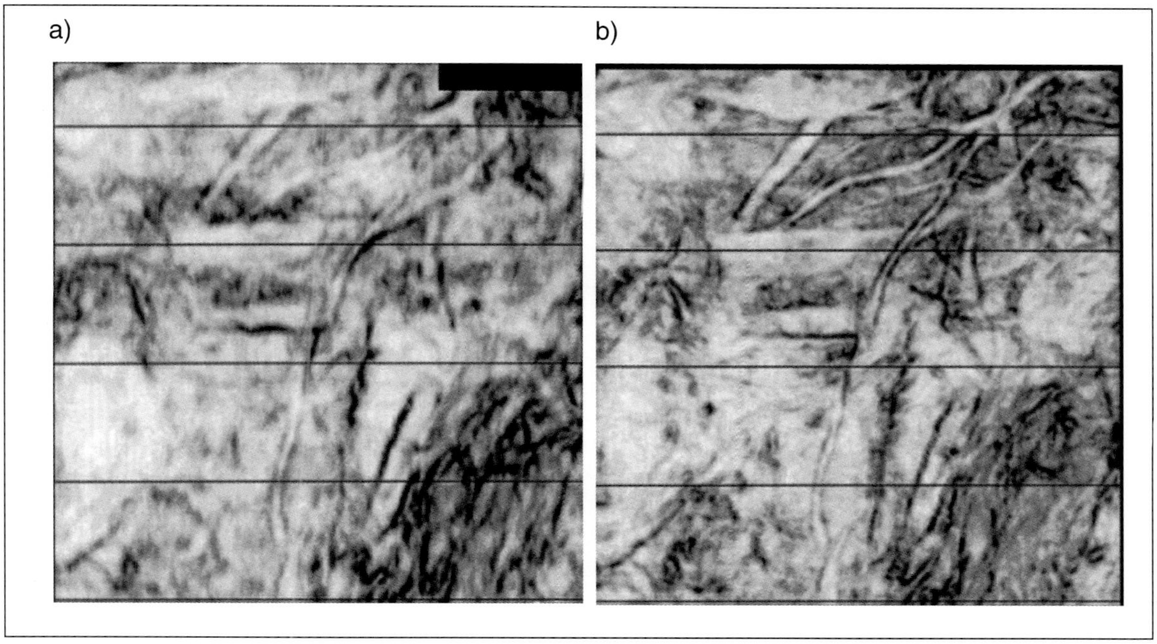

Figure 5.3. An example showing coherence after (a) an initial velocity model and (b) the final 3D velocity model for the same data. Note the marked improvement in lateral resolution provided by more careful velocity analysis.

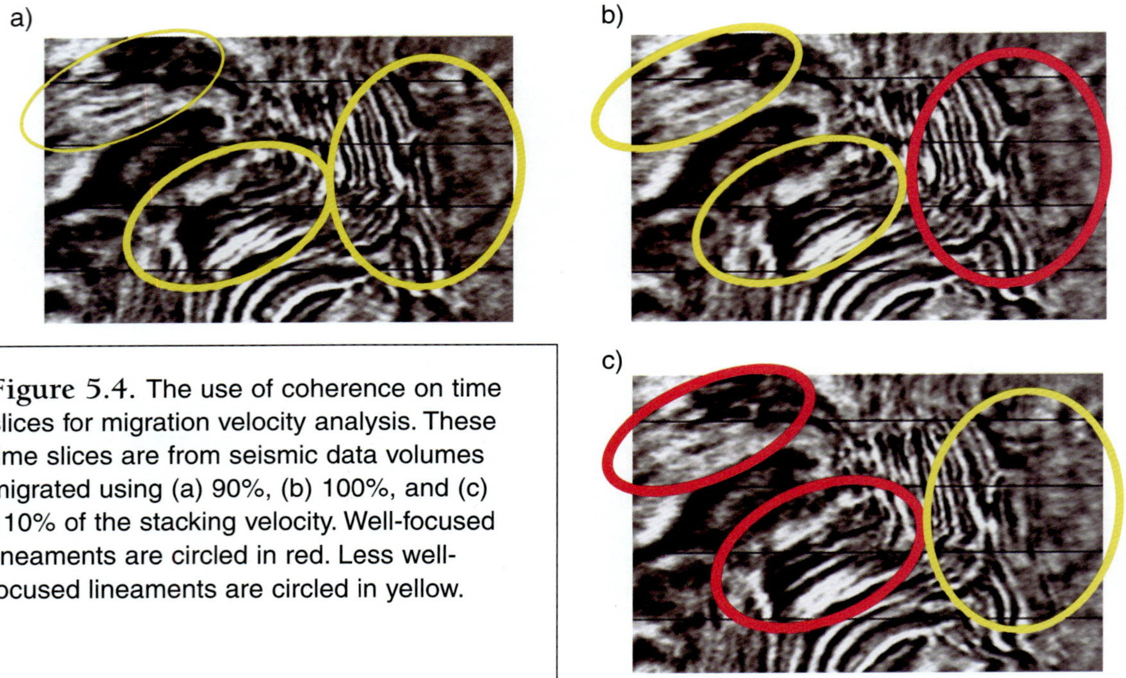

Figure 5.4. The use of coherence on time slices for migration velocity analysis. These time slices are from seismic data volumes migrated using (a) 90%, (b) 100%, and (c) 110% of the stacking velocity. Well-focused lineaments are circled in red. Less well-focused lineaments are circled in yellow.

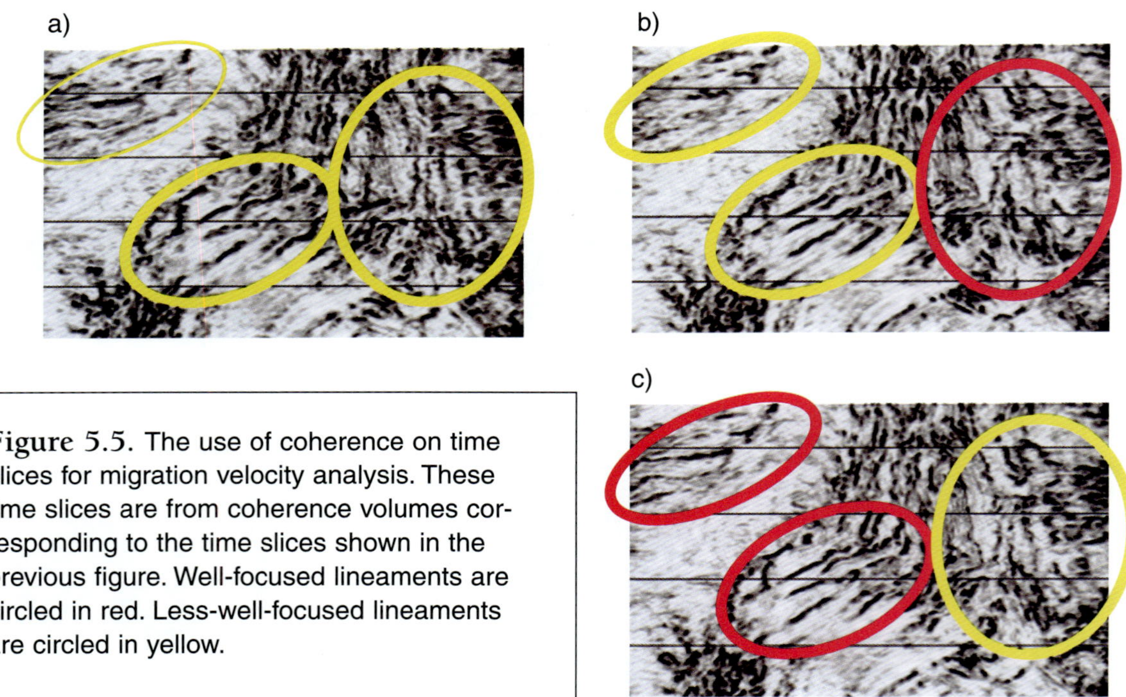

Figure 5.5. The use of coherence on time slices for migration velocity analysis. These time slices are from coherence volumes corresponding to the time slices shown in the previous figure. Well-focused lineaments are circled in red. Less-well-focused lineaments are circled in yellow.

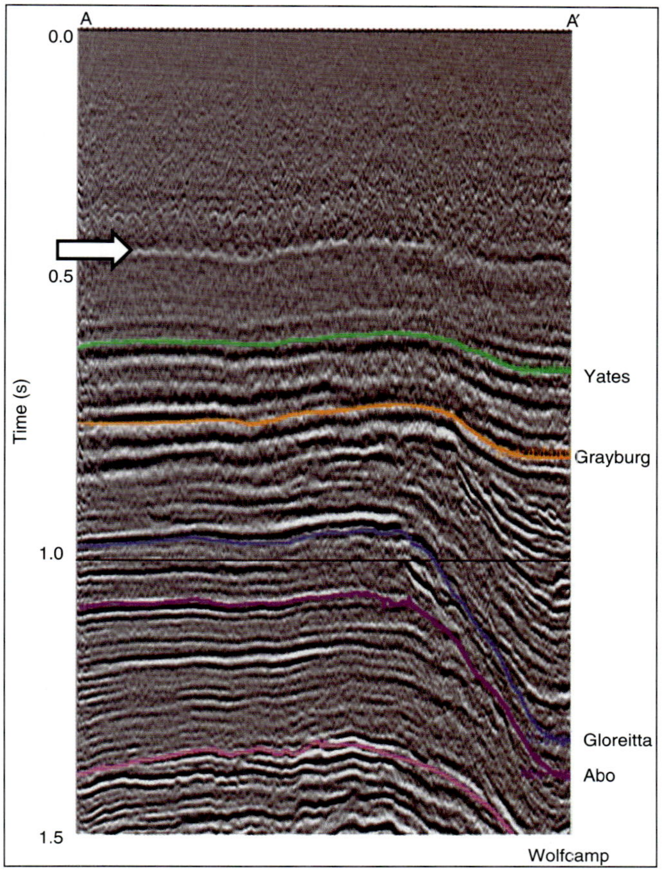

Figure 5.6. A vertical slice through conventional seismic data acquired over the Delaware Basin, New Mexico, U.S.A. The Yates and Grayburg strata are important producing horizons on the carbonate platform. The arrow indicates the appearance of an acquisition-footprint amplitude contamination at a horizon above the Yates. After Famini (2005).

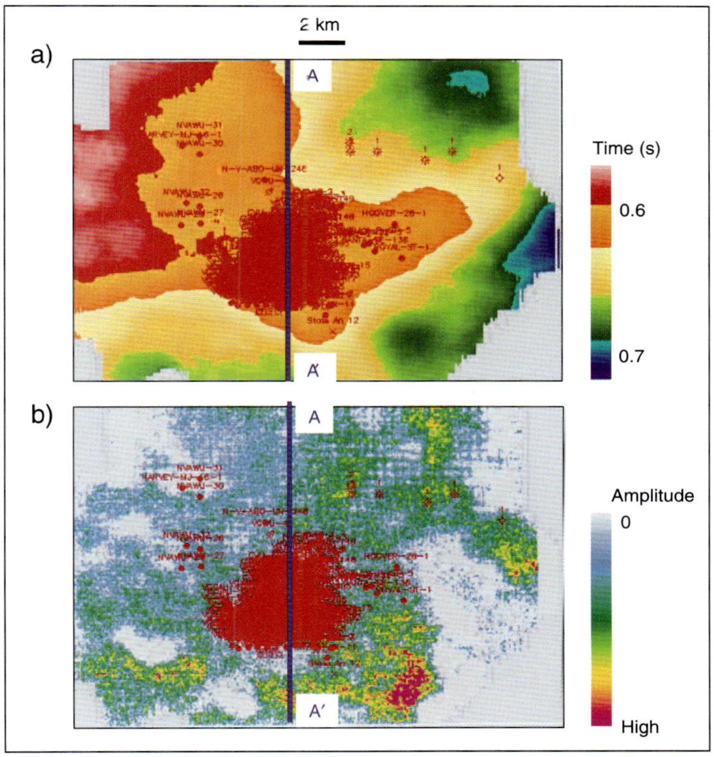

Figure 5.7. (a) A time-structure map and (b) an amplitude-horizon slice, both corresponding to the Yates horizon shown in the previous figure. An acquisition footprint contaminates the amplitude horizon slice. After Famini (2005).

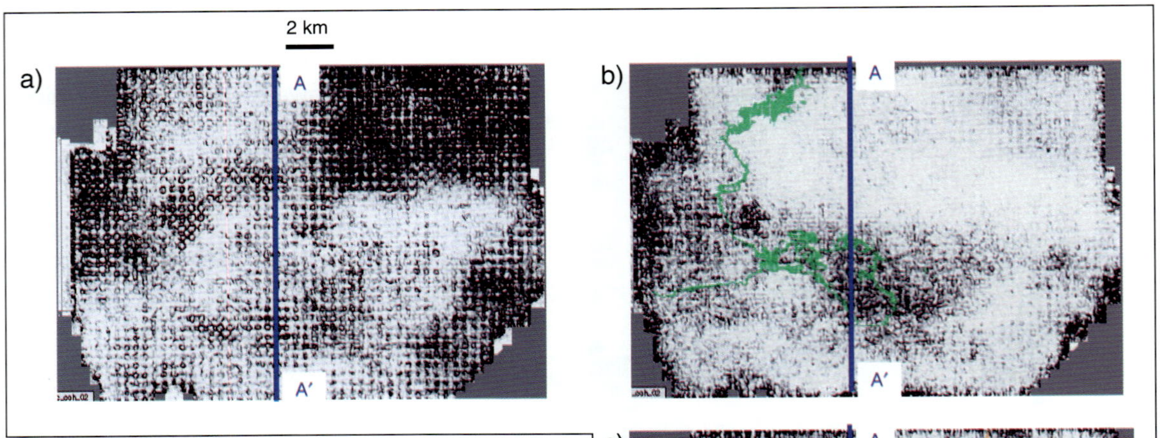

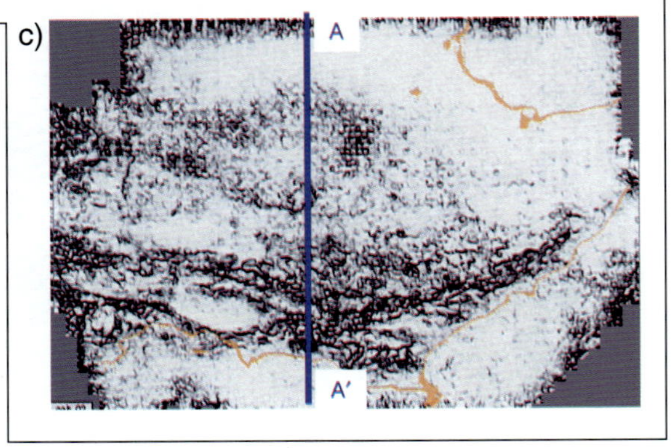

Figure 5.8. Time slices at (a) $t = 0.400$ s, (b) $t = 0.600$ s, and (c) $t = 0.800$ s through a coherence volume from a survey acquired over the Delaware Basin, New Mexico, U.S.A. Note how the acquisition footprint diminishes in deeper slices. Colors indicate horizons of interest cutting through the time slice. After Famini (2005).

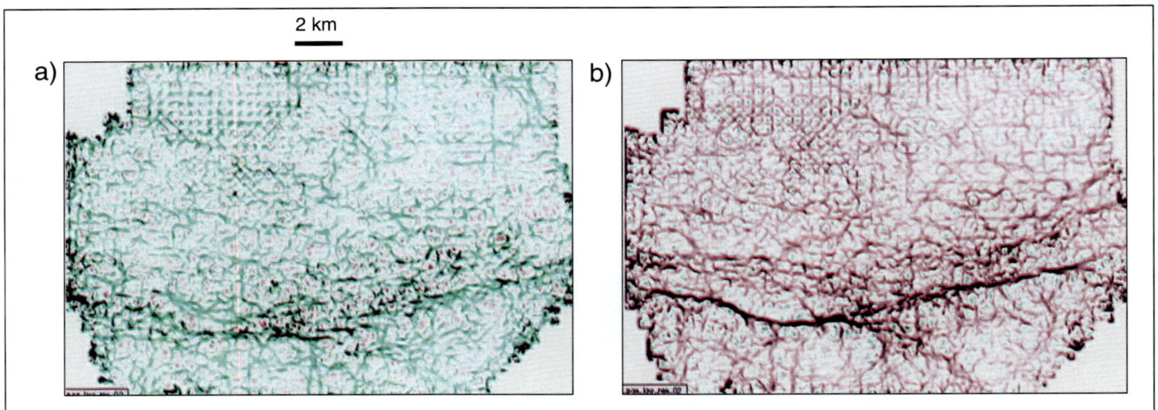

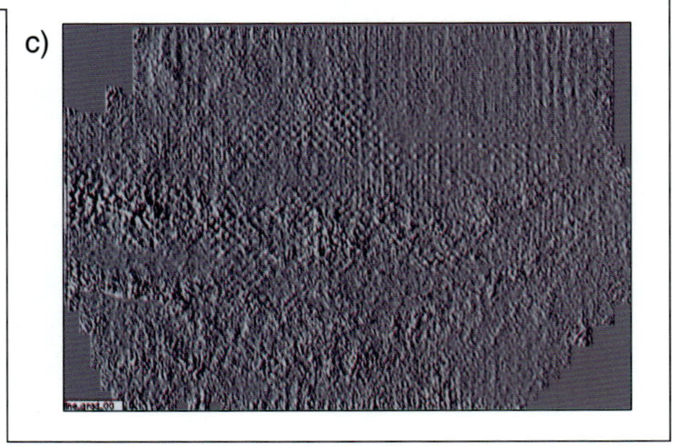

Figure 5.9. Time slices at $t = 0.800$ s through (a) the most-negative-curvature component, (b) the most-positive-curvature component, and (c) an east-west component of the energy-weighted coherent-amplitude-gradient volume for the Delaware Basin data shown in the previous figure. The acquisition footprint impacts all attributes, including curvature, as can be seen on the vertical section in Figure 5.6. After Famini (2005).

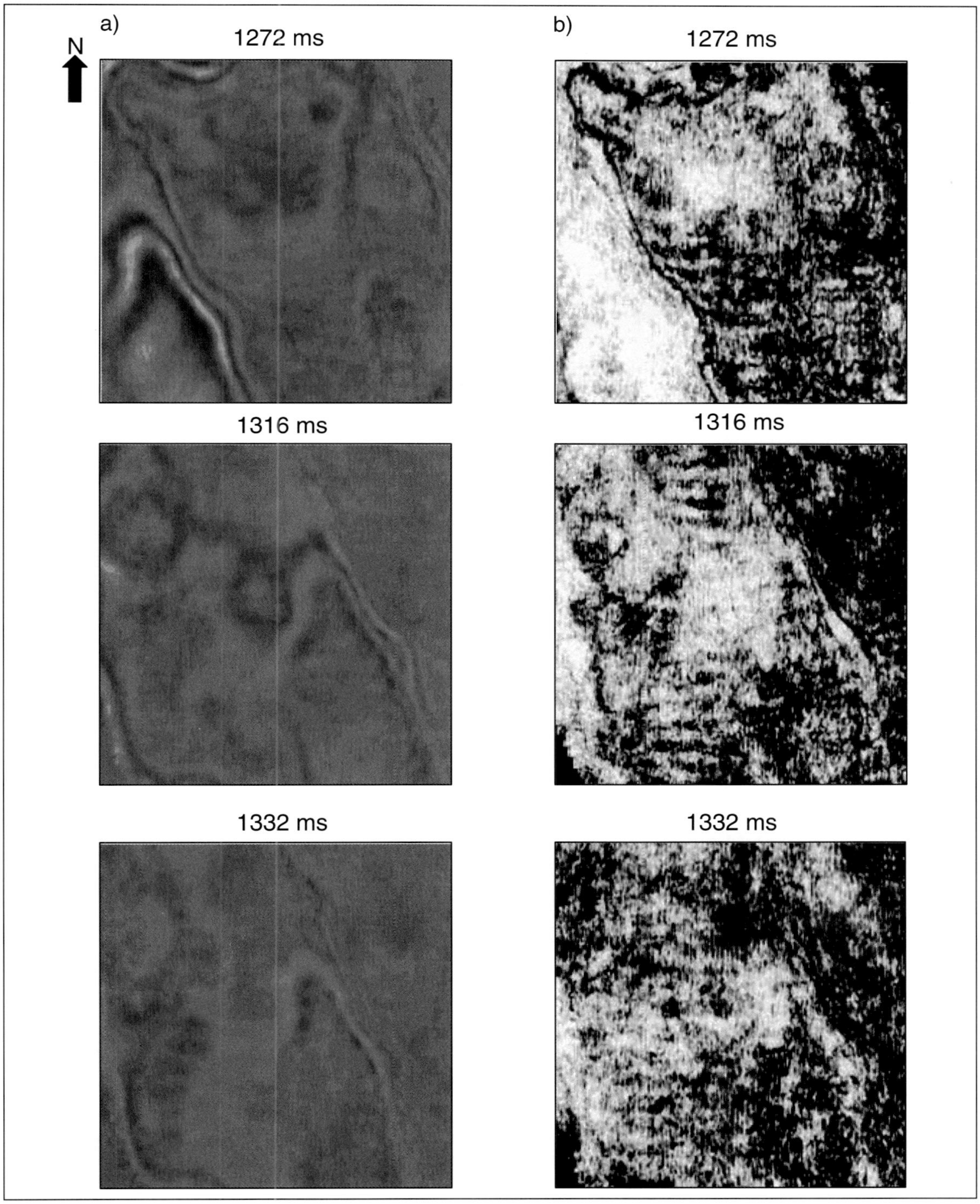

Figure 5.10. Time slices at t = 1.272, 1.316, and 1.332 s through (a) a seismic data volume and (b) the corresponding coherence volume. The acquisition footprint does not impede our ability to interpret the original seismic amplitude data, but it is exacerbated by the coherence calculation, in which we see wide east-west and narrow north-south artifacts. After Chopra (2002).

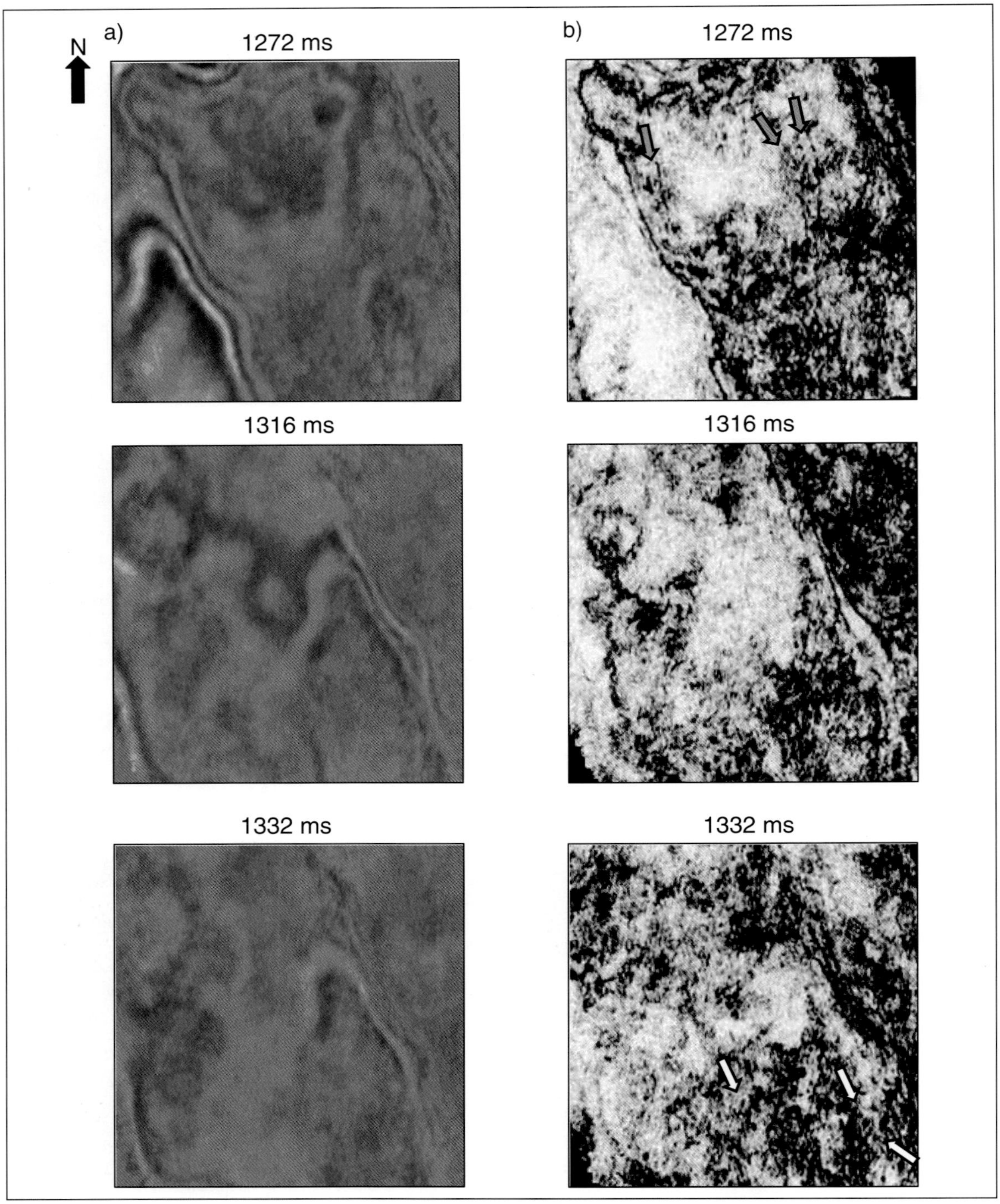

Figure 5.11. The same data slices as those shown in the previous figure, but displayed here after additional seismic processing (balancing, statics, and improved velocity analysis) to minimize the acquisition footprint. Although the impact on the original seismic data is minimal, we now can discern minor faults (indicated by gray arrows) and possible stratigraphic features (indicated by white arrows) that previously were masked by the acquisition footprint. After Chopra (2002).

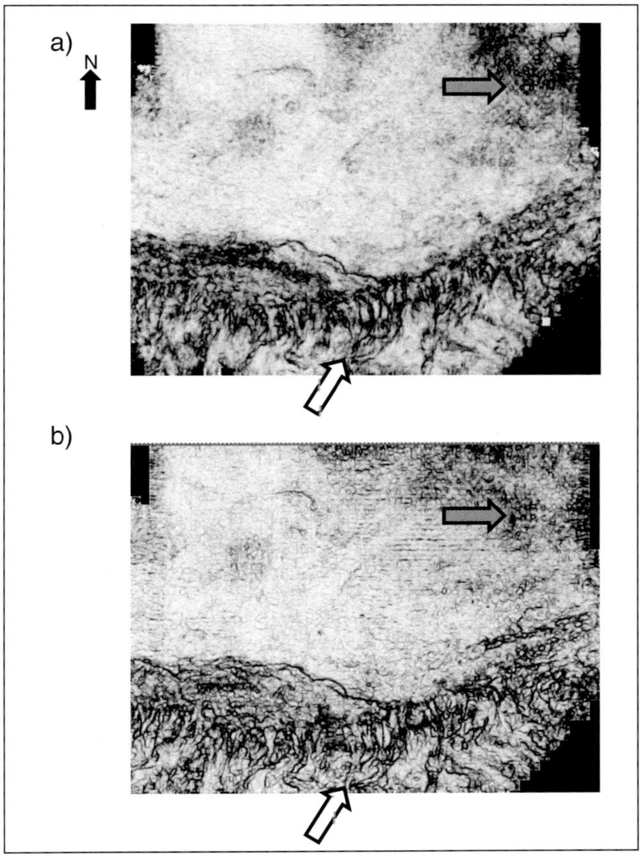

Figure 5.12. Comparison of a time slice at 1.0 s through coherence volumes generated for a survey from the Delaware Basin, New Mexico, U.S.A. The coherence slice from (a) original processing and (b) after careful reprocessing to improve lateral resolution. Note the trade-off between the increased acquisition footprint in the northeast corner, indicated by the gray arrow, and improved resolution in the Brushy Canyon slump features, indicated by the white arrow. After Famini (2005).

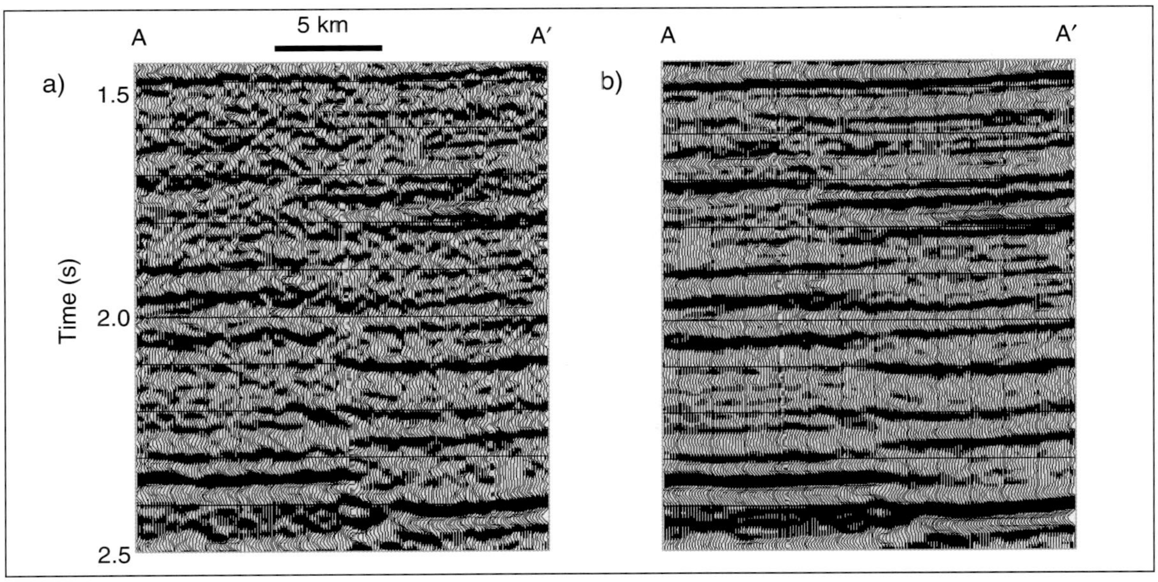

Figure 5.13. A seismic line (a) before and (b) after footprint removal using careful dip filtering. Note that the strong backscattered ground-roll noise has been suppressed.

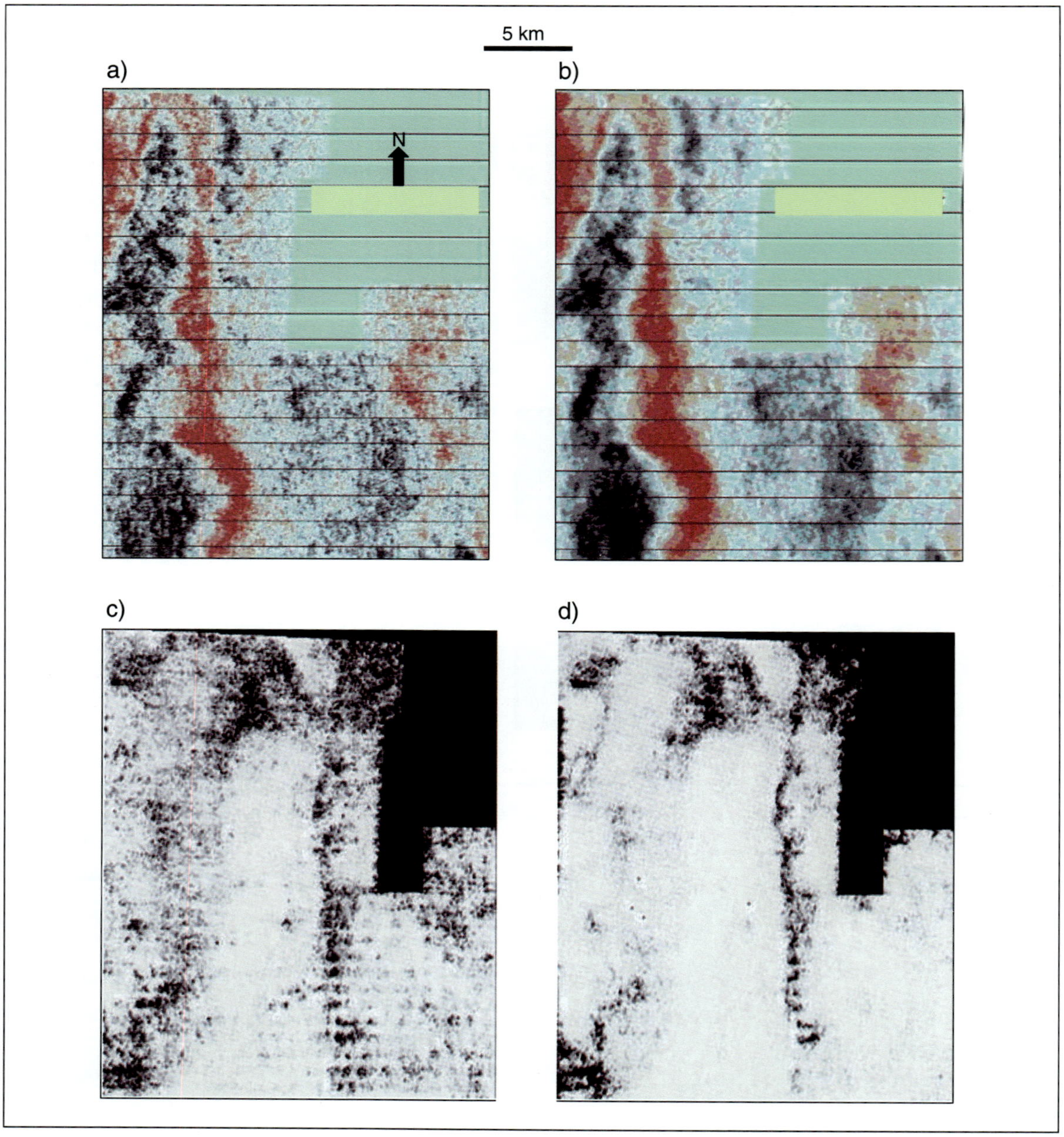

Figure 5.14. (a)-(b).Time slices at t = 1.000 s through the seismic data volume shown in the previous figure (a) before and (b) after footprint attenuation. (c)-(d).Time slices at t = 1.000 s through the corresponding coherence volumes (c) before and (d) after footprint attenuation. Note the strong east-west acquisition footprint resulting from backscattered ground roll in (a). The processor uses coherence images of the acquisition footprint to judge the effectiveness of alternative processing flows and parameters.

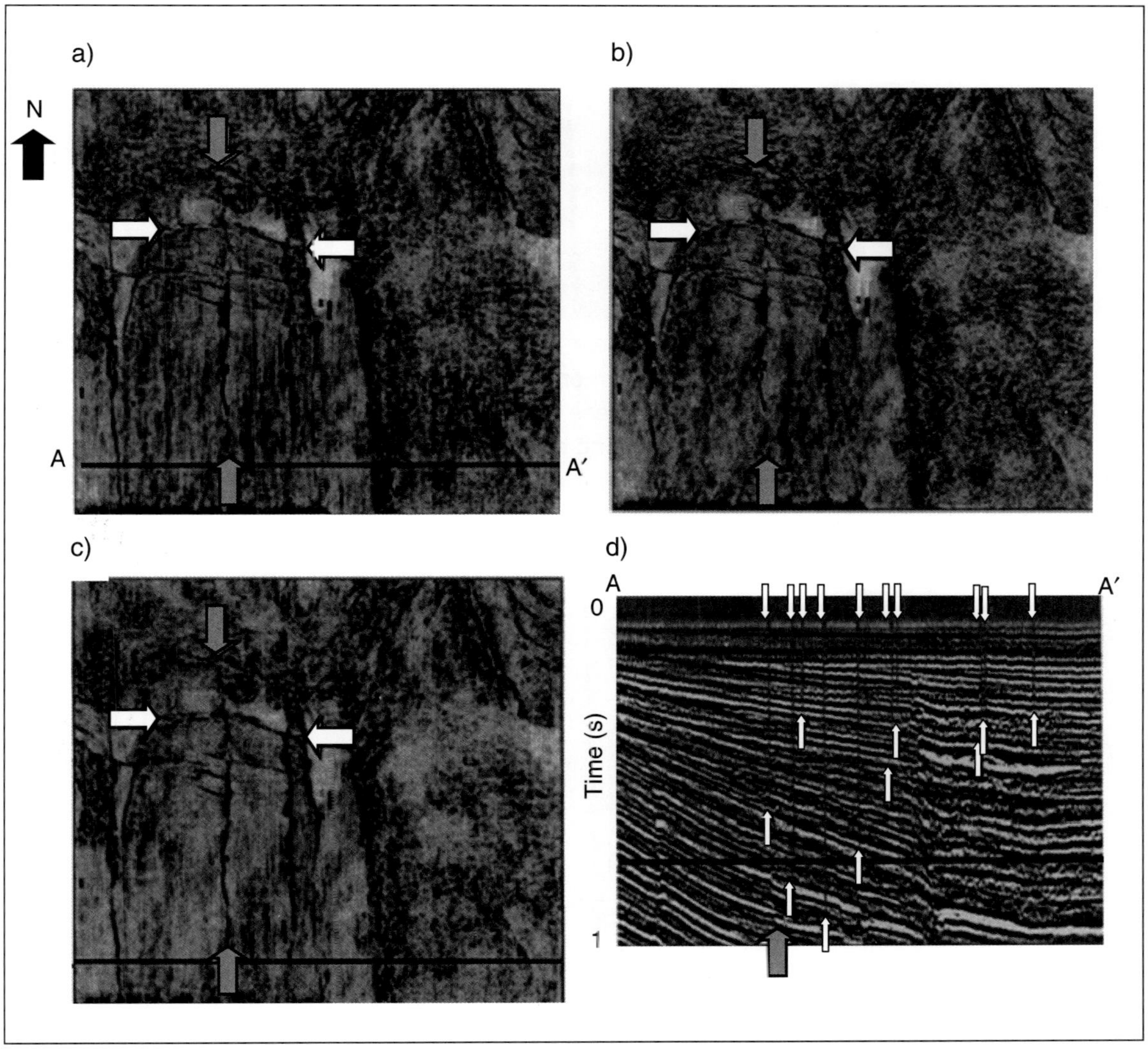

Figure 5.15. (a)-(c).Time slices at 0.75 s from a survey offshore Trinidad through coherence volumes generated from (a) original data, (b) data after k_x-k_y filtering, and (c) data after 3D dip filtering. (d) A vertical slice through the original seismic data. The small white arrows indicate areas of low fold resulting from cable feathering. Note that the k_x-k_y filter eliminates the north-south-trending faults indicated by gray arrows. After Marfurt et al. (1999).

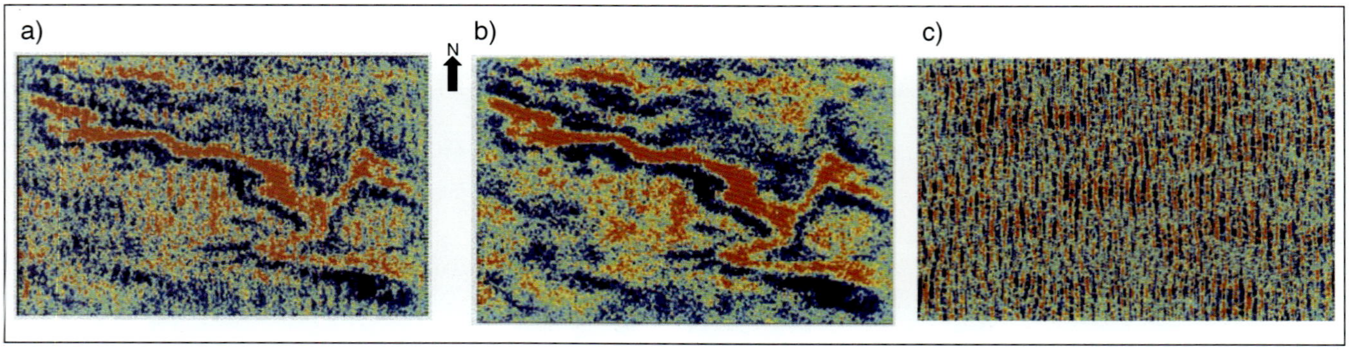

Figure 5.16. A time slice through a data volume from onshore Algeria, (a) showing a strong acquisition footprint and (b) after adaptive filtering designed to suppress the footprint. (c) Rejected noise obtained by subtracting the data shown in (b) from those in (a). After Drummond et al. (2001).

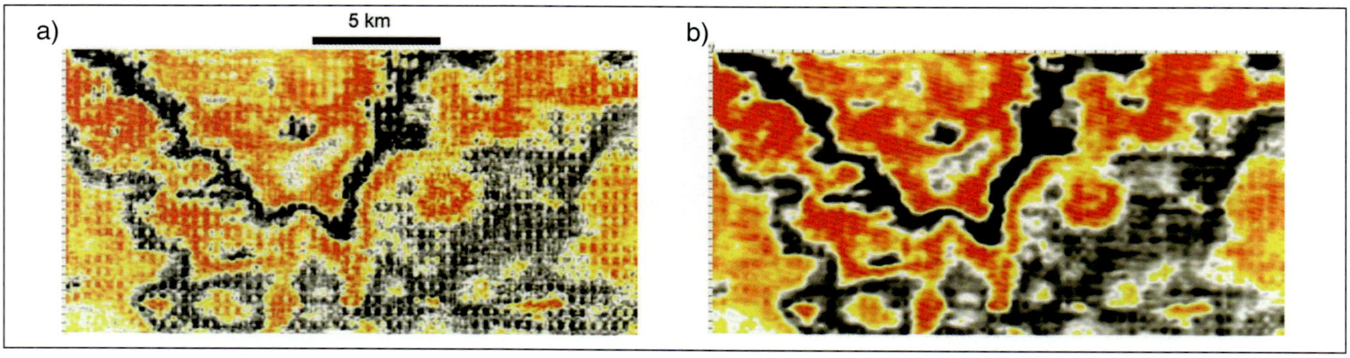

Figure 5.17. Suppression of acquisition footprint by using eigenvectors of the noise. Time slices at $t = 0.320$ s through the seismic data (a) before and (b) after footprint removal using an eigenvector-decomposition approach. Only the first few eigenvectors are used to describe the noise. Note that the strong north-south and east-west striping is suppressed. After al-Bannagi et al. (2005).

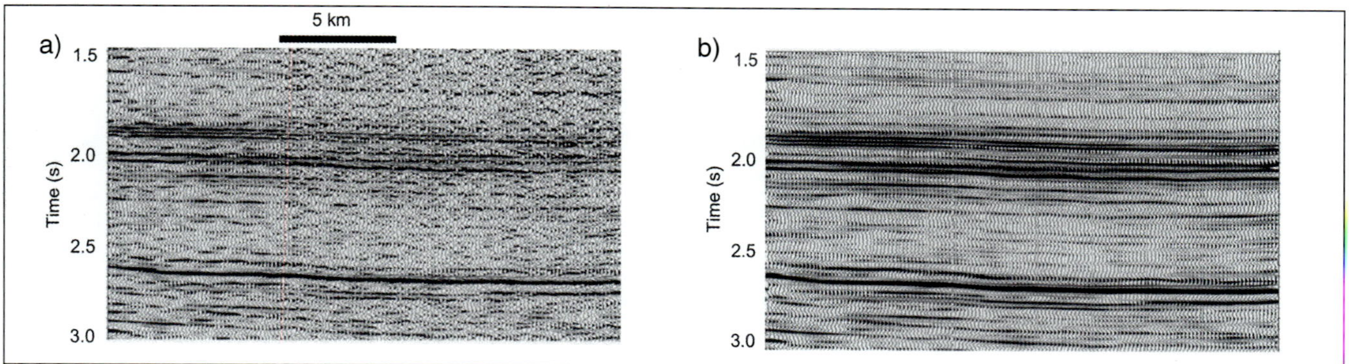

Figure 5.18. Suppression of an acquisition footprint by using eigenvectors of the noise. Vertical slices through the seismic data (a) before and (b) after footprint removal using an eigenvector-decomposition approach. Unlike the case with time slice shown in the previous figure, here the acquisition footprint appears less well organized — almost random — on the vertical section. After al-Bannagi et al. (2005).

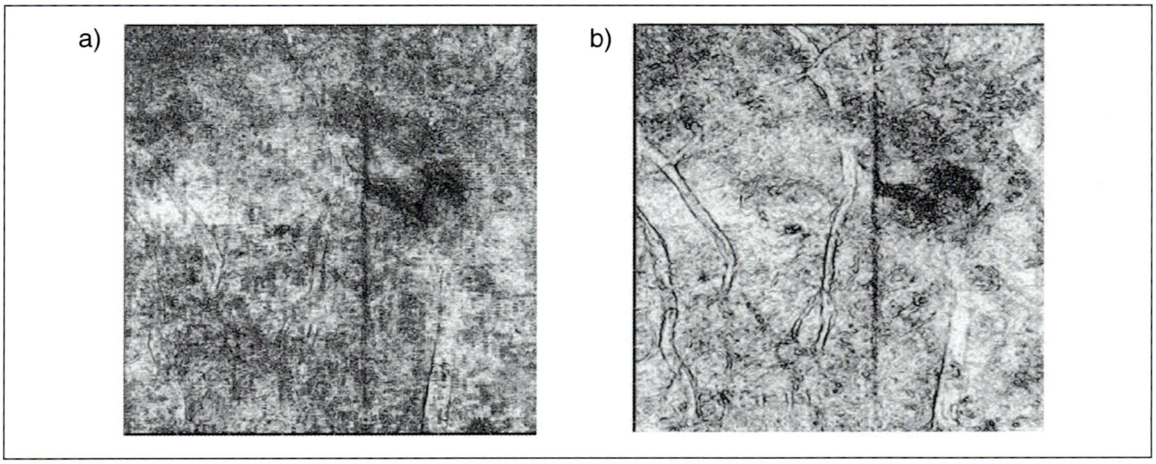

Figure 5.19. Suppression of an acquisition footprint by using eigenvectors of the noise. Time slices through edge-detection attribute volumes (with a lateral change in amplitude), computed here from the seismic data (a) before and (b) after footprint removal using a eigenvector-decomposition approach. Several distinct channels that were buried in the acquisition footprint in (a) can be seen in (b). After al-Bannagi et al. (2005).

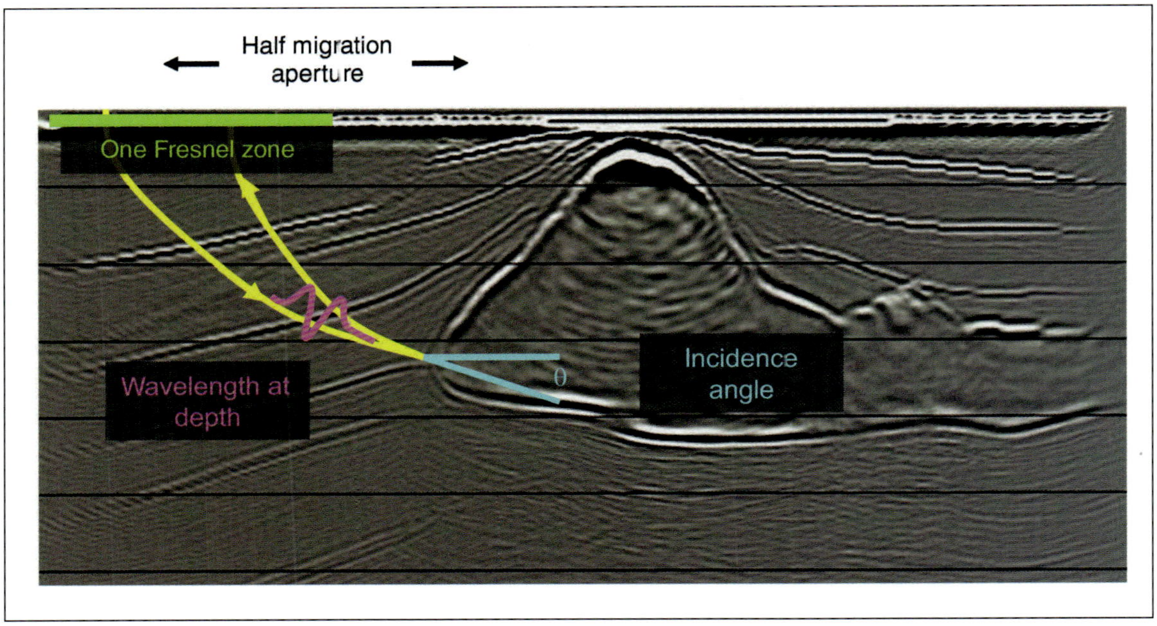

Figure 5.20. Lateral resolution as a function of depth and migration aperture, for a velocity function that has the form $v = v_0 + kz$, with values $v_0 = 3$ km/s and $k = 0.3$. The wavenumber increases more slowly with the size of the aperture for deeper reflectors, which indicates that for a fixed aperture, lateral resolution decreases with depth. Note that increasing the migration aperture much beyond the aperture being equal to depth results in a diminishing increase in horizontal wavenumber bandwidth. After Lazarevic (2004).

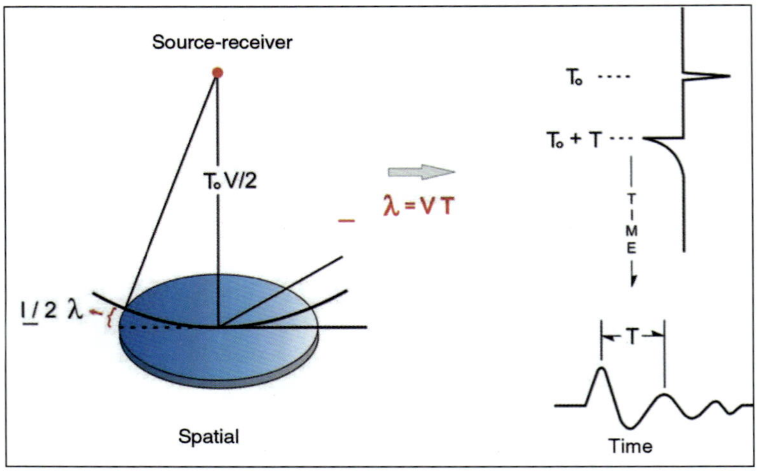

Figure 5.21. Description of the Fresnel zone in spatial and temporal domains. Here, λ is the dominant wavelength, *T* is the two-way traveltime to the target from the surface, and *V* is the average velocity of the overburden. After Hilterman (1982).

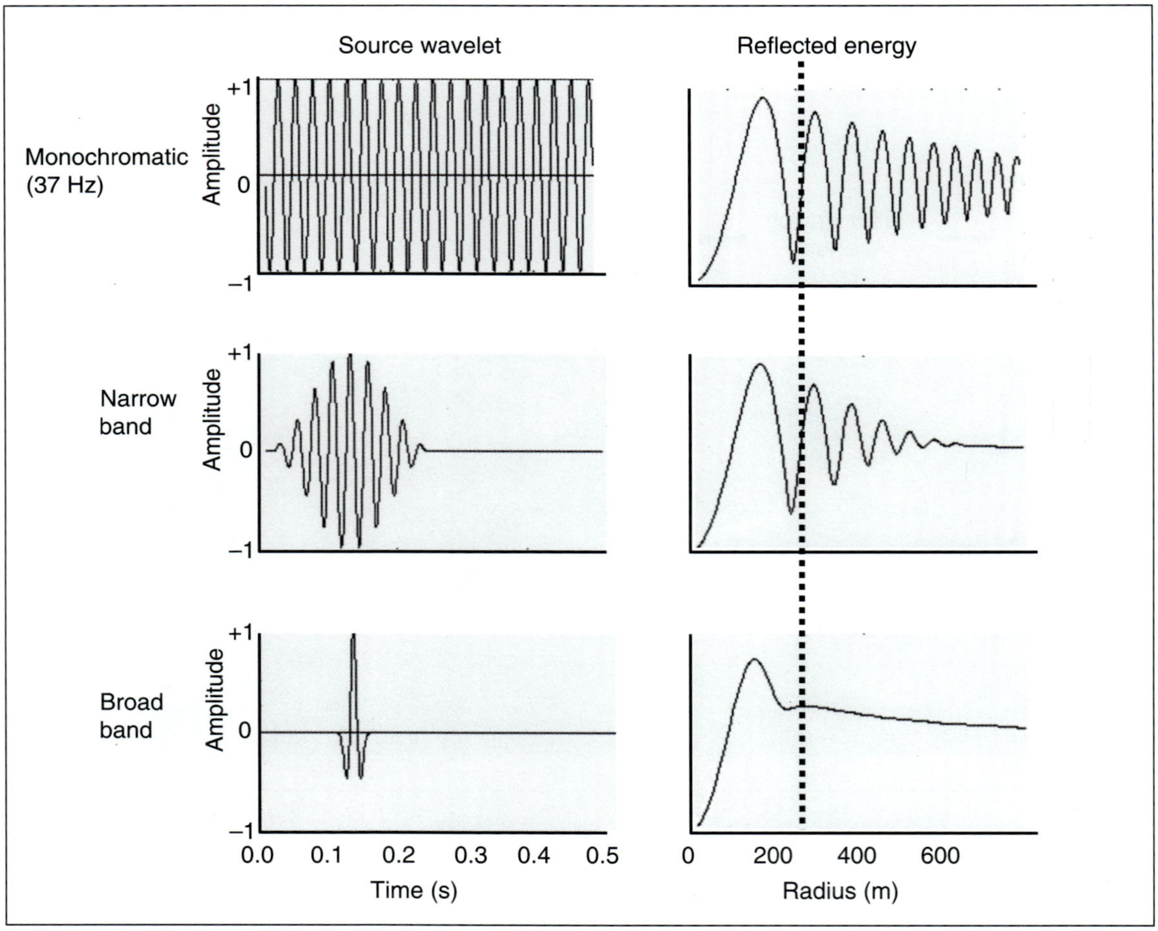

Figure 5.22. Reflected amplitude as a function of reflector radius (Fresnel zones) for monochromatic, narrow-band, and broadband sources. All three wavelets have a center frequency of 37.1 Hz. Reflector depth is 1000 m, and velocity is 2000 m/s. After Bruhl et al. (1996).

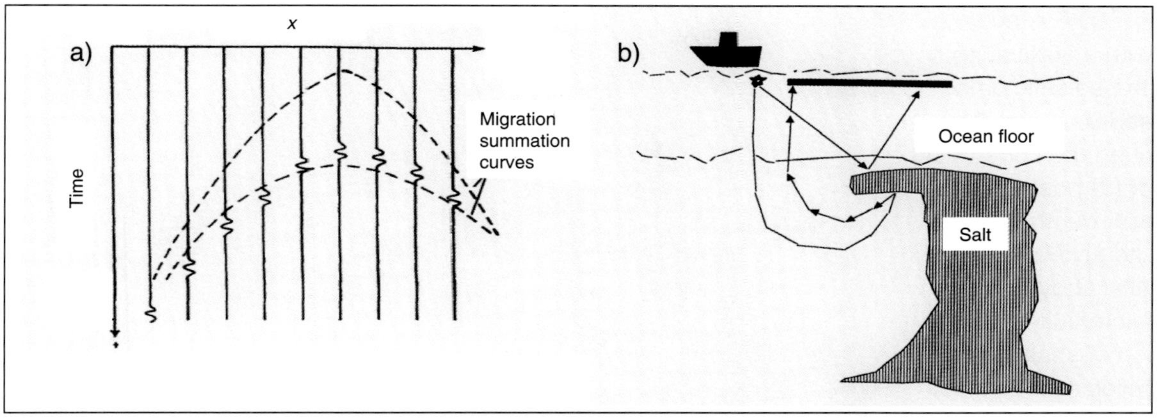

Figure 5.23. The impact of migration on resolution. (a) Correct migration velocities sum along an accurate diffraction hyperbola and thereby produce high-resolution images. After French (1975). (b) Migration also rotates our estimates of resolution so that frequency content controls lateral resolution and migration aperture controls vertical resolution along a steep salt flank. After Brown (1994).

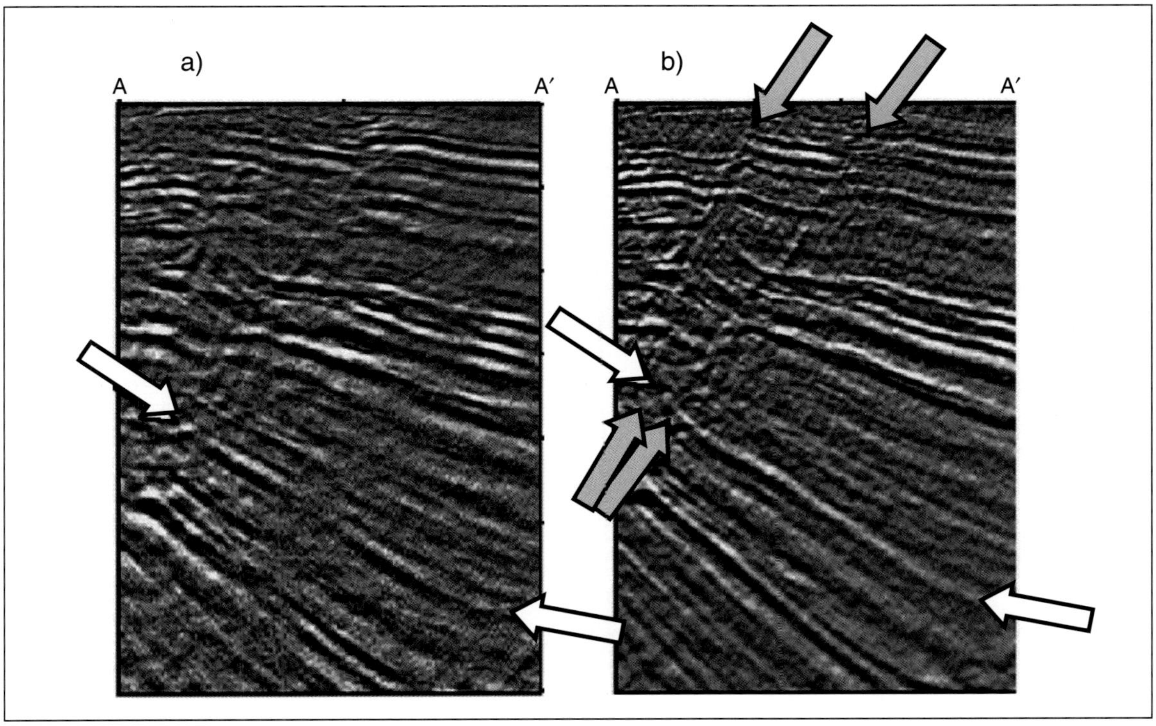

Figure 5.24. Vertical seismic images generated using (a) poststack and (b) prestack depth migration, for a survey from offshore Trinidad. Note the improved resolution of reflector terminations and fault-plane reflections, indicated by gray arrows. Also note white arrows, which indicate the greater continuity of the reflector. After Rietveld et al. (1999).

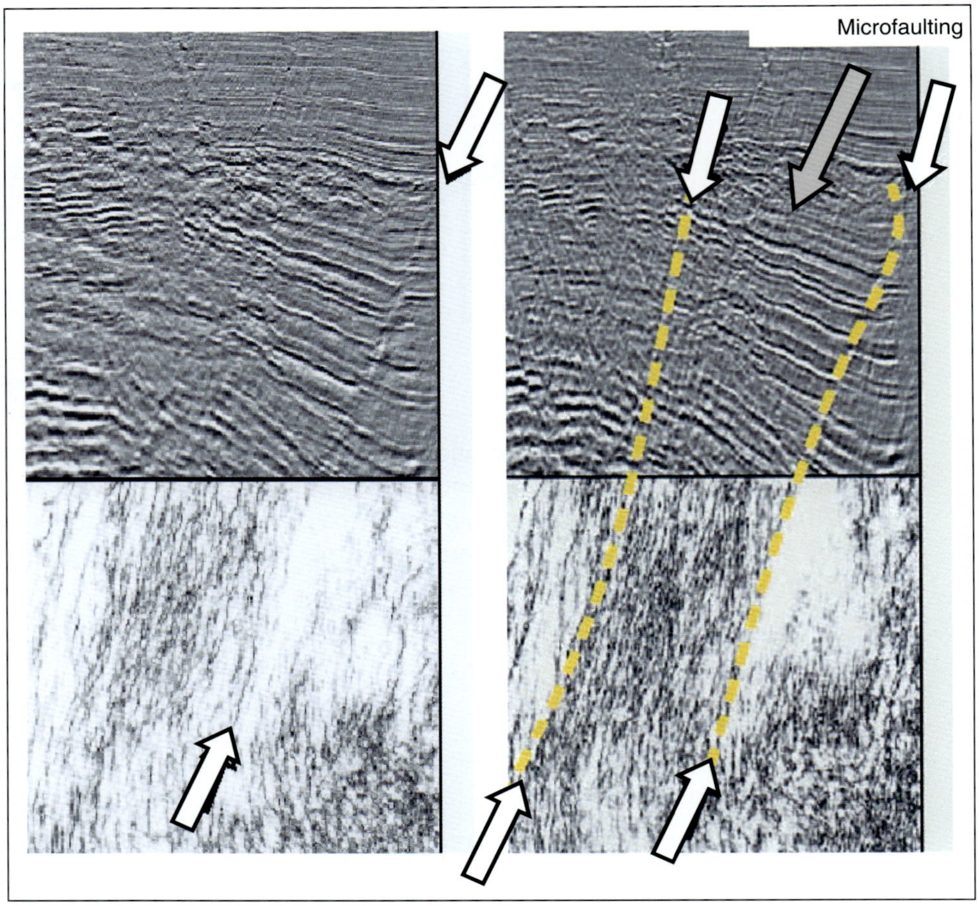

Figure 5.25. Folded images with a vertical slice through the seismic data volume and a horizontal slice through the coherence volume of (a) poststack and (b) prestack depth-migrated data. White arrows indicate major faults. Gray arrow indicates microfaulting, which is common in shale-dominated lithologies. Microfaulting appears as a broad incoherent zone on the depth slice. After Rietveld et al. (1999).

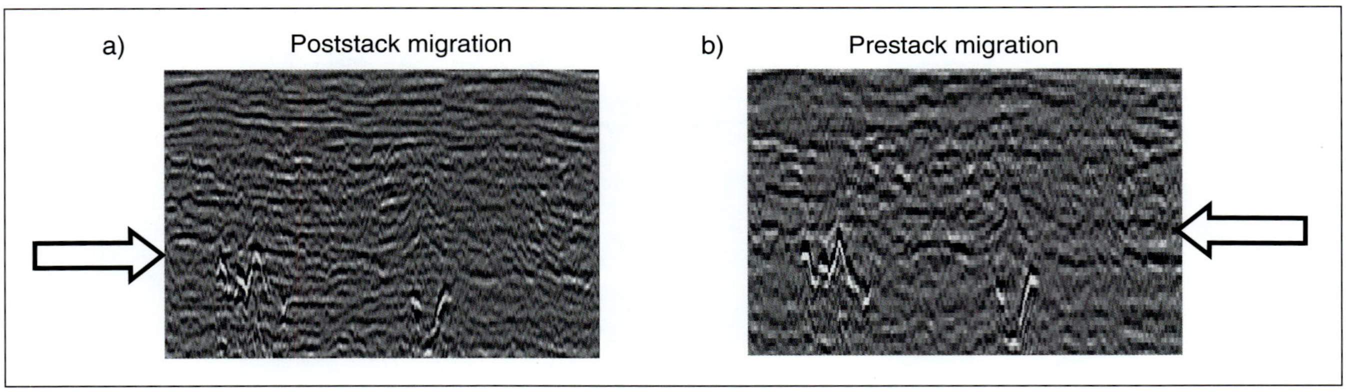

Figure 5.26. Vertical slices through a shale-dewatering sequence from offshore the Netherlands, taken from post-stack and prestack depth-migrated volumes. Minimal structural definition is present. After Rietveld et al. (1999).

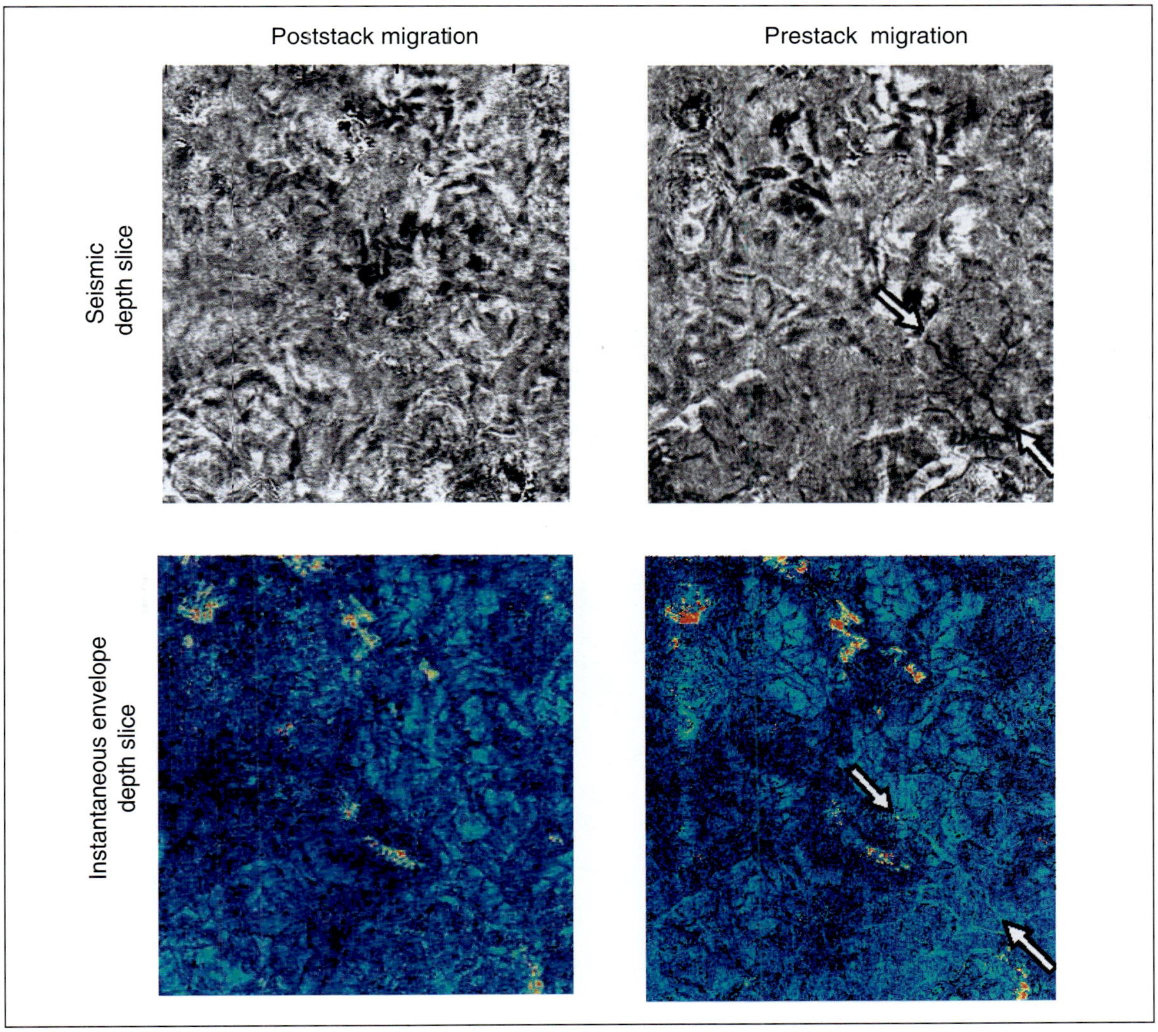

Figure 5.27. Depth slices at the level indicated by white arrows in the previous figure, shown here through the seismic data volume and an energy-envelope (reflection-strength) volume after poststack and prestack depth migration. Note the improved lateral resolution on the envelope slice after prestack depth migration. After Rietveld et al. (1999).

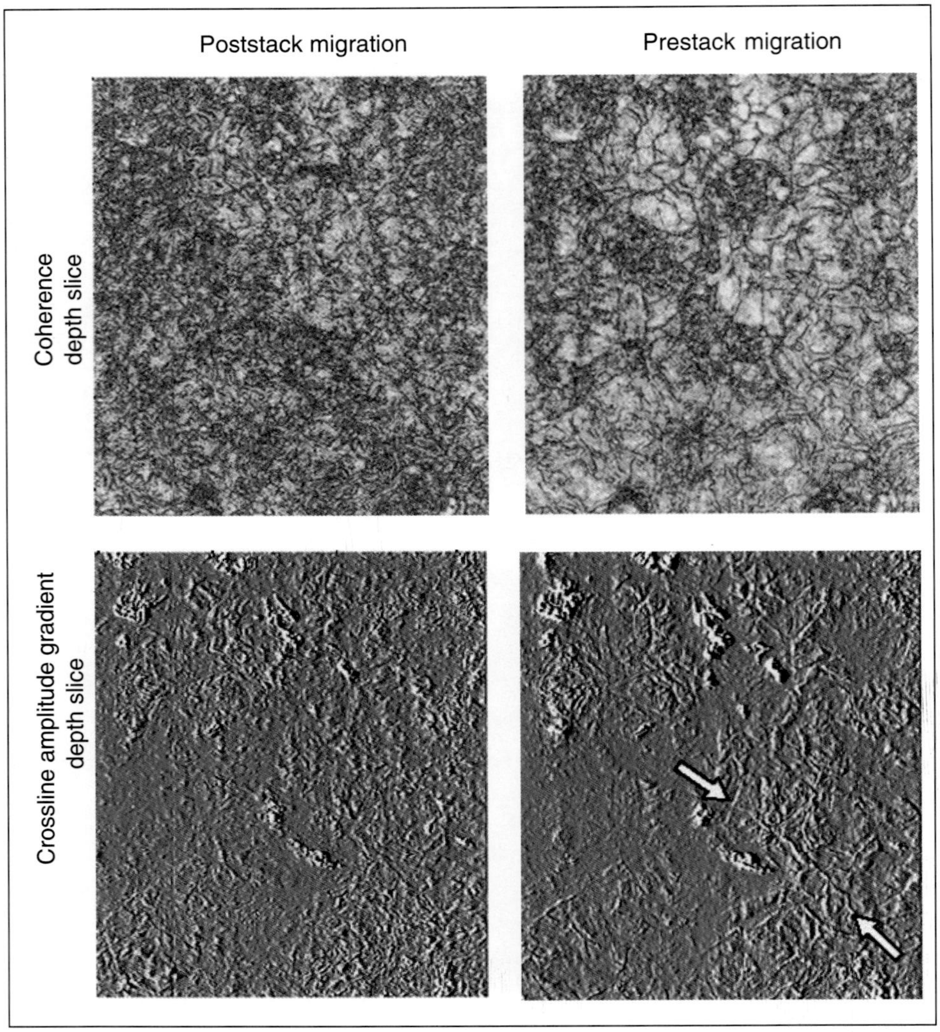

Figure 5.28. Depth slices at the level indicated by the white arrows in Figure 5.26 and corresponding to same depth level as in Figure 5.27, shown here through the coherence volume and the north-south component of the energy-weighted coherent-amplitude-gradient volume after poststack and prestack depth migration. The polygonal geometries associated with the dewatering process are resolved much better in the prestack depth-migration images. After Rietveld et al. (1999).

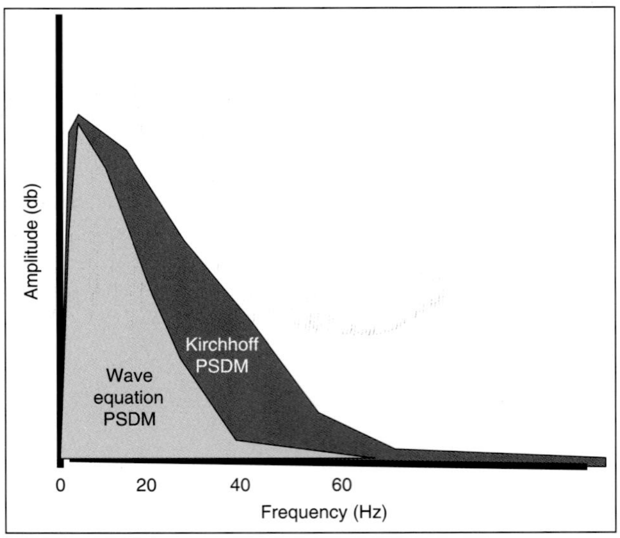

Figure 5.29. Frequency content of Kirchhoff and wave-equation prestack depth-migration images from a Yucatán Peninsula, Mexico, data volume corresponding to the next figure. The Kirchhoff image is at 20 db at 40 Hz, and the wave equation is at 20 db at 20 Hz. After Lazarevic (2004).

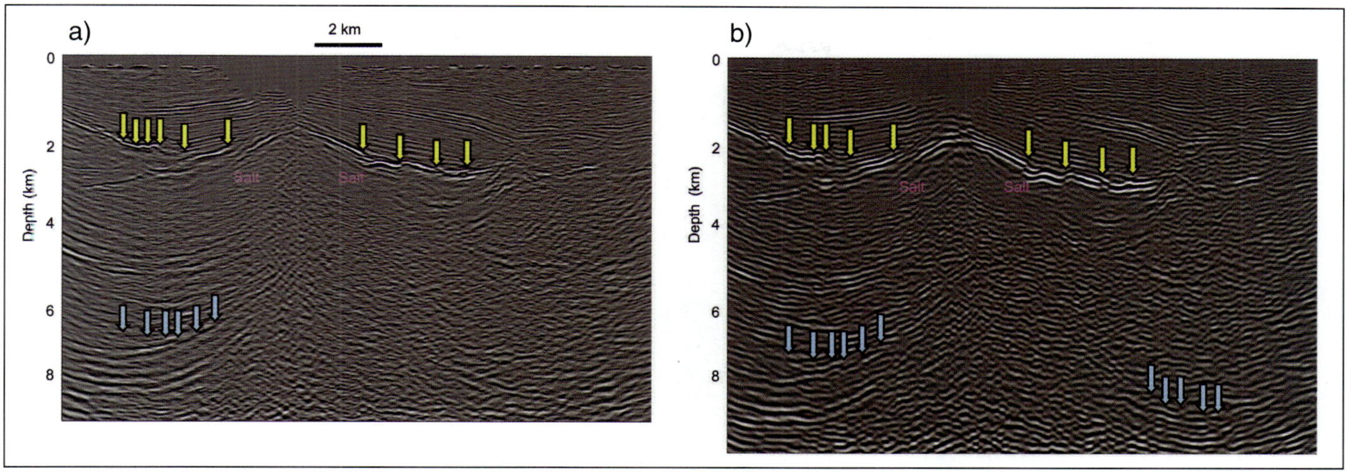

Figure 5.30. Vertical slices through (a) a Kirchhoff prestack depth-migration image and (b) a wave-equation prestack depth-migration image. Lateral discontinuities in the shallow horizon (yellow arrows) are defined better on Kirchhoff migration. At depth, the more accurate physics implemented by the multiarrival wave-equation-migration algorithm provides a better image, with improved continuity and lateral resolution (blue arrows), even though its frequency content is less. After Lazarevic (2004).

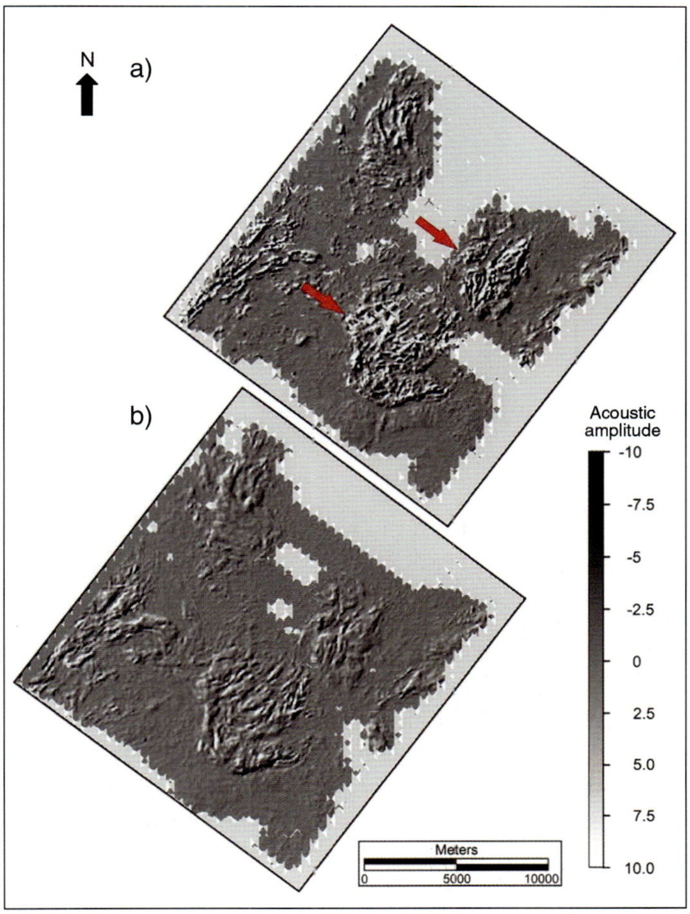

Figure 5.31. Shallow horizon slices, corresponding to yellow arrows in the previous figure, of the inline energy-weighted coherent-amplitude gradient on (a) a Kirchhoff prestack depth-migration image and (b) a wave-equation prestack depth-migration image. Note the improved lateral resolution of the Kirchhoff image, which shows indications of channels and other stratigraphic features. After Lazarevic (2004).

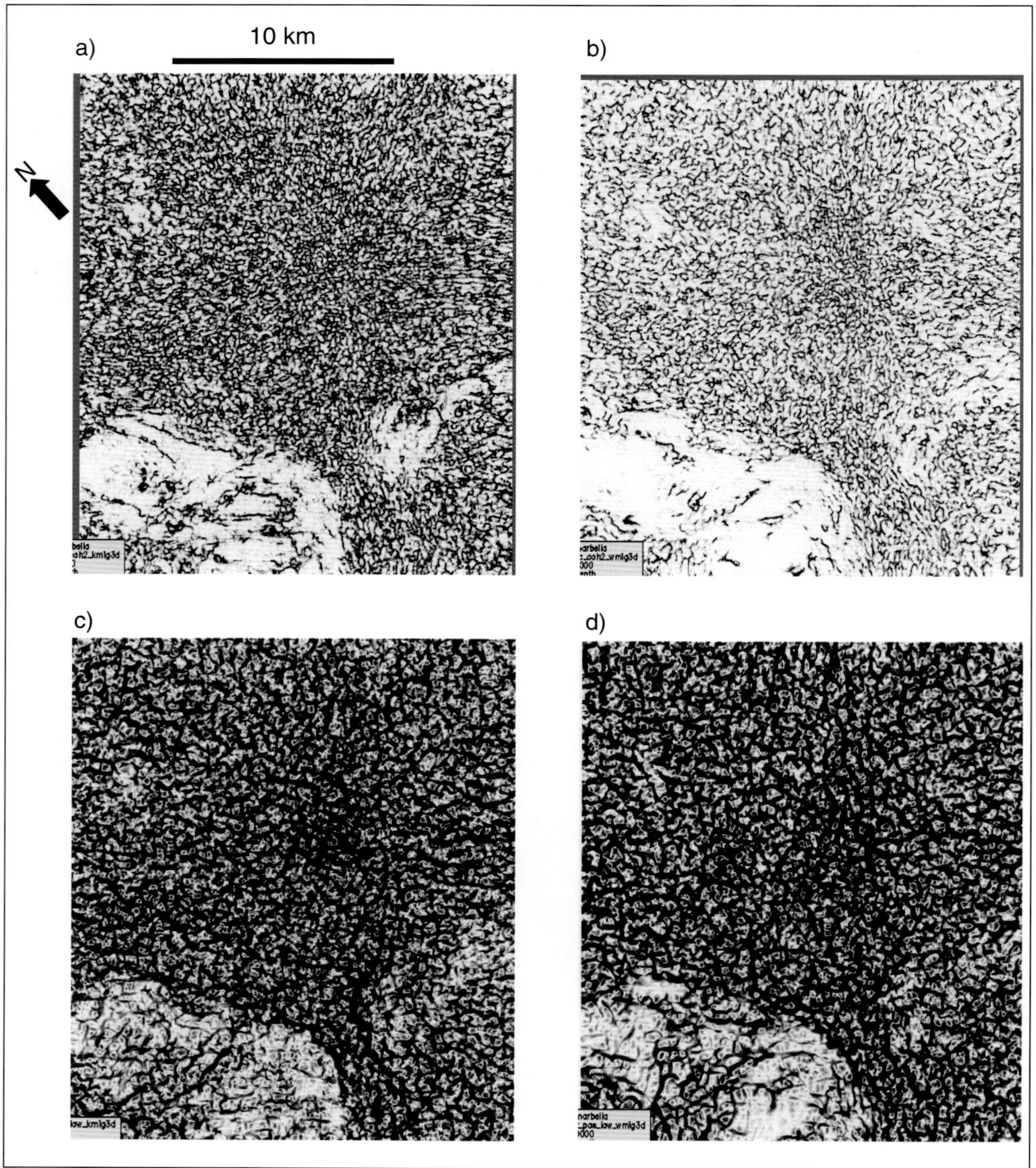

Figure 5.32. (a)-(b). Depth slices at $z = 8$ km through coherence volumes generated from (a) a Kirchhoff prestack depth-migration volume and (b) a wave-equation prestack depth-migration volume. (c)-(d). Corresponding depth slices through most-positive-curvature volumes from (c) a Kirchhoff prestack depth-migration volume and (d) a wave-equation prestack depth-migration volume. The poor attribute images outside the minibasin in the lower left corner are the result of overlying salt and carbonates. To state it succinctly, "fancy" attributes cannot help a poorly migrated image. After Lazarevic (2004).

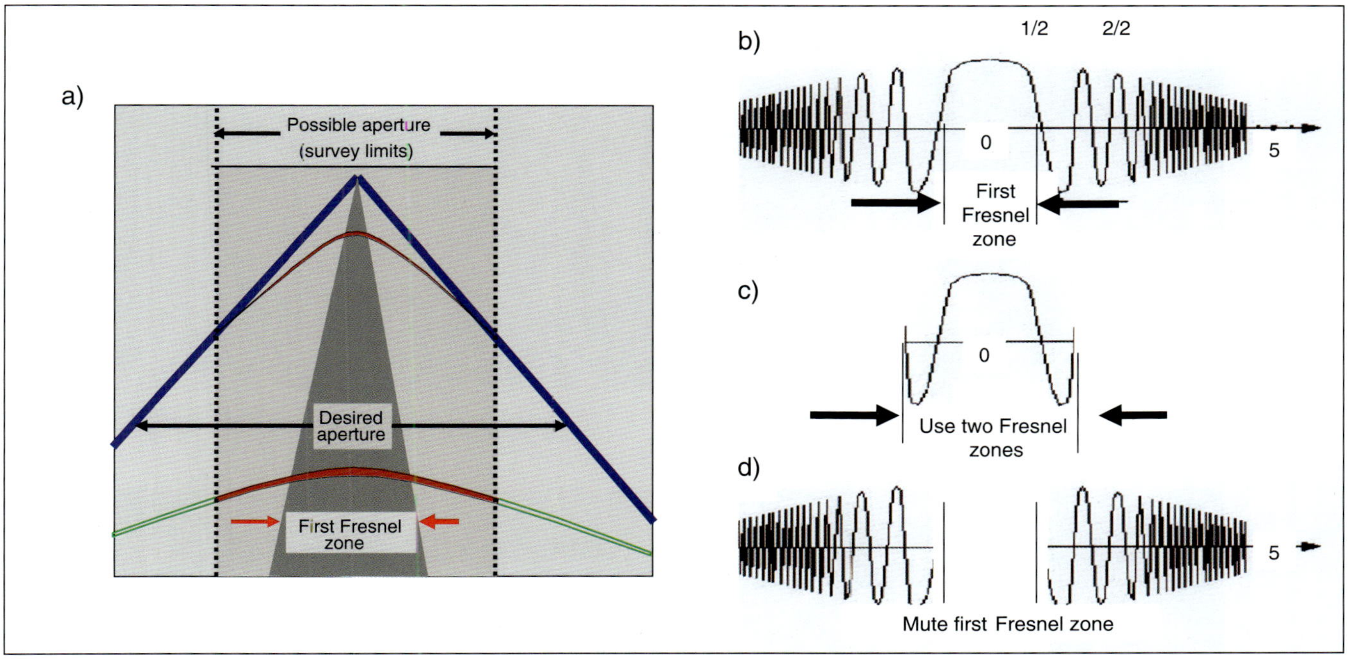

Figure 5.33. (a) Schematic diagram showing the size of the Fresnel zone at a given time level for a given velocity. Theoretically, in conventional migration, we wish to include as many Fresnel zones as permitted by the limits of acquisition and computation. (b) Each Fresnel zone is defined by an annulus (for constant velocity), in which all amplitudes are summed together with the same sign. (c) Typically, cost constraints, errors in velocity modeling, and operator aliasing lead us to limit the migration aperture to about two Fresnel zones. (d) To enhance scattering objects versus specular reflectors, we exclude the interior Fresnel zones from computation, as shown here. After Kozlov et al. (2004).

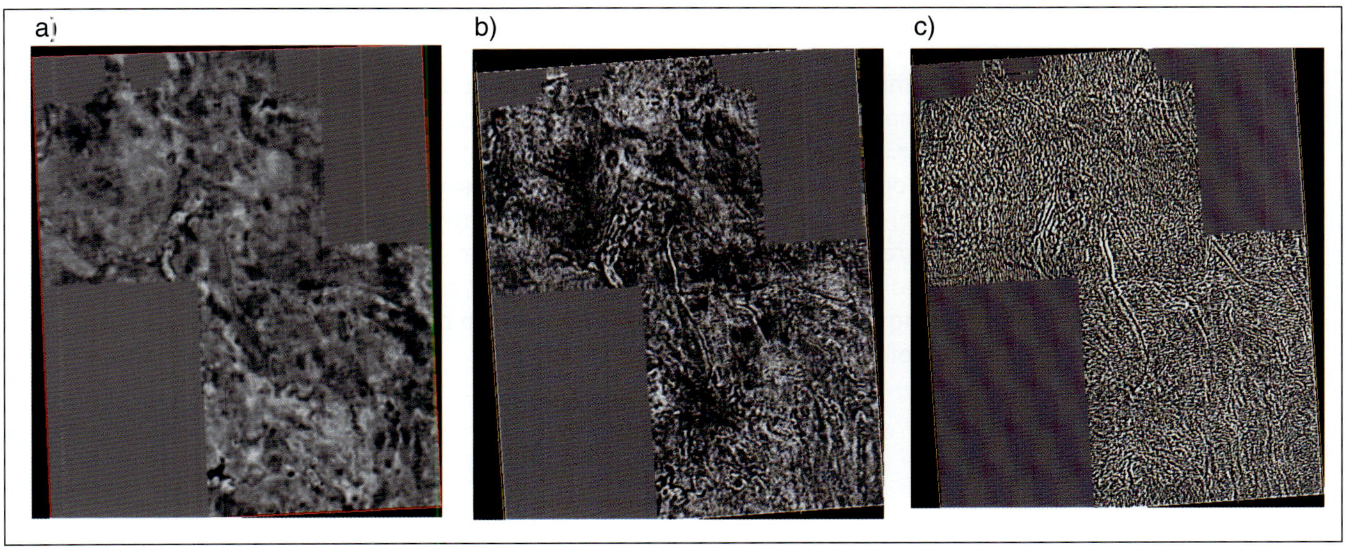

Figure 5.34. Depth slices at 3296 m through a survey from eastern Siberia using (a) all Fresnel zones, (b) all but the nearest Fresnel zones, and (c) only the farthest Fresnel zones. Note the improved illumination of stratigraphic discontinuities. After Kozlov et al. (2004).

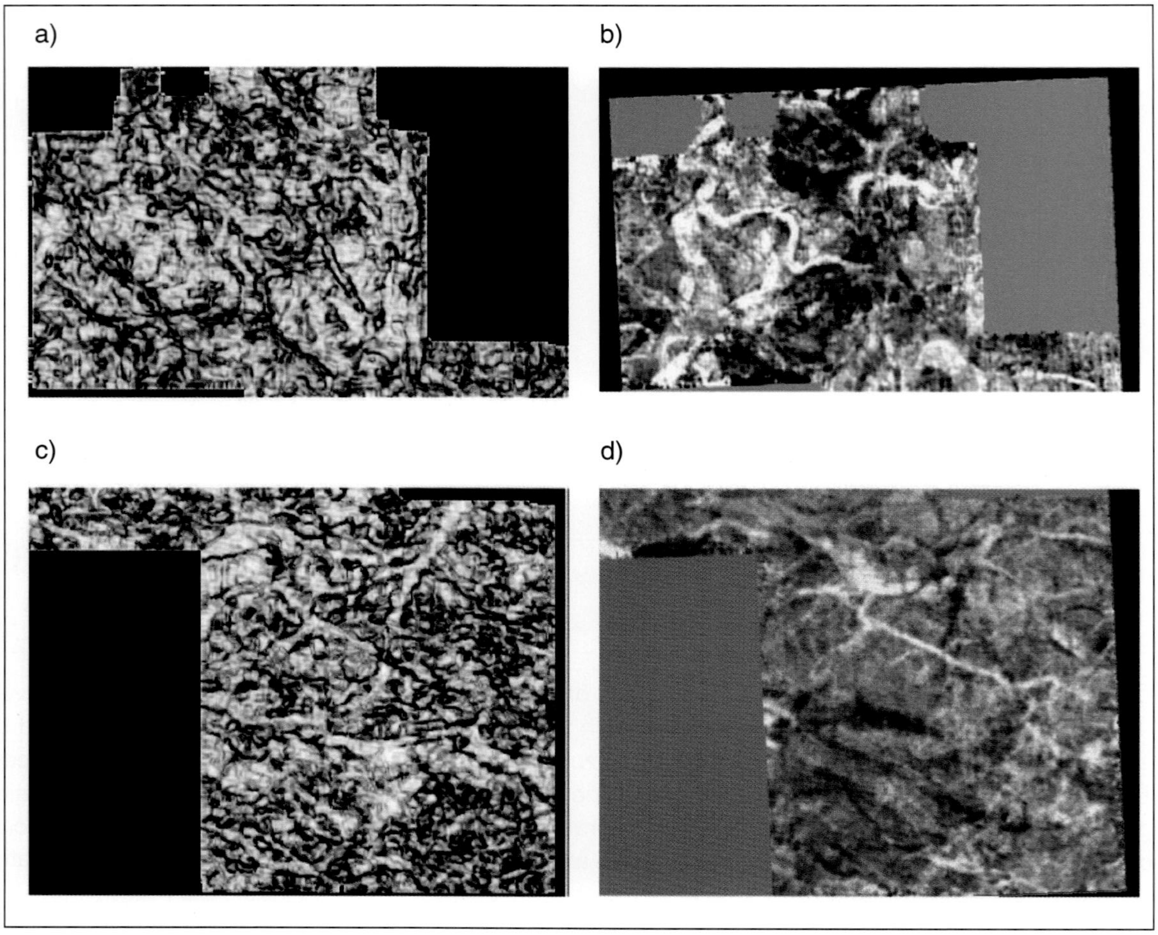

Figure 5.35. (a)-(b). Depth slices at 842 m in data from a survey acquired in eastern Siberia through (a) a coherence volume calculated from conventional prestack depth migration using the full migration aperture, encompassing all Fresnel zones, and (b) a "scattering-object-image" volume, obtained using the same depth-migration algorithm but limited to the outermost Fresnel zones. In this algorithm, the Kirchhoff migration diffraction summation operator excludes contributions from the crest of the operator where the width of the exclusion crest is measured in terms of Fresnel zone diameters. (c)-(d). Using data from the same survey, depth slices here are at 1762 m through (c) the coherence volume calculated from conventional prestack depth migration using the full migration aperture, encompassing all Fresnel zones, and (d) a scattering-object-image volume, obtained using the same depth-migration algorithm but limited to the outermost Fresnel zones. After Kozlov et al. (2004).

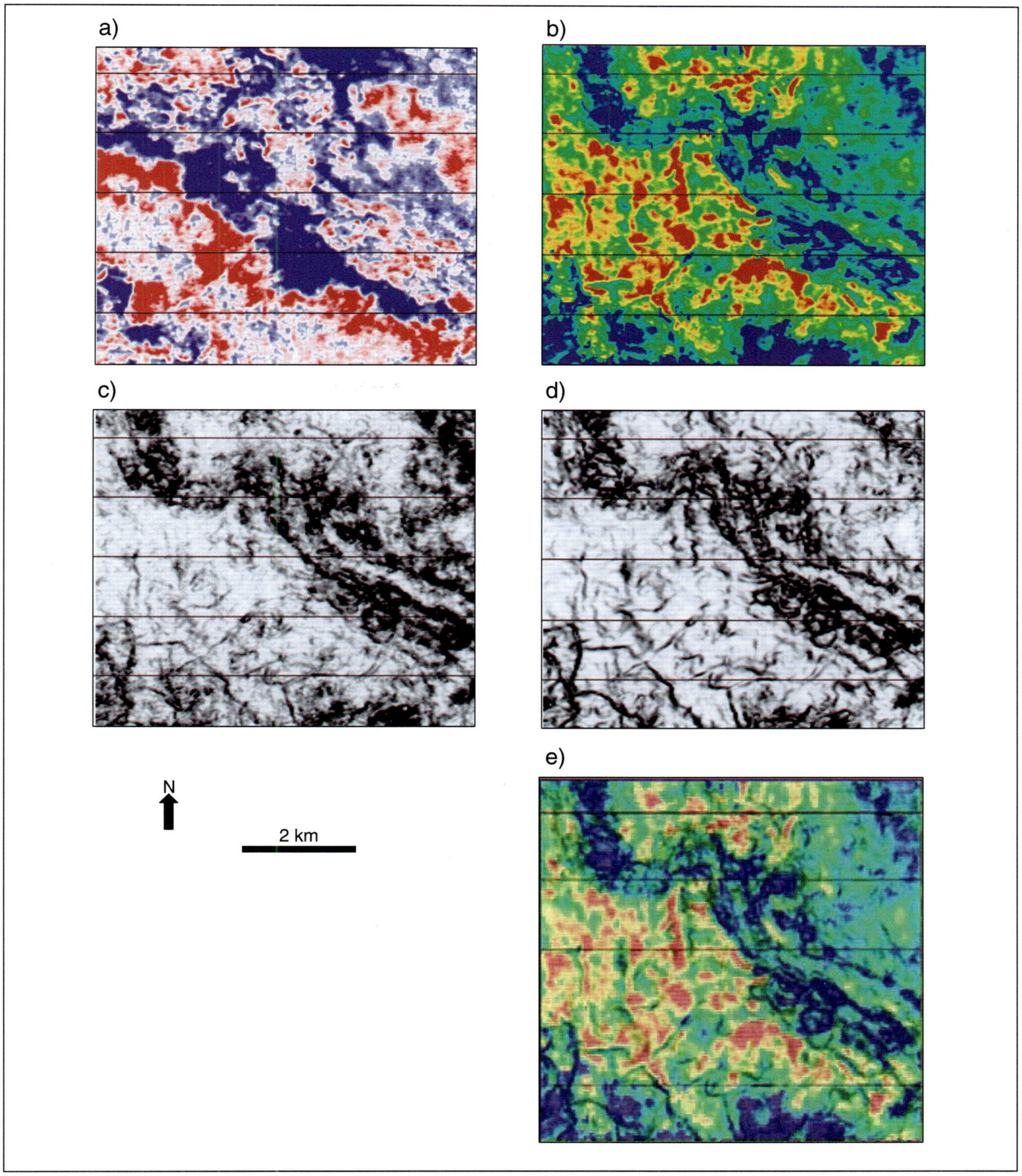

Figure 5.36. Time slices from a survey acquired in Alberta, Canada. (a) A seismic data volume, (b) an acoustic-impedance volume, (c) a coherence volume computed from the seismic data volume displayed in (a), and (d) a coherence volume computed from the acoustic-impedance volume displayed in (b). Note the improved resolution in (d). In contrast, image (c) shows that discontinuities occurring within a narrow stratigraphic window are blended with events above and below, through the narrower-bandwidth longer seismic wavelet. (e) The coherence volume shown in (d) is alpha-blended here with the impedance shown in (b).

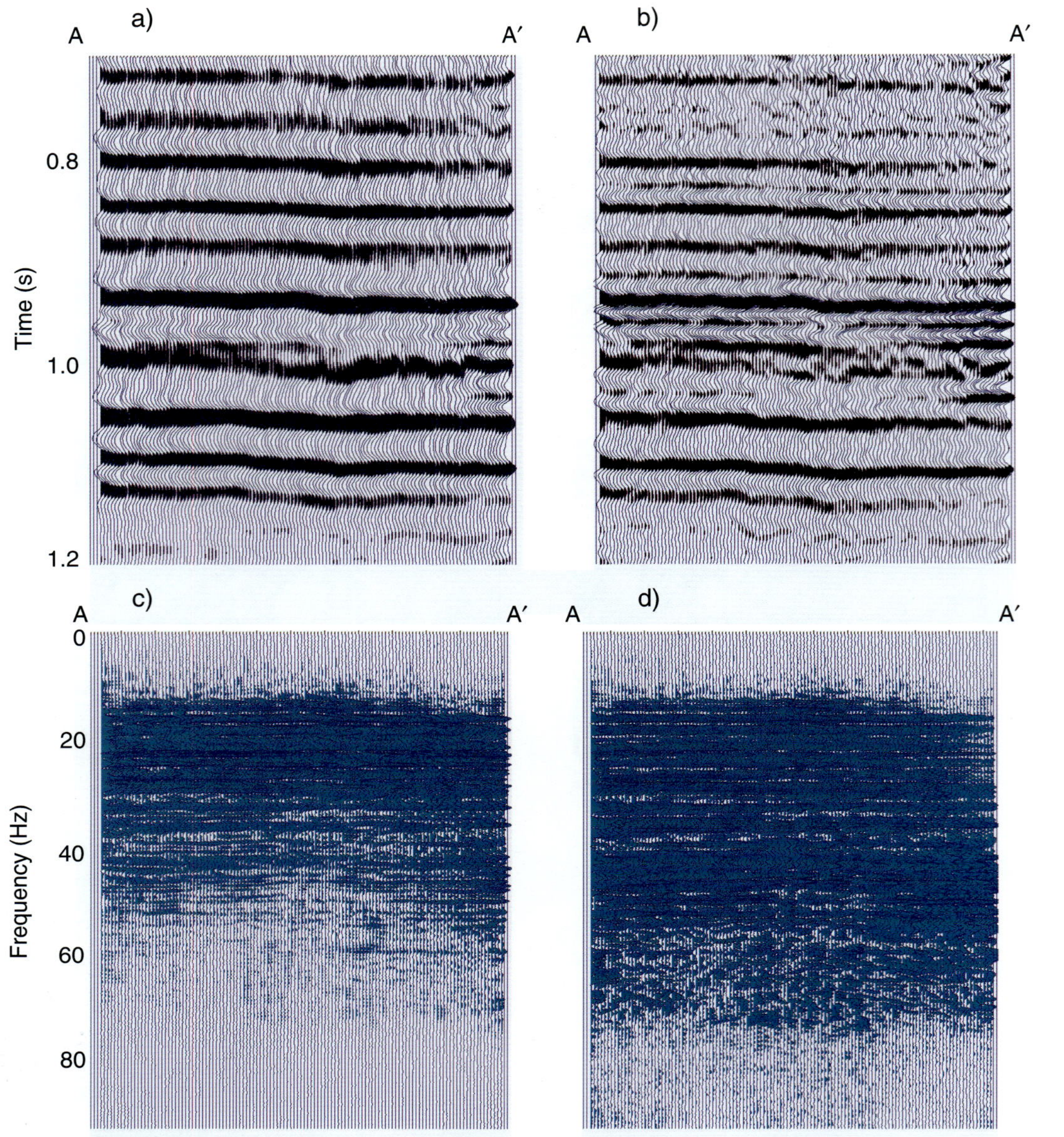

Figure 5.37. (a)-(b). Vertical slices through a seismic volume (a) before and (b) after VSP (vertical-seismic-profiling)-driven spectral enhancement. (c)-(d). Spectra of the same seismic line, shown (c) before and (d) after VSP-driven spectral enhancement. After Chopra et al. (2003).

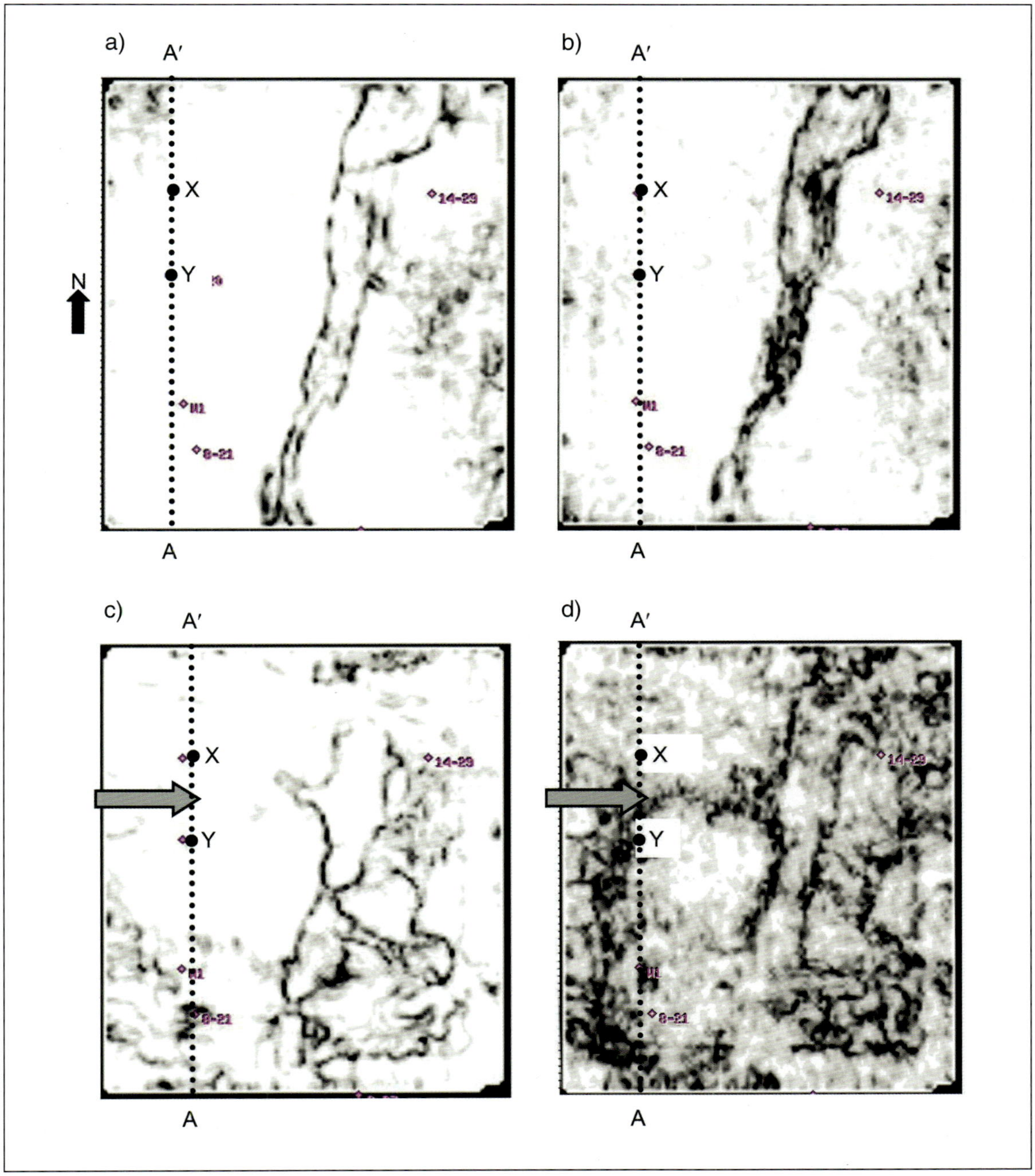

Figure 5.38. Time slices through coherence volumes generated from seismic data in the previous figure, shown here at (a) $t = 0.984$ s, (b) $t = 0.984$ s, (c) $t = 1.024$ s, and (d) $t = 1.024$ s. The arrow indicates a zone where no communication existed between wells X and Y. After Chopra et al. (2003).

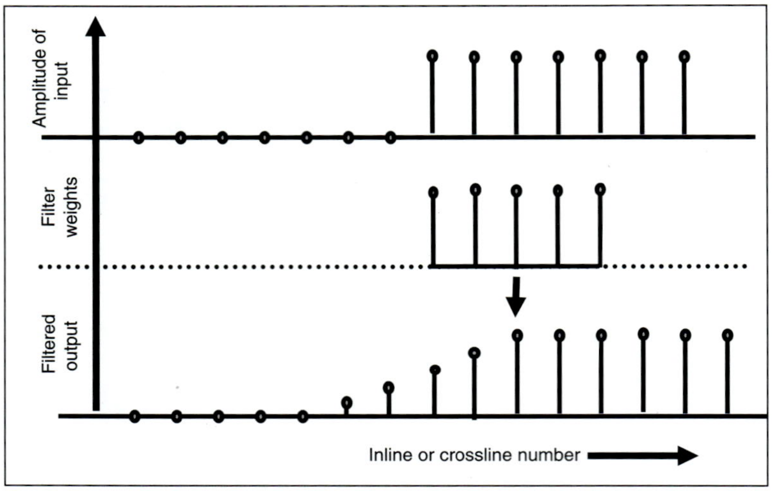

Figure 5.39. A cartoon showing implementation of a running sum filter or a mean filter which, with the median filter, commonly is applied by interpreters to maps of seismic traveltime, amplitude, and attributes. In this cartoon, the filter moves from left to right, with the output being the average of five adjacent traces that fall within the operator "gate." After Luo et al. (2001).

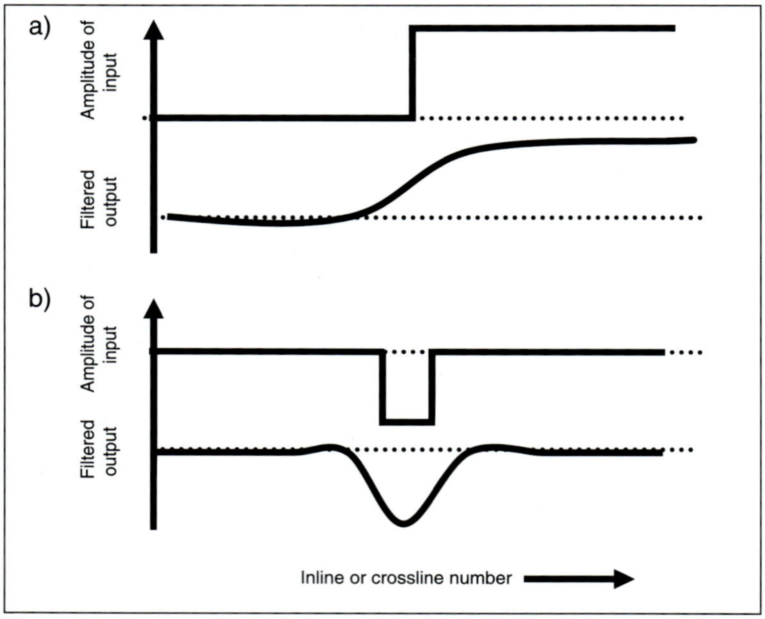

Figure 5.40. The result of applying a running sum filter or a mean filter to an idealized (a) fault and (b) channel. The discontinuity in amplitude (or dip or whatever component is being smoothed) is blurred, thereby decreasing lateral resolution. After Luo et al. (2001).

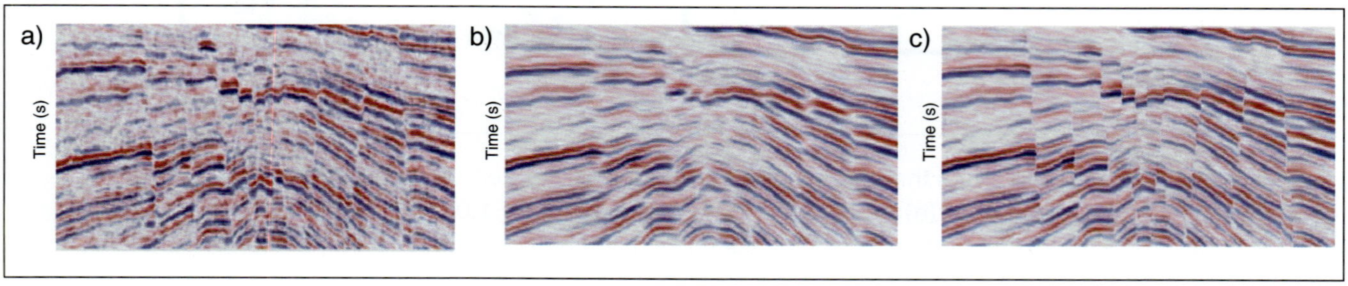

Figure 5.41. Seismic data (a) before and (b) after structure-oriented filtering (SOF) with no edge preservation and (c) after structurally oriented filtering using an anisotropic diffusion operator. Note how the fault edges are preserved in (c). After Hoecker and Fehmers (2002).

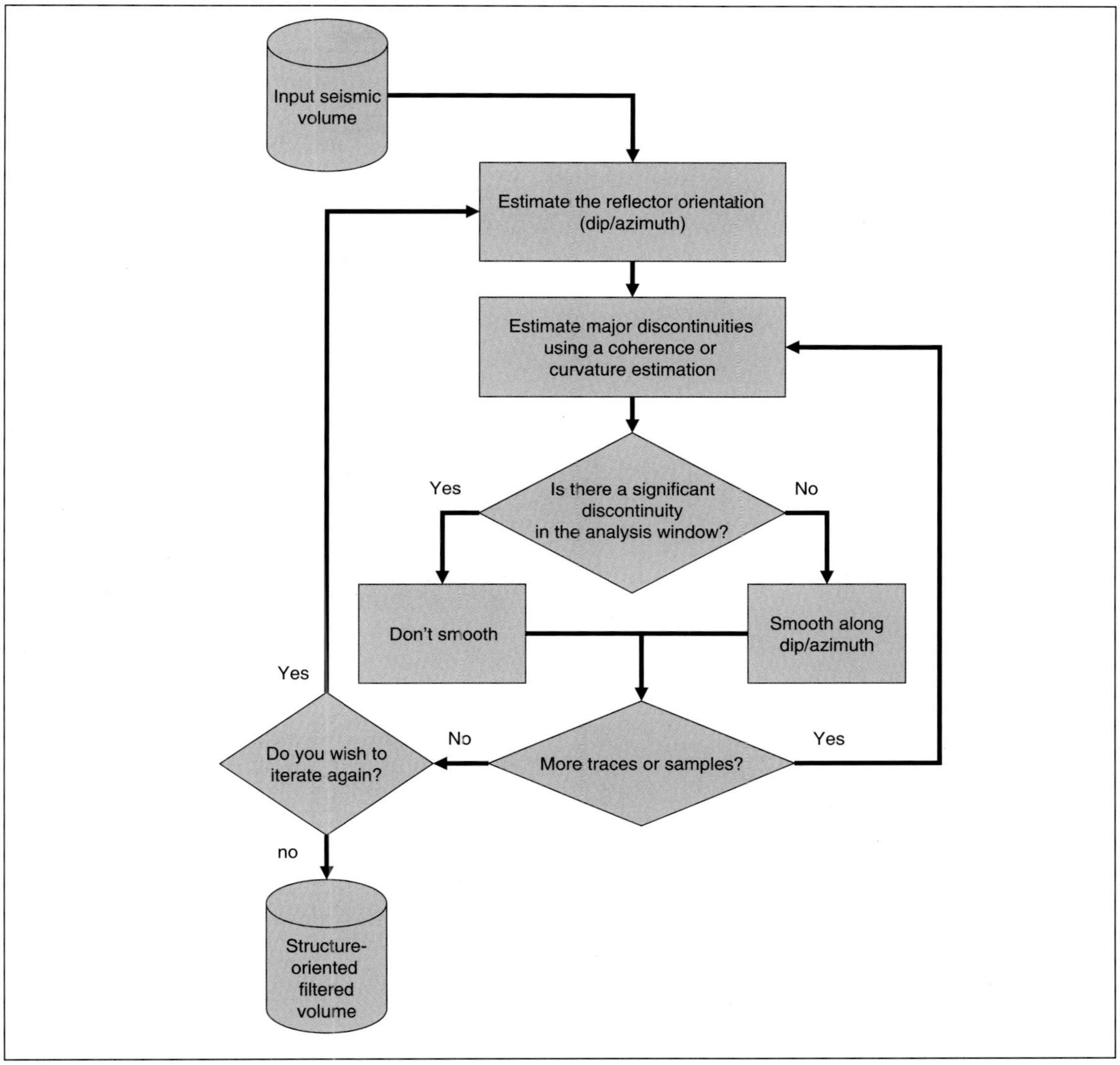

Figure 5.42. A simplified flowchart of Hoecker and Fehmers' (2002) structure-oriented filtering algorithm. In actual implementation, the amount of smoothing varies between the two "yes" and "no" extremes shown in the figure.

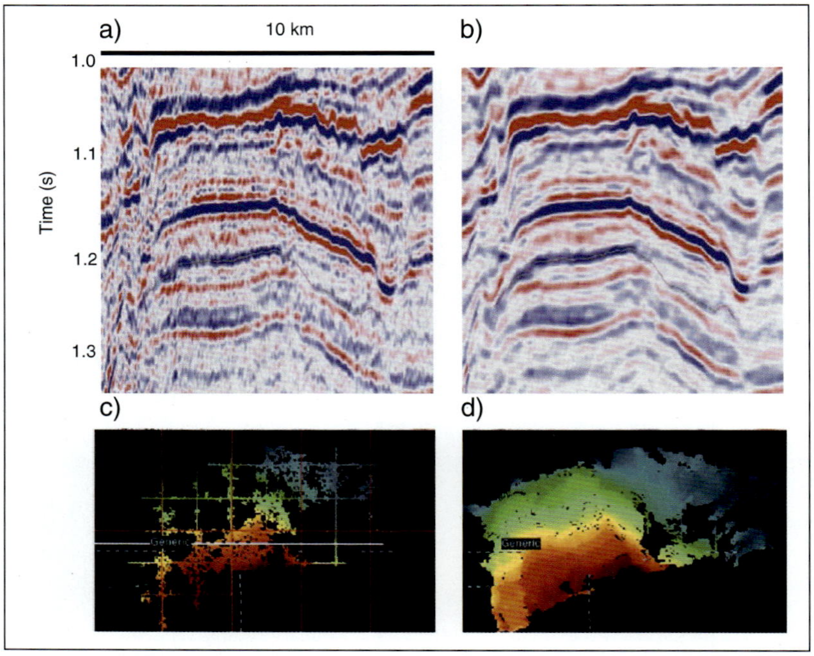

Figure 5.43. The influence of structure-oriented filtering on autotrackers. A vertical slice (a) before and (b) after structure-oriented fltering. An autotracked horizon (c) before and (d) after structure-oriented filtering using an anisotropic diffusion equation. The autotracker fails to pick the complete horizon when it is applied to noisy original data, but it succeeds when applied to filtered data. After Hoecker and Fehmers (2002).

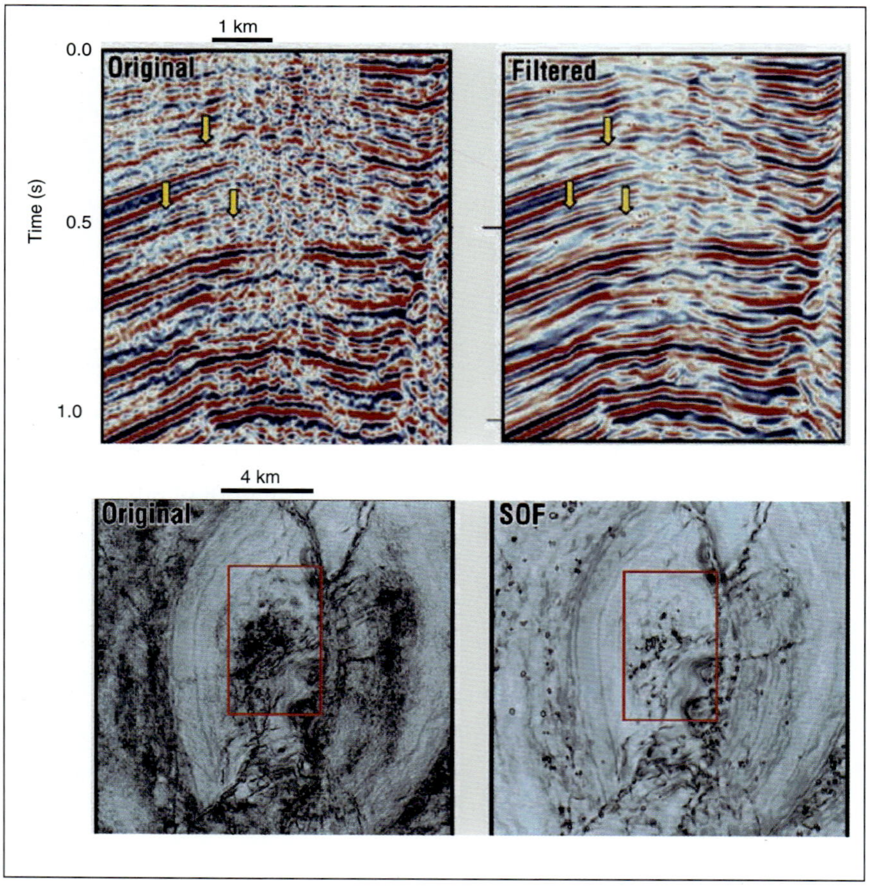

Figure 5.44. The influence of structure-oriented filtering on geometric attributes for a survey from Oman. (a)-(b). Vertical slices through the seismic volume (a) before and (b) after structure-oriented filtering. (c)-(d). Time slices at $t = 0.420$ s through the semblance estimate of coherence (c) before and (d) after structure-oriented filtering. After Masaferro et al. (2003).

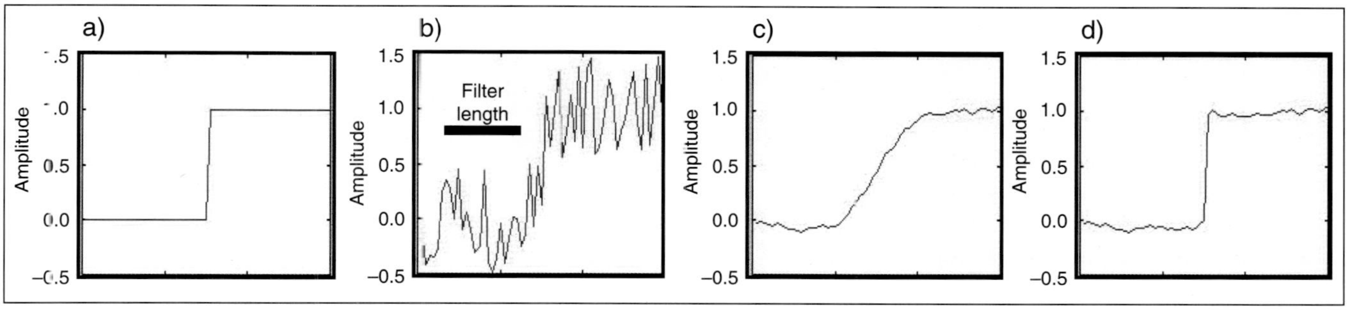

Figure 5.45. A simple example showing multiwindow (Kuwahara) filtering. An idealized amplitude discontinuity (a) without and (b) with additive noise. (c) The result of filtering the noisy signal in (b) with a 21-point running-window mean filter and (d) with a 21-point multiwindow Kuwahara filter. After Luo et al. (2002).

Figure 5.46. (a) An amplitude time slice at t = 1.026 s from a survey acquired in Saudi Arabia. (b) The corresponding time slice through an edge-detection volume generated using the long-wavelength Hilbert transform edge-detection algorithm described in Unit 3. (c) The corresponding time slice through the long-wavelength Hilbert transform edge-detection algorithm applied to the seismic data after multiwindow Kuwahara "edge-preserving smoothing." Arrows indicate a narrow channel that meanders through the survey. After Luo et al. (2002).

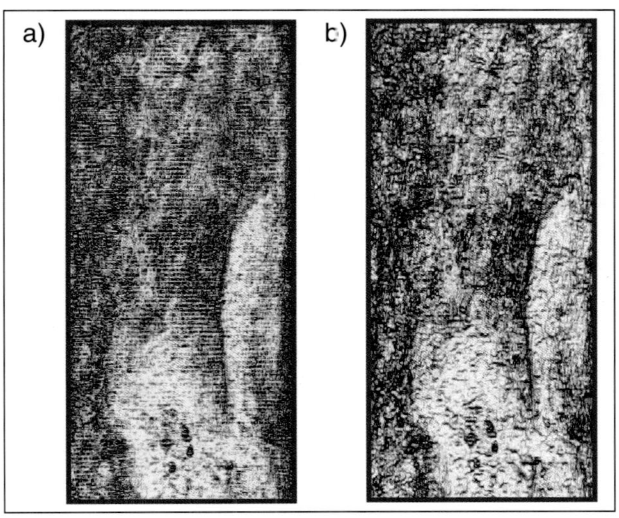

Figure 5.47. The influence of edge-preserving smoothing on an acquisition footprint from a survey acquired in Saudi Arabia. A time slice through edge-detection attribute volumes using the long-wavelength Hilbert transform edge-detection algorithm described in Unit 3 and generated (a) before and (b) after Kuwahara multiwindow edge-preserving smoothing. After Luo et al. (2002).

Figure 5.48. (a)-(b). Time slices at t = 0.800 s and (c)-(d) at 1.040 s through principal-component coherence volumes generated for a survey acquired over the Central Basin Platform, Texas, U.S.A. Slices (a) and (c) were calculated from original time-migrated data, and (b) and (d) were calculated from the same data after two passes of structure-oriented filtering using a multiwindow principal-component algorithm. Seismic data are courtesy of OXY.

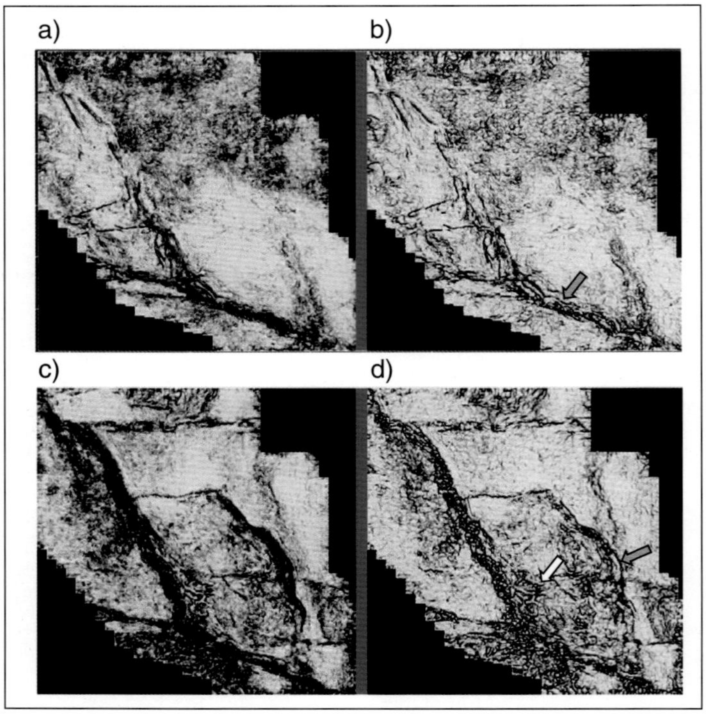

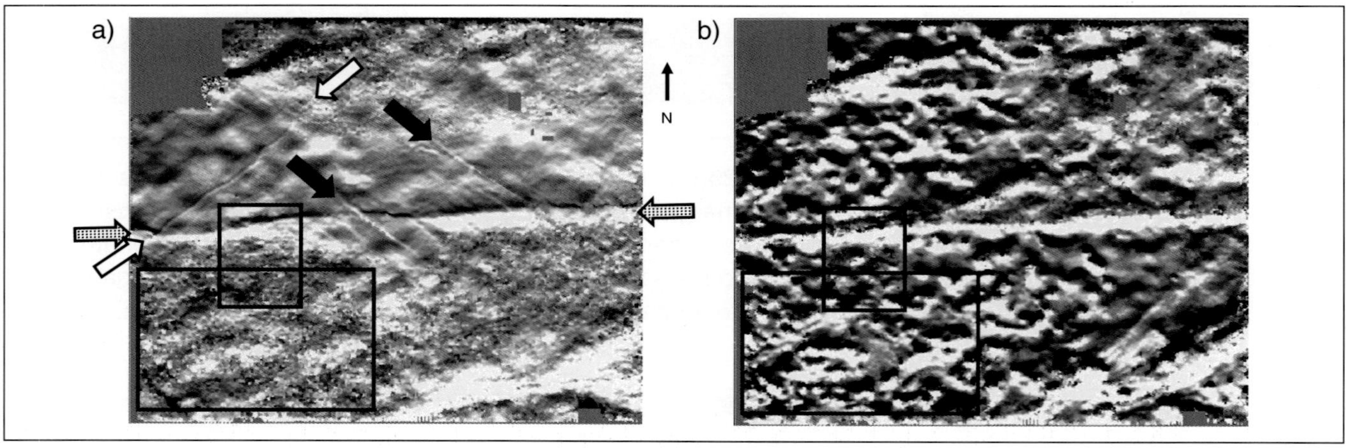

Figure 5.49. Time slices at t = (a) 0.800 and (b) 1.100 s through a north-south dip volume from a survey in the Fort Worth Basin, Texas, U.S.A. Speckled arrows indicate a broad east-west strike-slip fault. Black arrows indicate thinner northwest-southeast lineaments. White arrows indicate a northeast-southwest channel, which is visible because of differential compaction. Rectangles indicate two older, smaller surveys that were incorporated into a larger, merged survey. All three surveys were reprocessed as a unit, so differences in data quality are the result of acquisition rather than of processing. After al-Dossary and Marfurt (2003).

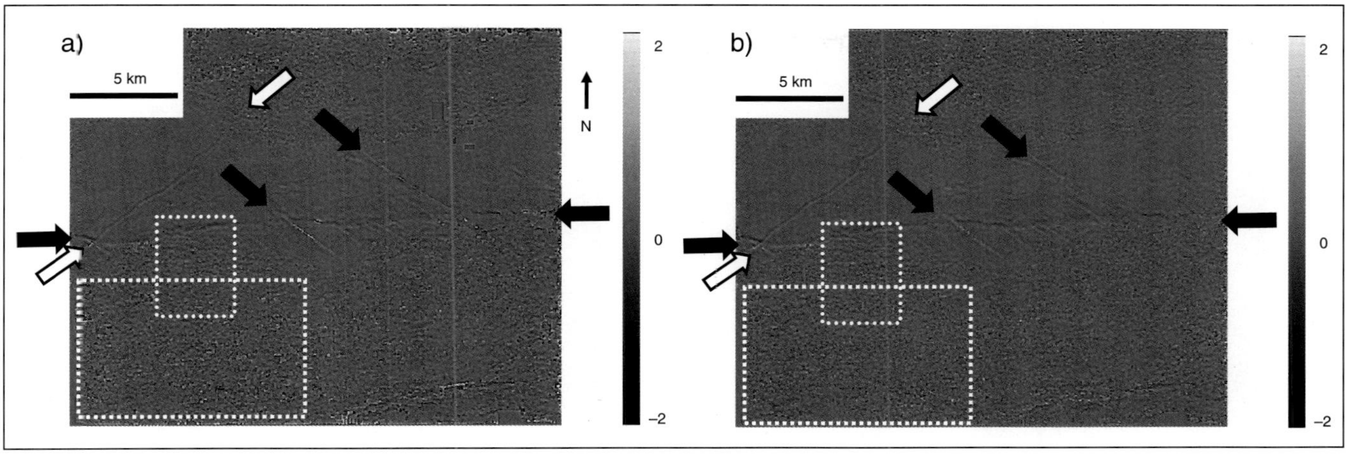

Figure 5.50. The rejected "noise" resulting from applying two passes of (a) a mean filter and (b) a multistage median modified trimmed mean (MSMMTM) filter to the image shown in part (a) of the previous figure and subtracting them from the original. Note that the mean filter rejects the short-wavelength components of the lineaments. The MSMMTM filter rejects a smaller amount of random noise than the mean filter does. The MSMMTM also rejects more long-wavelength components of the signal, thereby indicating that it has a "sharpening" component typical of median filters. After al-Dossary and Marfurt (2003).

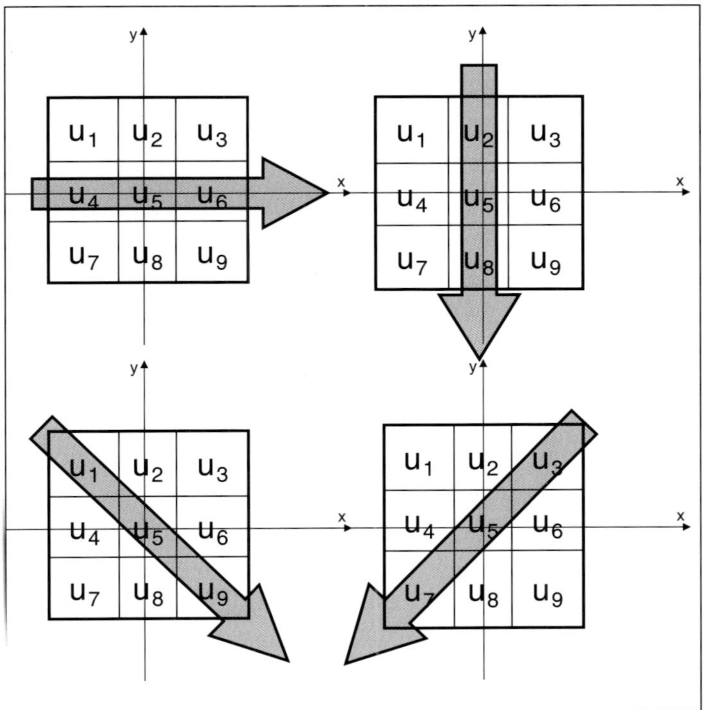

Figure 5.51. A multistage median filter (MSM) consists of four median filters one seismic bin wide. First we obtain the median of u4, u5, and u6. Then we obtain the median of u2, u5, and u8. We now have two numbers, both of which are medians. Include in this collection the value u5. Now we have three numbers. Take the median of those three numbers, which we will refer to as M13. Repeat this procedure for medians calculated on the diagonals. Call that median M24. Then take the median of M13 and M24 along with the value of u5. We call this final (multistage) median, u_{MSM}. After al-Dossary and Marfurt (2003).

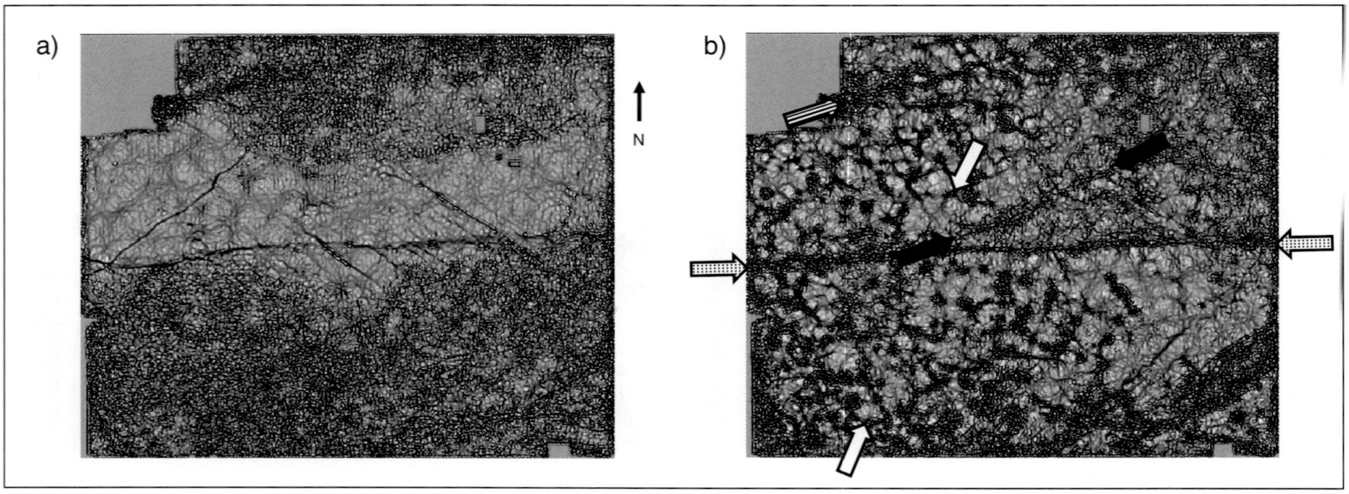

Figure 5.52. Time slices through the short-wavelength (α = 1.0), most-negative-curvature volume, k_{neg}, at t = (a) 0.800 and (b) 1.100 s. The most negative curvature was calculated from the unfiltered vector dip, the north-south component of which is shown in Figure 5.49. After al-Dossary and Marfurt (2003).

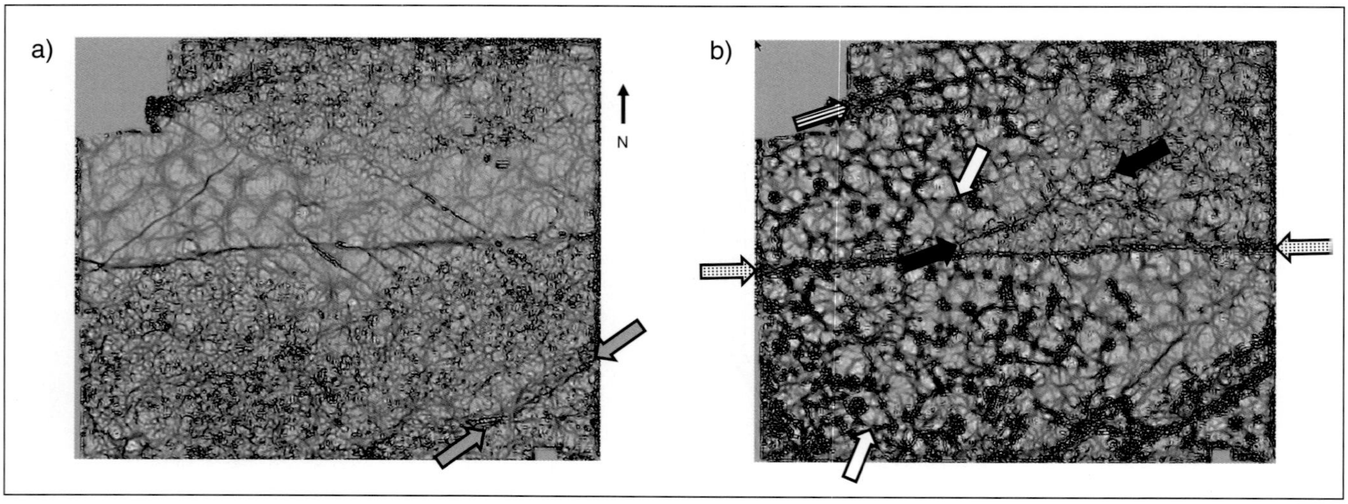

Figure 5.53. Time slices through the short-wavelength (α = 1.0), most-negative-curvature volume, k_{neg}, at t = (a) 0.800 and (b) t = 1.100 s, calculated from the north-south and east-west components of vector dip after two passes of MSMMTM filtering using a 3 × 3 window. The fault edge indicated by the gray arrows in (a) shows up more clearly than it did in Figure 5.52a. The Reidel shear zone (black arrows) branching off the strike-slip fault (speckled arrows) in (b) shows up as a clear lineament rather than as the incoherent zone seen in Figure 5.52b. Likewise, a north-northeast to south-southwest lineament connecting collapse features (white arrow) is more visible. Although the fault zone indicated by the striped arrow is still diffuse, it is significantly clearer here than it was in Figure 5.52b. After al-Dossary and Marfurt (2003).

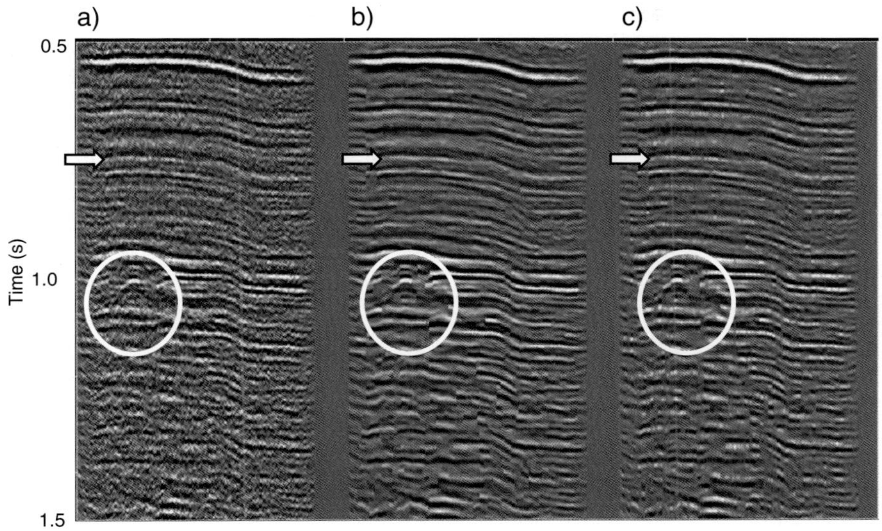

Figure 5.54. Pitfalls of structure-oriented filtering. (a) A vertical slice through original seismic data from a survey acquired in the Illinois Basin, U.S.A. (b)-(c). Vertical slices through the same data (b) after two passes of structure-oriented multiwindow filtering using a mean filter and (c) after two passes of structure-oriented multiwindow filtering using a principal-component filter. Note the conflicting dips of the diffraction tails circled in (a) and the underlying horizontal reflectors. After the filtering in (b) and (c), the tails of this diffraction have been eliminated so that the residual diffraction apex might be misinterepreted to be a carbonate buildup. Seismic data are courtesy of Continental Resources of Illinois, Inc.

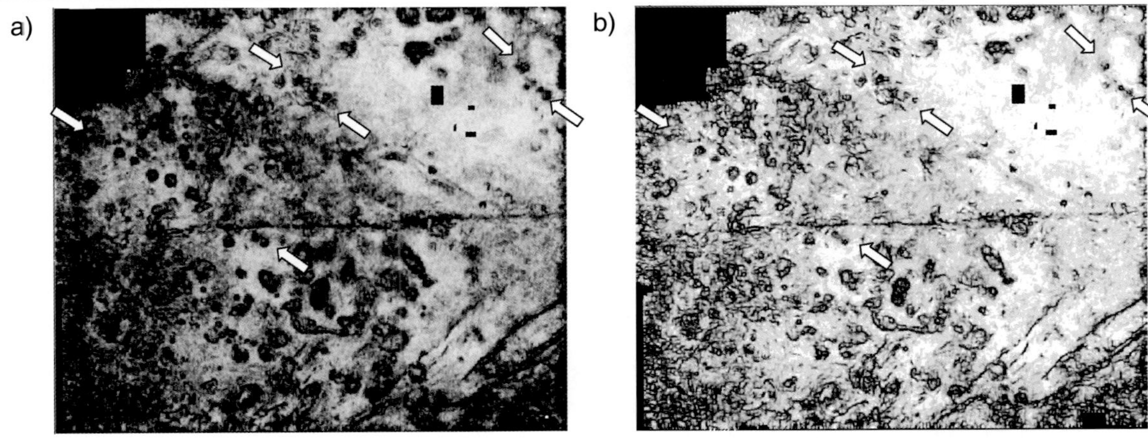

Figure 5.55. Time slices at $t = 1.100$ s through coherence volumes generated for a survey acquired over the Fort Worth Basin, Texas, U.S.A. Coherence calculated (a) from the original seismic data and (b) after two passes of structure-oriented filtering. The effect of structure-oriented filtering has been to diminish the "incoherent" appearance of the aligned collapse features indicated by arrows. Although structure-oriented filtering generally produces sharper attribute images of faults and channels, not all geologic features of interest (such as the collapse features shown here) are represented well by edges. Seismic data are courtesy of Devon Energy.

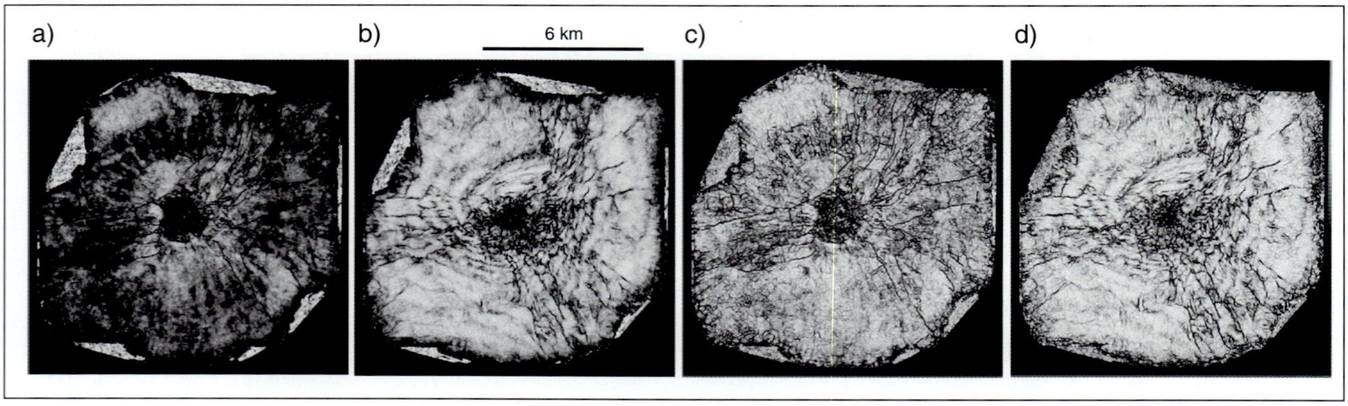

Figure 5.56. Time slices at t = (a) 0.600 and (b) 1.500 s through coherence volumes calculated from a prestack time-migrated seismic survey over Vinton Dome, Louisiana, U.S.A. (c)-(d). The corresponding coherence time slices generated after seismic data had undergone structure-oriented filtering. After Marfurt et al. (2002). Seismic data are courtesy of OPEX.

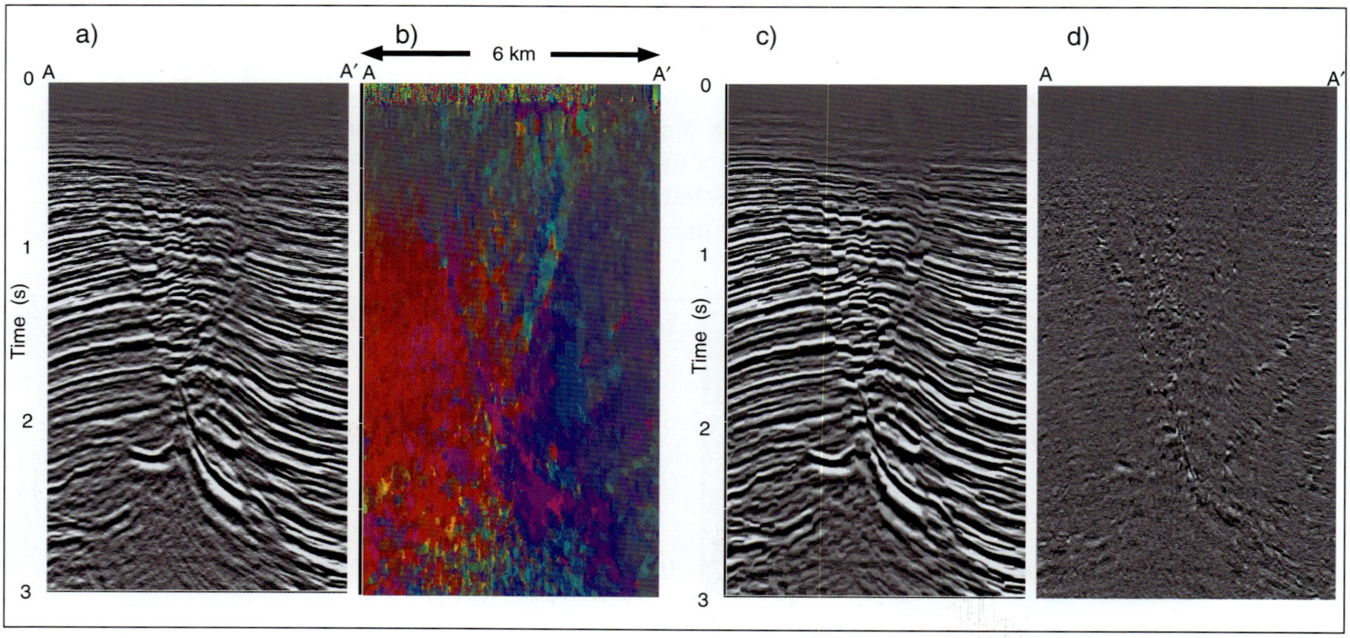

Figure 5.57. A vertical slice (a) through a prestack time-migrated data volume, (b) through the corresponding dip and azimuth cube, and (c) after two passes of structure-oriented filtering using a multiwindow principal-component algorithm. (d) The difference, or rejected "noise," between the data shown in (c) and in (a). The color wheel and location of line AA′ are shown in Figure 5.59. After Marfurt et al. (2002).

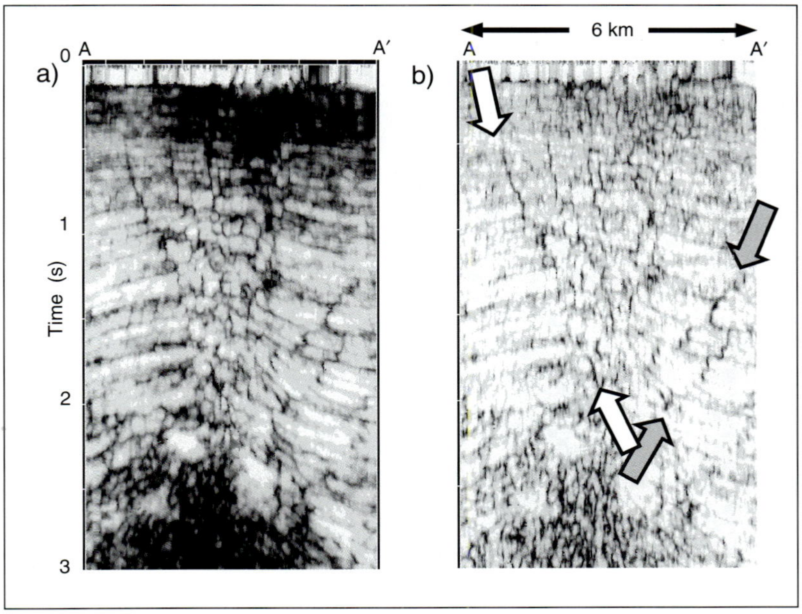

Figure 5.58. Vertical sections through coherence volumes corresponding to Figure 5.57a and 5.57c. Note that although structure-oriented filtering reduces noise in the shallow section (the black area), it does not realign poorly imaged faults. Such misalignment generates the "wormy" faults seen on the coherence image in Figure 5.56d. Although misalignment of faults is the result of poor seismic imaging, faults do not always need to align. In Unit 4, we presented examples of faults misaligned because of brittle deformation of carbonates embedded in plastically deformed shales.

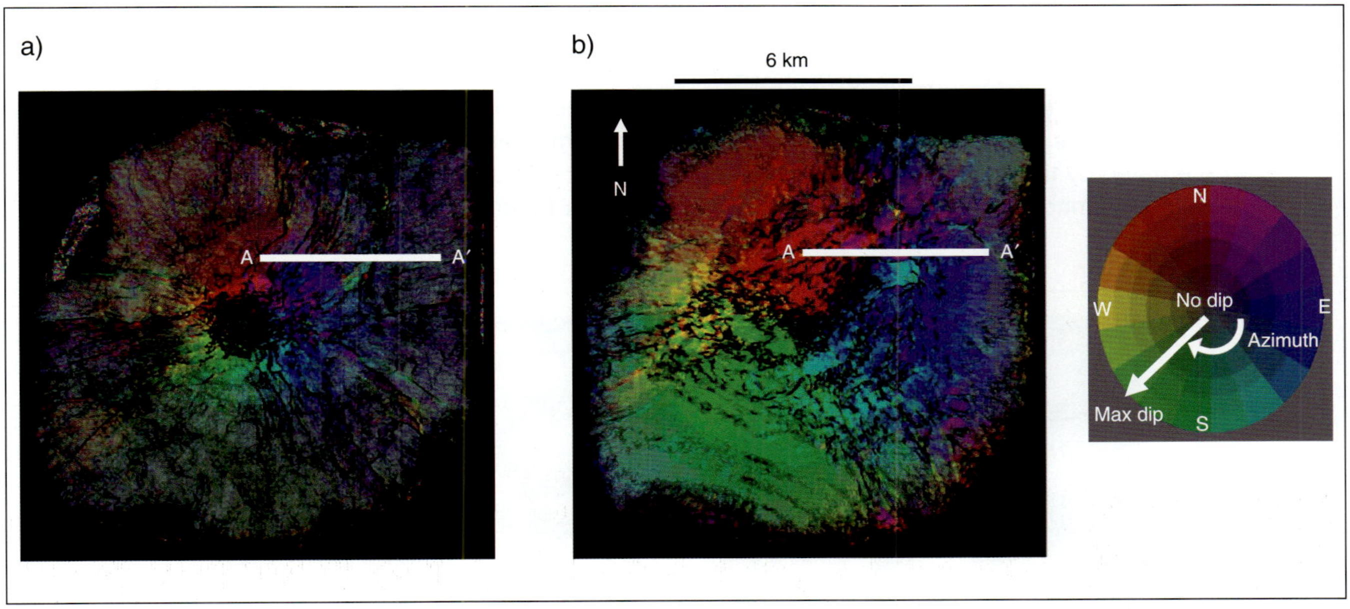

Figure 5.59. Time slices at t = (a) 0.750 and (b) 2.250 s through a dip/azimuth/coherence composite volume, using the techniques described in Unit 1. Note the change in azimuth from north (red) to east (blue), intersected by line AA′. Such a change indicates a rotation of the sediments around a complicated fault zone. After Marfurt et al. (2002).

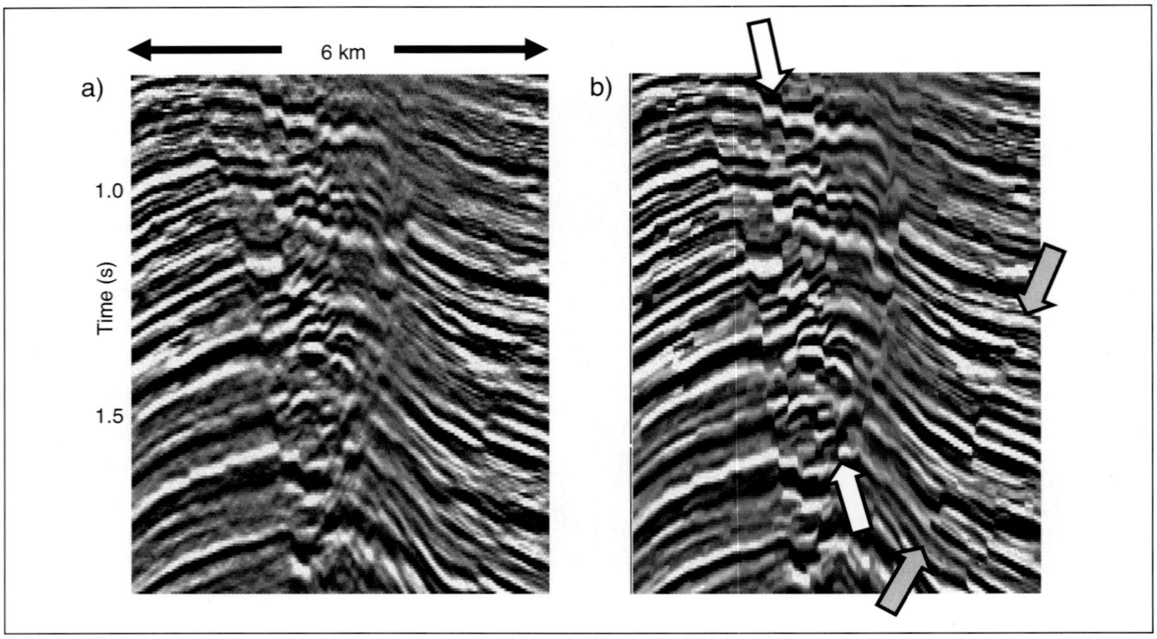

Figure 5.60. Close-up views of data (a) before and (b) after two passes of structure-oriented filtering, shown respectively in Figure 5.57a and 5.57c. Note that although the edges are indeed sharper in (b), they do not line up along smooth faults. Because the geology consists of easily deformed sands and shales, we attribute this misalignment to an inaccurate velocity model, exacerbated by the conflicting reflector dip shown in Figure 5.59. These two factors result in a geologically unacceptable interfingering of reflectors. After Marfurt et al. (2002).

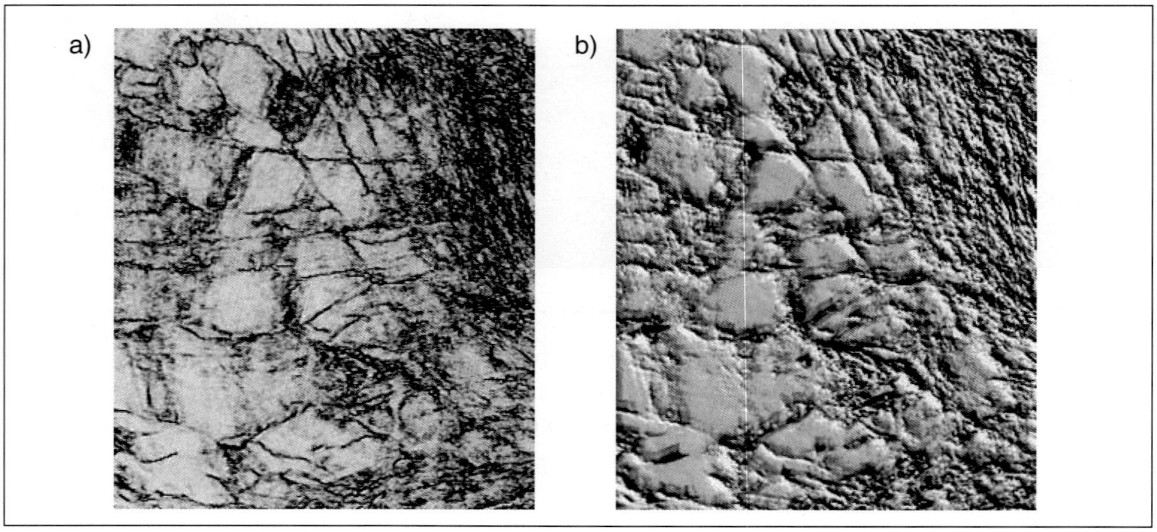

Figure 5.61. Time slices of (a) a coherence volume and (b) a relief-enhanced coherence volume that eliminates many of the low-coherence values associated with horizontal stratigraphy.

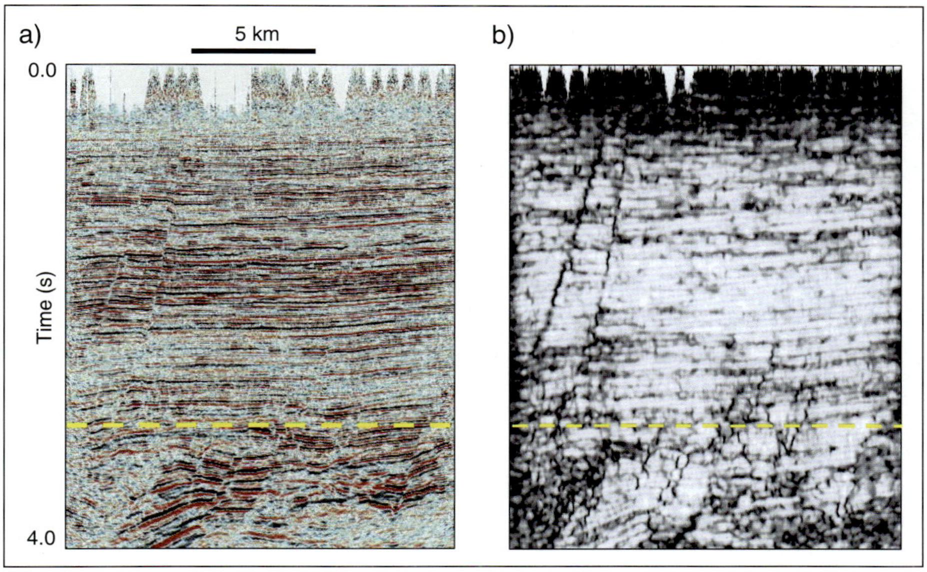

Figure 5.62. Vertical slices through (a) a seismic volume and (b) the corresponding discontinuity volume for a survey acquired in southern Louisiana, U.S.A. Data are courtesy of Seitel, and attribute analysis is courtesy of Landmark Graphics. After Barnes (2005).

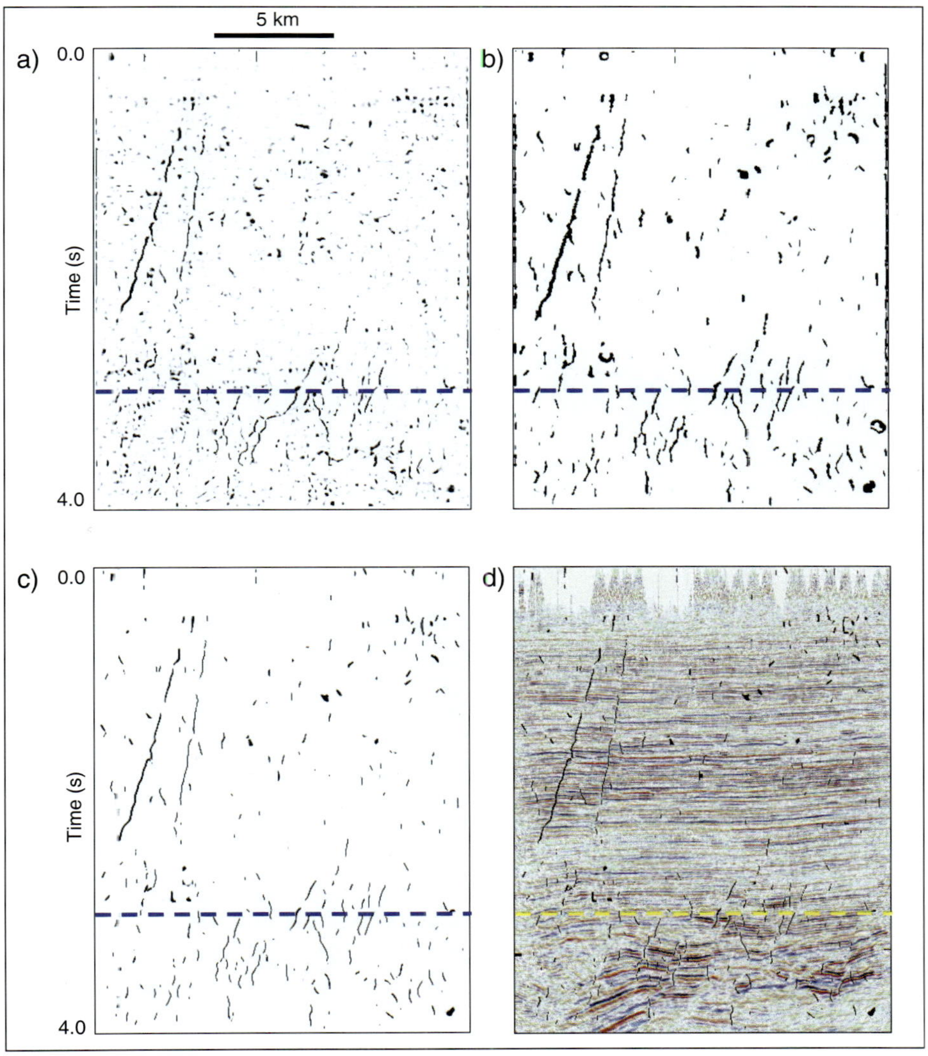

Figure 5.63. The results of successive filtering of the discontinuity volume shown on the vertical slice displayed in Figure 5.64b, after (a) suppression of horizontal discontinuities, (b) image dilation, and (c) image erosion. (d) A composite image of discontinuity and seismic data. After Barnes (2005).

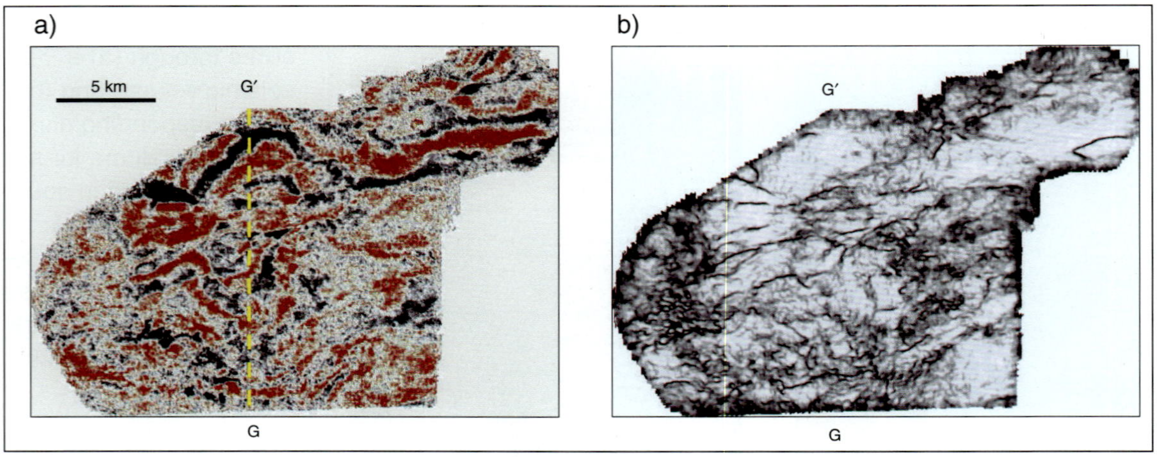

Figure 5.64. (a) A time slice at $t = 3.000$ s through the seismic data and (b) corresponding time slice through the discontinuity volume computed from the survey shown in Figure 5.62. Data are courtesy of Seitel, and attribute analysis is courtesy of Landmark Graphics. After Barnes (2005).

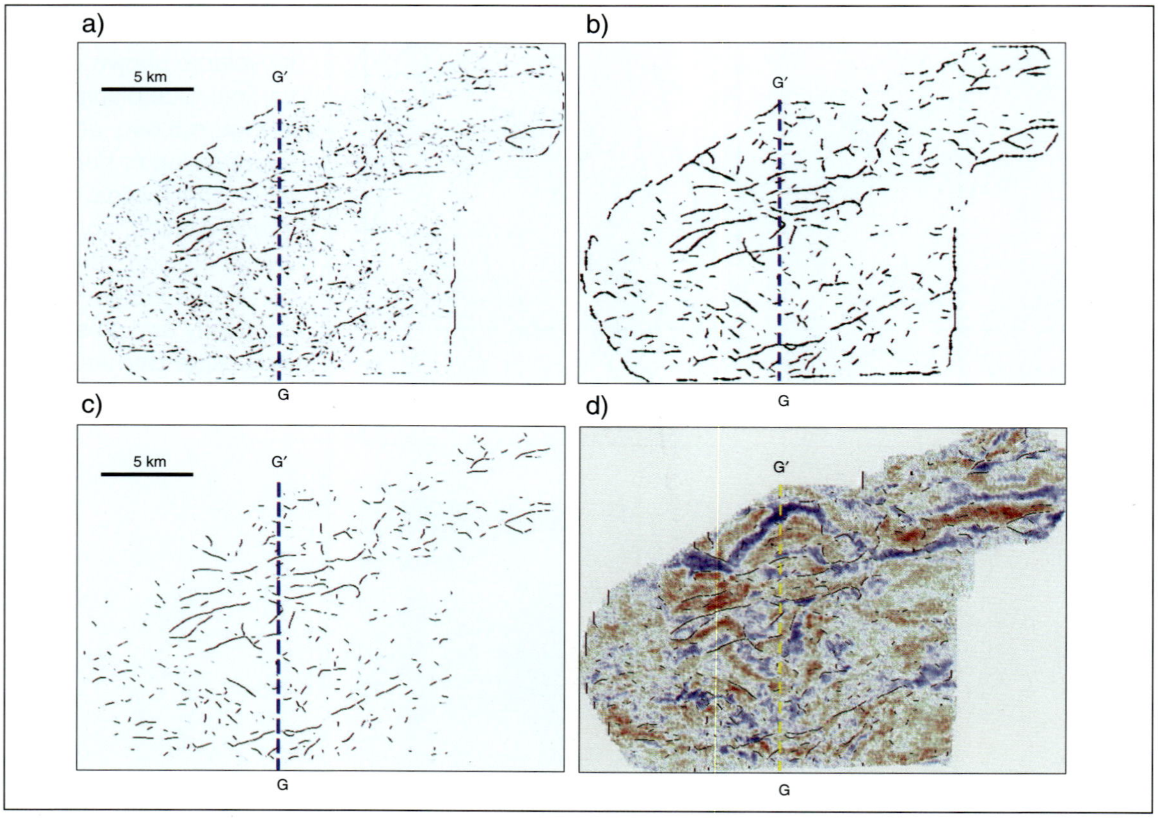

Figure 5.65. The results of successive filtering of the discontinuity volume shown on the time slice at $t = 3.000$ s in Figure 5.63b, after (a) suppression of horizontal discontinuities, (b) image dilation, and (c) image erosion. (d) A composite image of discontinuity and seismic data. After Barnes (2005).

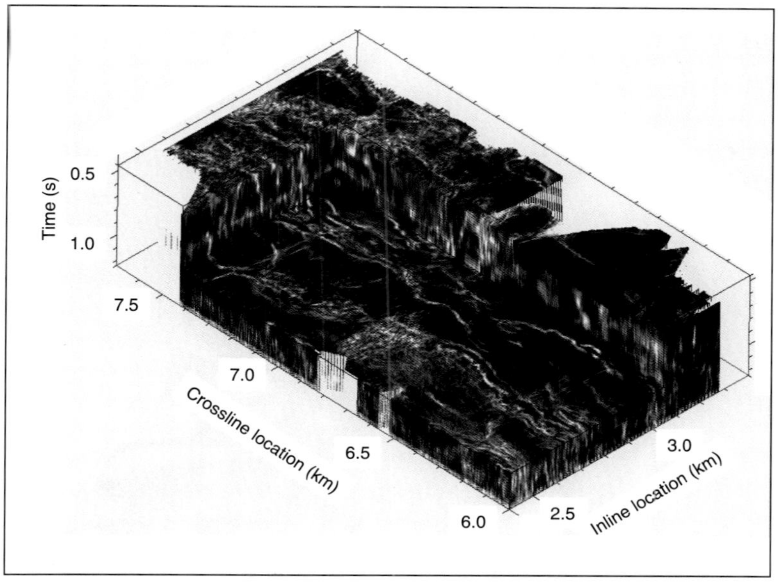

Figure 5.66. A coherence volume generated for the Wytch Farm survey, U. K., that will be used as input to an automatic fault-extraction algorithm. Data are courtesy of BP, Premier Oil, ONEPM, Kerr-McGee Oil (U. K.), and Talisman North Sea. After Dorn (2005).

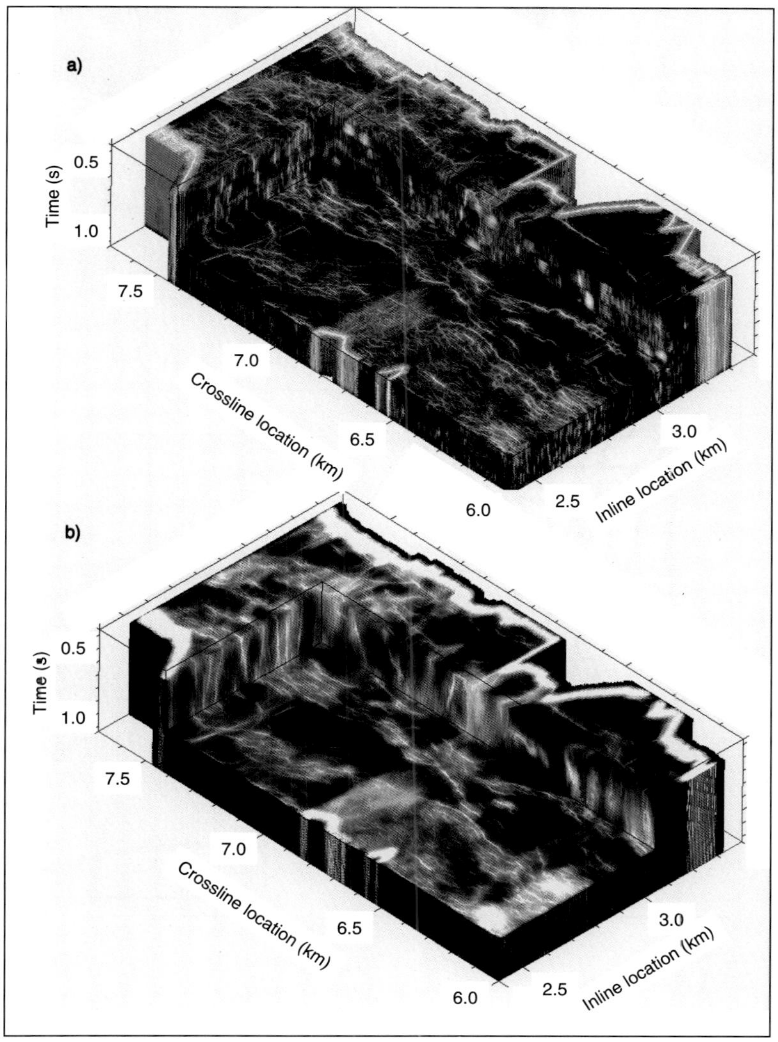

Figure 5.67. (a) A lineament-enhanced and (b) a fault-enhanced volume, both corresponding to the coherence volume in the previous figure. After Dorn (2005).

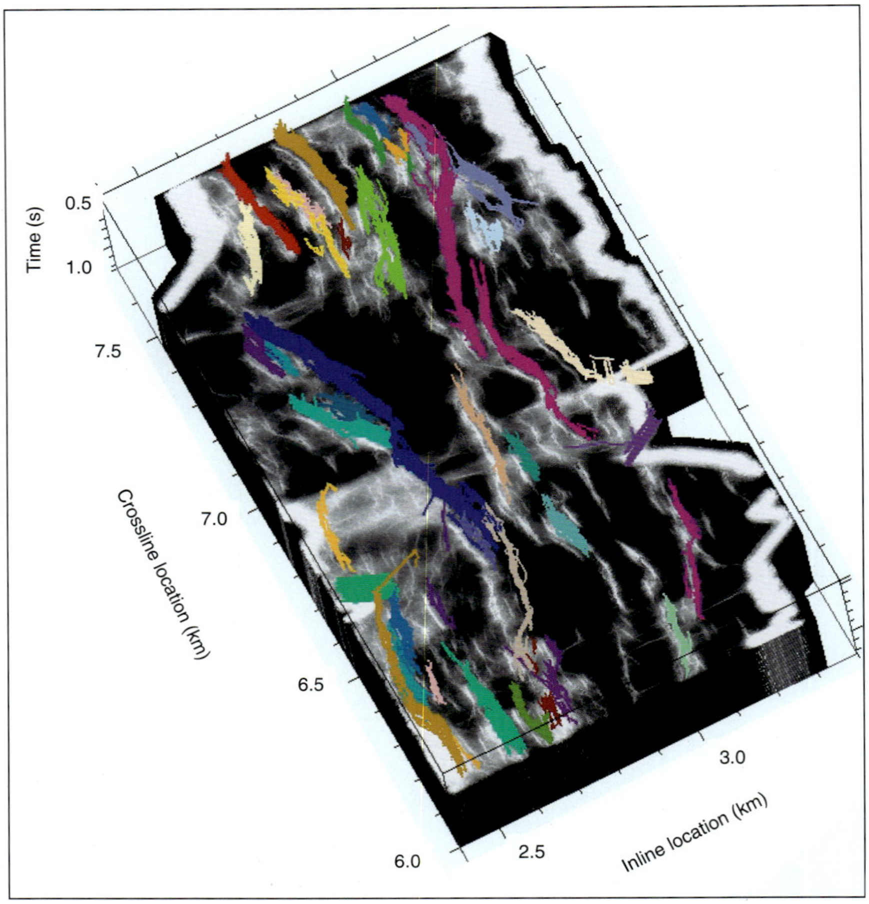

Figure 5.68. The 40 largest faults, of 133 faults extracted using the technique described in the previous two figures. To preserve image clarity, the remaining 83 faults are not shown. After Dorn (2005).

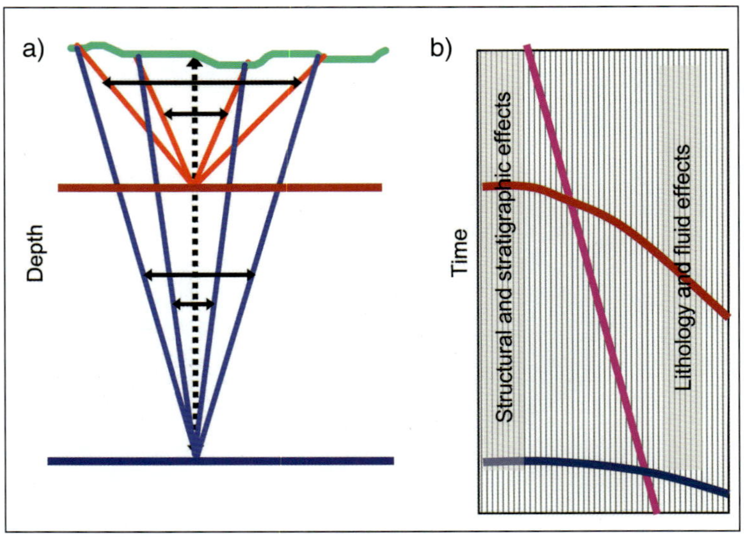

Figure 5.69. Idealized (a) geometry and (b) seismic data corresponding to a common-midpoint (CMP) gather. Often, structural and stratigraphic effects are seen best on near-angle seismic volumes, and lithofluid effects are seen best on far-angle seismic images.

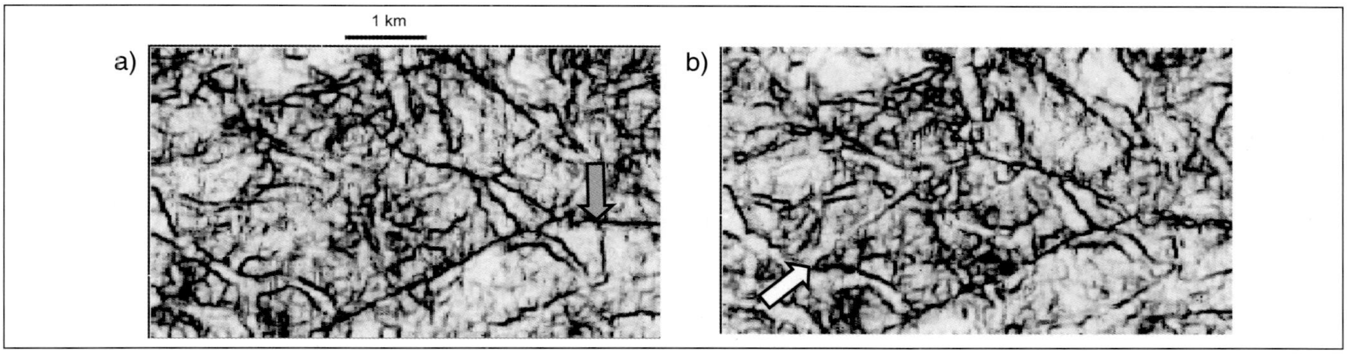

Figure 5.70. Time slices at $t = 1.218$ s through coherence volumes generated from (a) a far-angle and (b) a near-angle migrated stack. The fault indicated by the gray arrow is illuminated better on the far-angle stack, whereas the channel indicated by the white arrow is illuminated better on the near-angle stack. Data are courtesy of Fairfield.

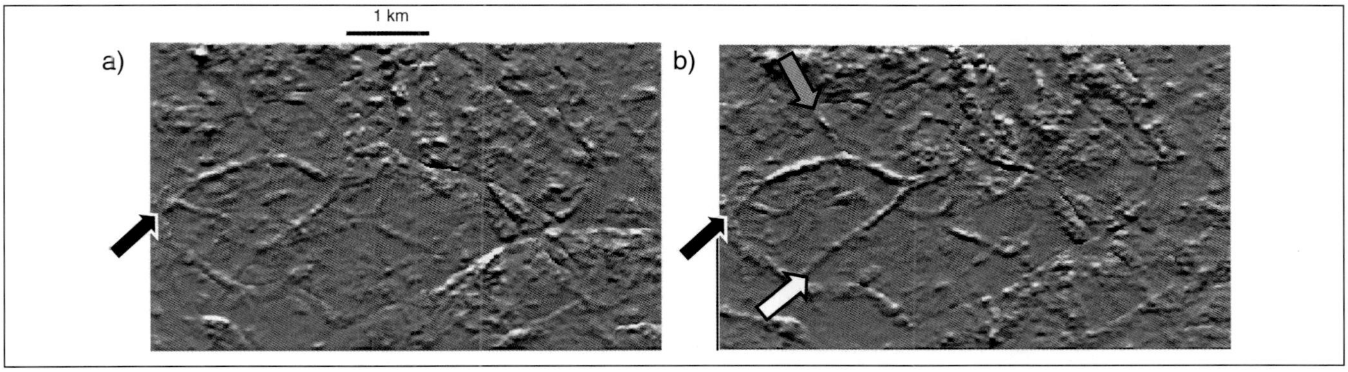

Figure 5.71. Time slices, at $t = 1.218$ s, through north-south energy-weighted coherent-amplitude-gradient volumes generated from (a) a far-angle migrated stack and (b) a near-angle migrated stack. The channel indicated by the black arrow is well illuminated on both the near-angle stack and the far-angle stack. The fault indicated by the gray arrow is illuminated better on the far-angle stack, whereas the channel indicated by the white arrow is illuminated better on the near-angle stack. The data are courtesy of Fairfield.

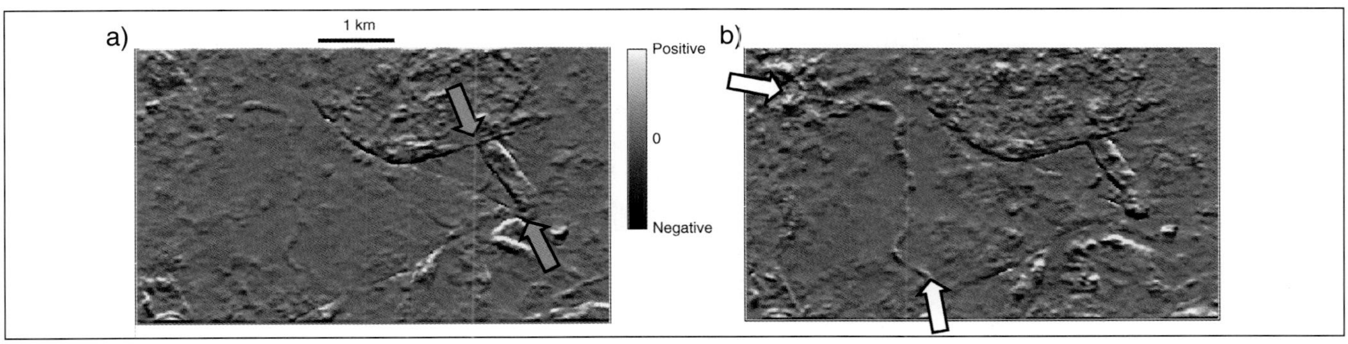

Figure 5.72. Time slices at $t = 1.482$ s through north-south energy-weighted coherent-amplitude-gradient volumes generated from (a) a far-angle and (b) a near-angle migrated stack. The fault indicated by the gray arrow is illuminated better on the far-angle stack, whereas the channel indicated by the white arrow is illuminated better on the near-angle stack. Data are courtesy of Fairfield.

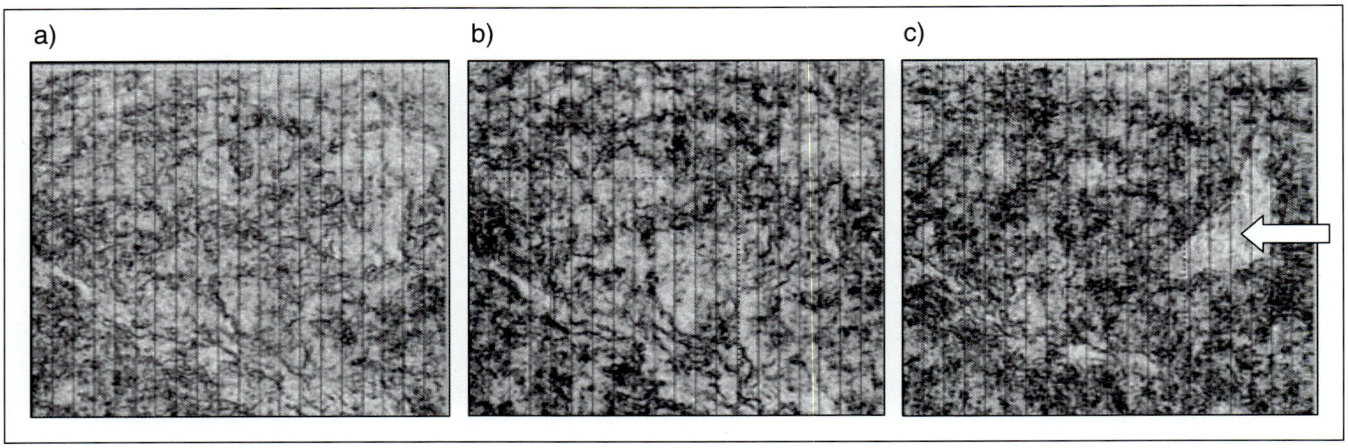

Figure 5.73. Coherence slices through a lower Tertiary gas reservoir for a survey acquired in the North Sea. Coherence was computed from the (a) full-offset, (b) near-offset, and (c) far-offset seismic volumes. The gas response for this AVO class can be seen clearly on the coherence slice of the far-offset volume, as indicated by the arrow in (c), but a gas response cannot be seen on the full-offset or near-offset volume.

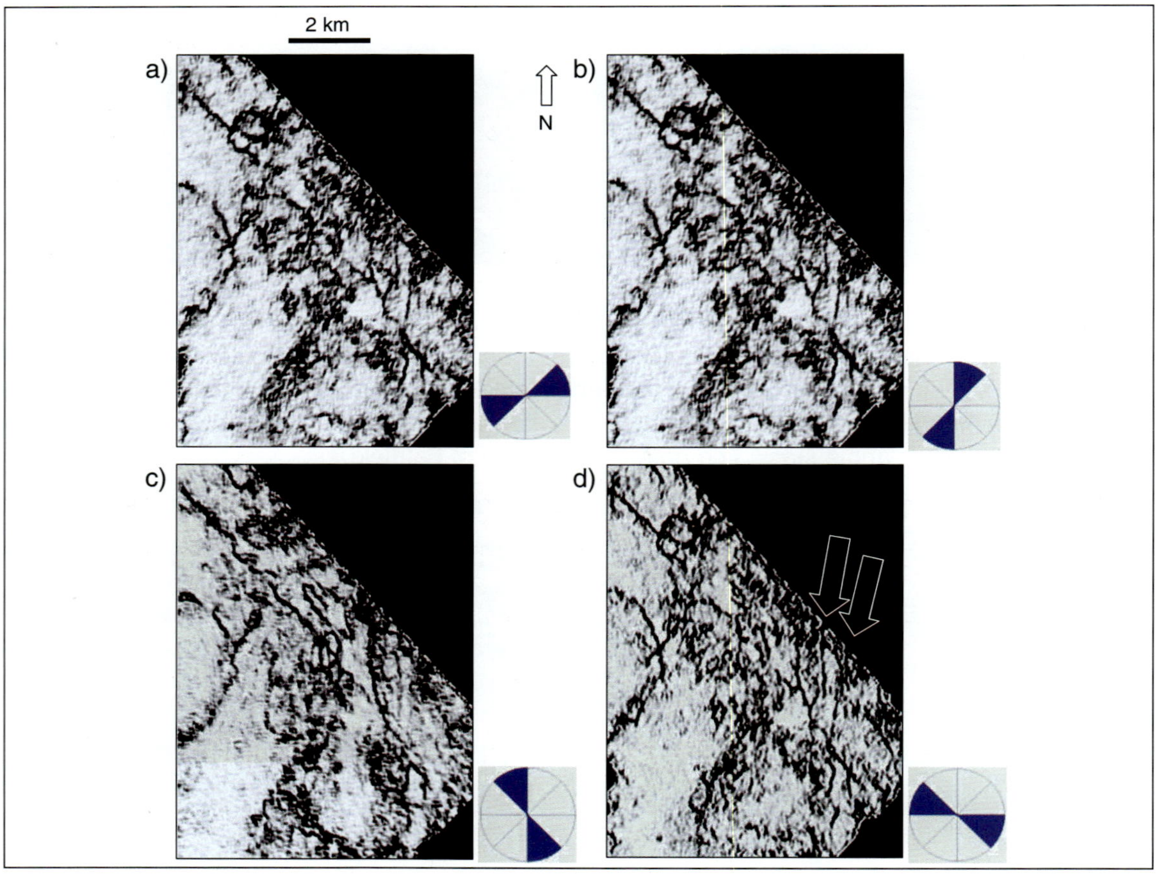

Figure 5.74. Time slices through coherence volumes generated on azimuthally limited seismic data volumes after image sharpening, as described earlier in this unit. The solid blue wedge in each image denotes azimuths.

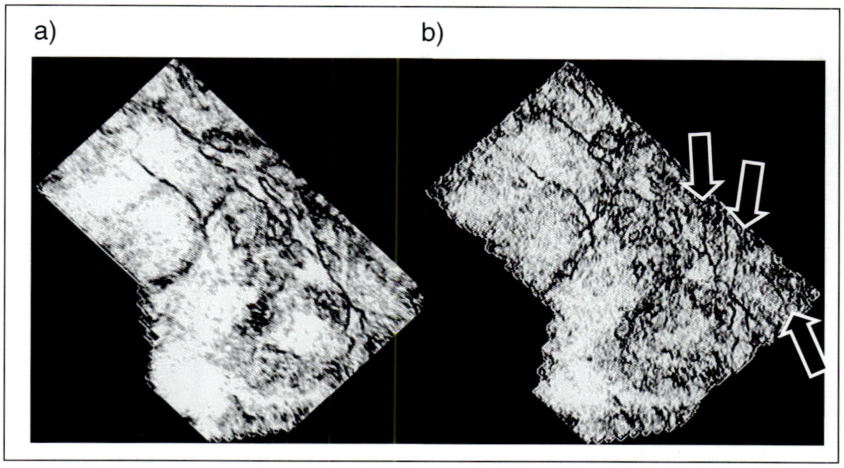

Figure 5.75. The impact of azimuthal binning on coherence. A coherence image run (a) on all azimuths and (b) on azimuths restricted to 45°. Images are courtesy of Core Lab.

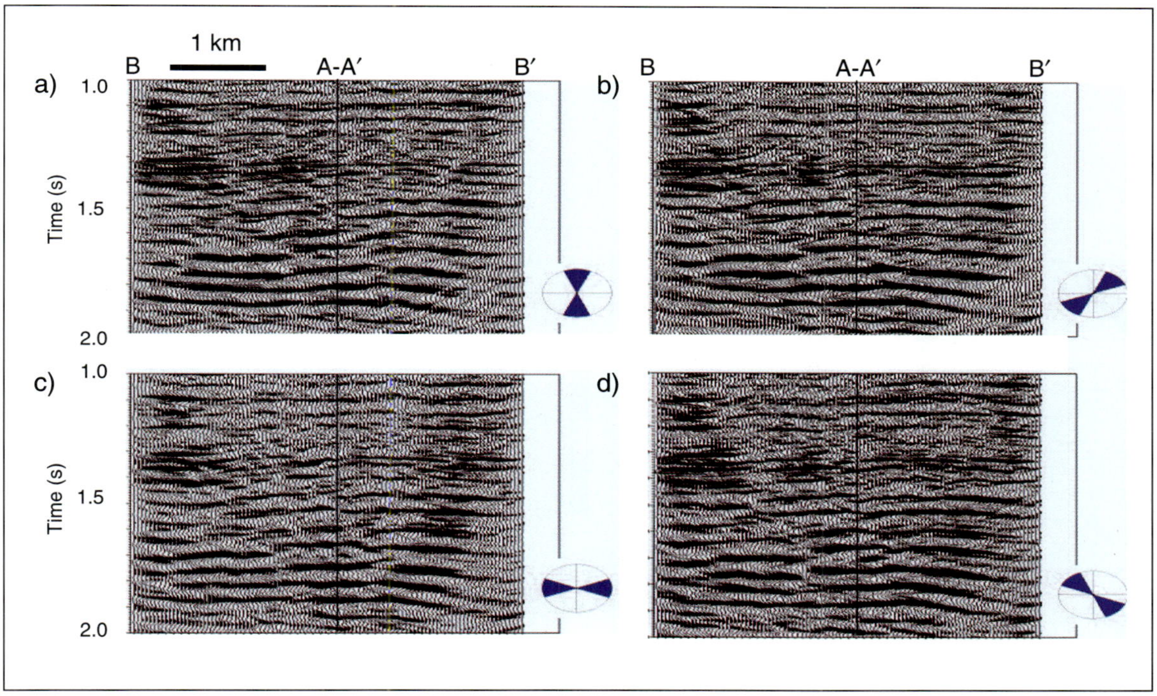

Figure 5.76. Line BB′ from four azimuth-limited seismic volumes at about (a) 0° and 180°, (b) 45° and 225°, (c) 90° and 270°, and (d) 135° and 315°.

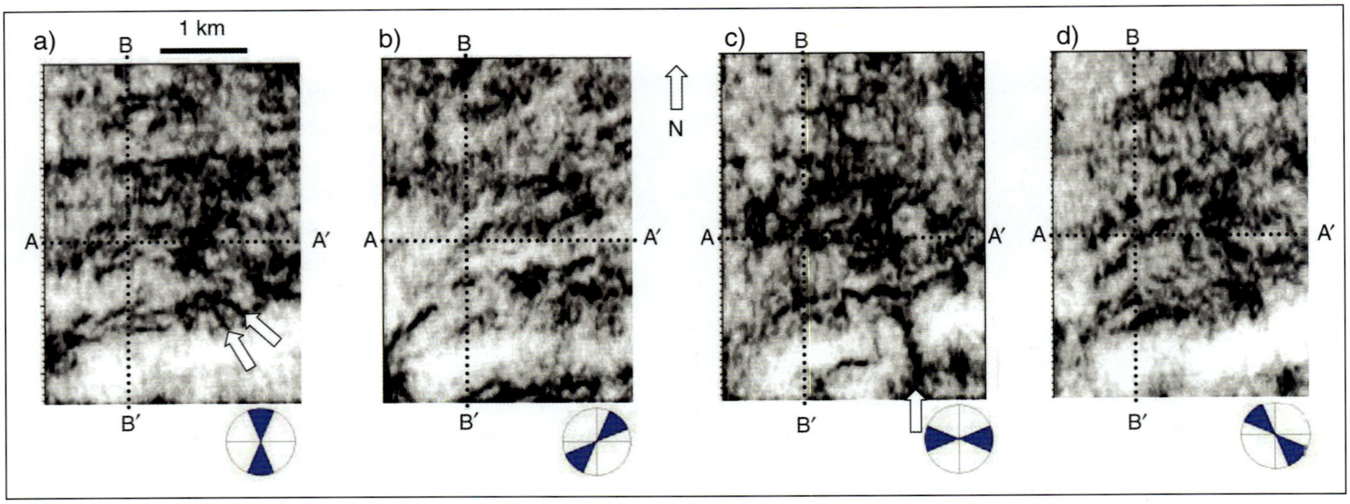

Figure 5.77. Times slices at t = 1.312 s through coherence volumes computed from azimuth-limited seismic volumes shown in the previous figure at about (a) 0° and 180°, (b) 45° and 225°, (c) 90° and 270°, and (d) 135° and 315°.

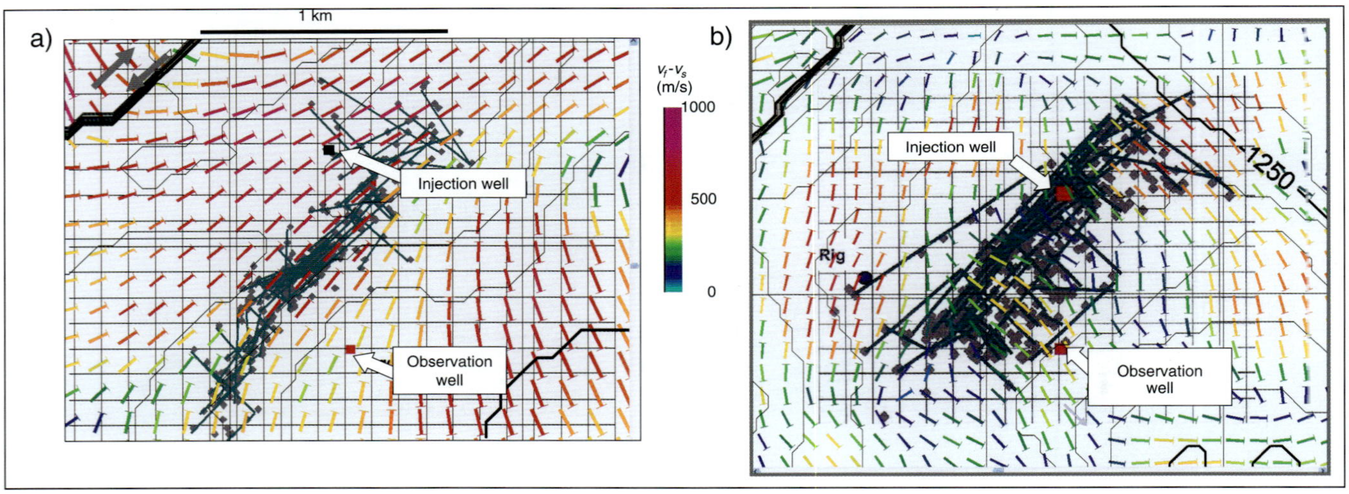

Figure 5.78. Velocity anisotropy at the reservoir level, plotted as icons. The azimuth and length of the icons reflect the azimuth and magnitude of the fast velocity, v_f. The color of the icon is proportional to the difference between the fast and slow velocity values, $v_f - v_s$. Black lines indicate fractures induced by hydrofracturing, mapped here using microseisms. (a) An injection well that resulted in moderate production resulting from fractures that have predominantly only one azmuth, aligned with strong anisotropy (red arrows). (b) An injection well that resulted in excellent production resulting from an intersecting network of fractures, associated with weak anisotropy (green and blue arrows). After Simon (2005). Data are courtesy of Devon Energy.

Figure 5.79. P-wave anisotropy, overplotted onto maximum curvature. Note how the azimuth of the fast velocity, v_f, lines up with the ridges and valleys seen in the curvature. After Simon (2005). Data are courtesy of Devon Energy.

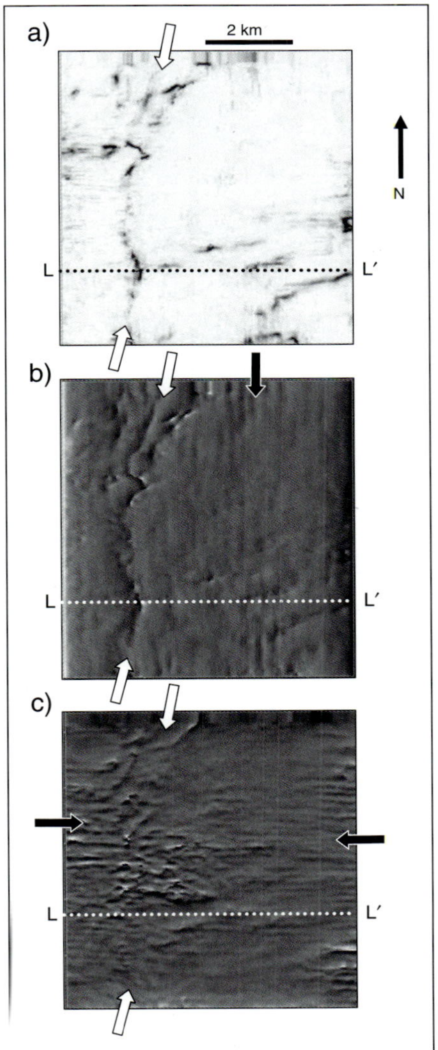

Figure 5.80. Time slices near the Escandalosa horizon, Barinas Basin, Venezuela, through attributes generated from the vertical-component P-P data volume through (a) a coherence volume, (b) an east-west component of the energy-weighted coherent-amplitude gradient, and (c) a north-south component of the energy-weighted coherent-amplitude gradient. Black arrows indicate an acquisition footprint. White arrows indicate a reverse fault. After D'Agosto-Palladino (2003). Data are courtesy of PdVSA.

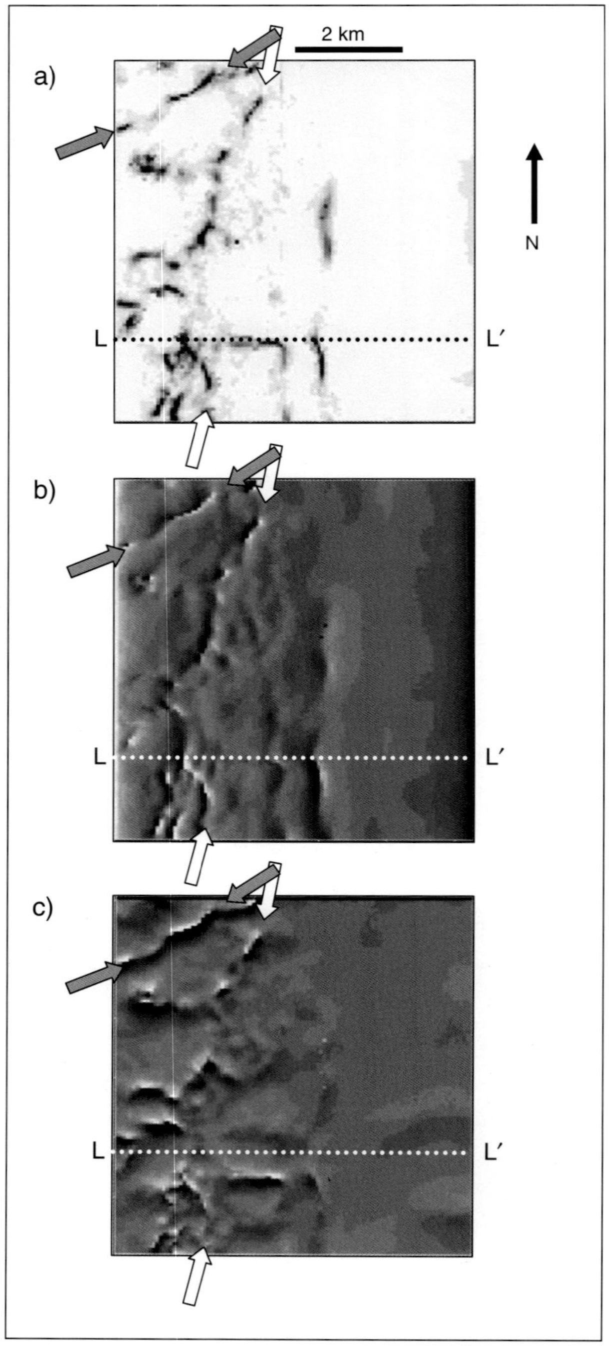

Figure 5.81. Time slices near the Escandalosa horizon through attributes generated from the radial-component P-S data volume through (a) a coherence volume, (b) an east-west component of the energy-weighted coherent-amplitude gradient, and (c) a north-south component of the energy-weighted coherent-amplitude gradient. White arrows indicate the reverse vault shown in the previous figure. Gray arrows indicate a northeast-southwest-trending fault not seenon P-P images. After D'Agosto-Pallodino, 2003. Data are courtesy of PdVSA.

References

Adeogba, A. A., T. R. McHargue, and S. A. Graham, 2005, Transient fan architecture and depositional controls from near-surface 3-D seismic data, Niger Delta continental slope: AAPG Bulletin, **89**, 627–643.

Al-Bannagi, M. S., K. Fang, P. G. Kelamis, and G. S. Douglass, 2005, Acquisition footprint suppression via the truncated SVD technique: Case studies from Saudi Arabia: The Leading Edge, **24**, 832–834.

al-Dossary, S., and K. J. Marfurt, 2003, Improved 3-D seismic edge detection filter applied to Vinton Dome, Louisiana: 73rd Annual International Meeting, SEG, Expanded Abstracts, 2370–2372.

Anstey, N., 2005, Attributes in color: The early years: CSEG Recorder, **30**, March, 12–15.

Bahorich, M. S., and S. L. Farmer, 1995, 3-D seismic coherency for faults and stratigraphic features: The Leading Edge, **14**, 1053–1058.

Bahorich, M.S., J. A. Lopez, N. L. Haskell, S. E. Nissen, and A. Poole, 1995, Stratigraphic and structural interpretation with 3-D coherence: 65th Annual International Meeting, SEG, Expanded Abstracts, 97–100.

Bahorich, M., A. Motsch, K. Laughlin, and G. Partyka, 2002, Amplitude responses image reservoir: Hart's E & P, January, 59–61.

Bakker, P., 2003, Image structure analysis for seismic interpretation: Ph.D. thesis, Technische Universiteit Delft.

Barnes, A. E., 2000a, Weighted average seismic attributes: Geophysics, **65**, 275–285.

———, 2000b, Attributes for automated seismic facies analysis: 70th Annual International Meeting, SEG, Expanded Abstracts, 553–556.

———, 2002, What a relief shade can be: AAPG Explorer, **8**, http://www.aapg.org/explorer/geophysical_corner/2002/08gpc.cfm (accessed June 22, 2005).

———, 2005, Fractal analysis of fault attributes derived from seismic discontinuity data: 67th Annual Conference and Technical Exhibition, EAGE, Extended Abstracts, P318.

Bergbauer, S., T. Mukerji, and P. Hennings, 2003, Improving curvature analyses of deformed horizons using scale-dependent filtering techniques: AAPG Bulletin, **87**, 1255–1272.

Biles, N. E., A. E. Hannan, G. A. Jamieson, A. Krueger, D. L. Shelander, and F. Snyder, 2003, Regional evaluation and hydrocarbon potential of the deep section of the Gulf of Mexico offshore Louisiana continental shelf from modern 3D seismic data: 73rd Annual International Meeting, SEG, Expanded Abstracts, 414–417.

Blumentritt, C. H., E. C. Sullivan, and K. J. Marfurt, 2003, Channel detection using seismic attributes on the Central Basin Platform, west Texas: 73rd Annual International Meeting, SEG, Expanded Abstracts, 466–469.

Brown, R. L., 1994, Image quality depends on your point of view: The Leading Edge, **13**, 669–673.

Bruhl, M., G. J. Vermeer, and M. Keihn, 1996, Fresnel zone for broadband data: Geophysics, **61**, 600–604.

Bubb, J. N., and W. G. Hatledid, 1978, Seismic stratigraphy and global changes of sea level, Part 10: Seismic recognition of carbonate buildups: AAPG Bulletin, **62**, 772–791.

Carr, T. R., S. Bhattacharya, and R. Pancake, 2000, Horizontal drilling in Kansas: A case history: http://www.kgs.ku.edu/Workshops/Horz2000/R640/ (accessed July 26, 2005).

Castagna, J. P., S. Sun, and R. W. Siegfried, 2003, Instantaneous spectral analysis: Detection of low-frequency shadows associated with hydrocarbons: The Leading Edge, **22**, 120–127.

Chopra, S., 2001, Adding the coherence dimension to 3D seismic data: CSEG Recorder, **26**, no. 1, 5–8.

———, 2002, Coherence cube and beyond: First Break, **20**, 27–33.

Chopra, S., A. Alexeev, and V. Sudhakar, 2003, High-frequency restoration of surface seismic data: The Leading Edge, **22**, 730–738.

Chopra, S., and I. McConnell, 2004, Using interwell chemical tracers and the coherence cube to understand reservoir communication, Oil and Gas Journal, May 17, 37–42.

D'Agosto-Palladino, C., 2003, Birefringence analysis at Borburata Field for fracture characterization: M.S. thesis, University of Houston.

Deptuck, M. E., 2003, Post-rift geology of the Jeanne d'Arc Basin, with a focus on the architecture and evolution of early Paleogene submarine fans, and insights from modern deep-water systems: Ph.D. thesis, Dalhousie University, Halifax, Nova Scotia, Canada.

Dorn, G. A., H. E. James, and L. Evins, 2005, Automatic fault extraction (AFE) in 3-D seismic data: CSEG National Convention, 247–250.

Drummond, J., J. Ryan, and R. Kasmi, 2001, Adapting to noisy 3-D data: Enhancing Algerian giant field development through strategic planning of 3-D seismic in Berkine Basin: The Leading Edge, **20**, 718–728.

Famini, J., 2005, Impact of seismic processing on seismic attribute analysis: Application to a survey from the Delaware Basin, NM, USA: M.S. thesis, University of Houston.

French, W. S., 1975, Computer migration of oblique seismic reflection profiles: Geophysics, **40,** 961–980.

Gallagher, J. W., and R. Heggland, 1995, Shallow gas evaluations based on conventional 3-D seismic data: 65th Annual International Meeting, SEG, Expanded Abstracts, 518–520.

Gao, D., 2003, Volume texture extraction for 3-D seismic visualization and interpretation: Geophysics, **68,** 1294–1302.

Gersztenkorn, A., and K. J. Marfurt, 1996, Eigenstructure based coherence computations: 66th Annual International Meeting, SEG, Expanded Abstracts, 328–331.

Gersztenkorn, A., J. A. Sharp, and K. J. Marfurt, 1999, Delineation of tectonic features offshore Trinidad using 3-D seismic coherence: The Leading Edge, **18**, 1000–1008.

Goloshubin, G., V. Korneev, and V. Vingalov, 2002, Seismic low-frequency effects from oil-saturated reservoir zones: 72nd Annual International Meeting, SEG, Expanded Abstracts, 1813–1816.

Hart, B. S., R. A. Pearson, and G. C. Rawling, 2002, 3-D seismic horizon-based approaches to fracture-swarm sweet spot definition in tight-gas reservoirs: The Leading Edge, **21**, 28–35.

Haskell, N., S. E. Nissen, M. J. S. Hughes, J. Grindhaug, S. Dhanani, R. P. Heath, J. D. Kantorowicz, L. Antrim, M. Cubanski, R. Nataraj, M. Schilly, and S. Wigger, 1999, Delineation of geological drilling hazards using 3-D seismic attributes: The Leading Edge, **18**, 373–382.

Haskell, N. L., S. E. Nissen, J. A. Lopez, and M. S. Bahorich, 1995, 3-D seismic coherency and the imaging of sedimentological features: 65th Annual International Meeting, SEG, Expanded Abstracts, 1532–1534.

Heggland, R., 2004, Definition of geohazards in exploration 3-D seismic data using attributes and neural-network analysis: AAPG Bulletin, **88**, 857–868.

Hilterman, F. J., 1982, Interpretative lessons from three-dimensional modeling: Geophysics, **47**, 784–808.

Hoecker, C., and G. Fehmers, 2002, Fast structural interpretation with structure-oriented filtering: The Leading Edge, **21**, 238–243.

Hughes, M. J., S. Dhanani, R. K. Frimpong, M. Gainski, N. L. Haskell, R. P. Heath, J. D. Kantorowicz, P. M. Maguire, and S. E. Nissen, 1999, Applications of the coherency cube in the UKCS, *in* A. J. Fleet and S. A. R. Bouldy, eds., Petroleum Geology of Northwest Europe: Proceedings of the 5th Conference: Geological Society of London, 1299–1305.

Johann, P., G. Ragignin, and M. Spinola, 2003, Spectral decomposition reveals geological hidden features in the amplitude maps from a deep-water reservoir in the Campos Basin: 73rd Annual International Meeting, SEG, Expanded Abstracts, 1740–1743.

Kerans, C., 1990, Depositional systems and karst geology of the Ellenburger Group (Lower Ordovician), subsurface west Texas: University of Texas at Austin, Bureau of Economic Geology Report of Investigations, **193**.

Knobloch, C., 1982, Pitfalls and merits of interpreting color displays of geophysical data: 52nd Annual International Meeting, SEG, Expanded Abstracts, 112.

Kozlov, E., N. Barasky, E. Korolev, A. Antonenko, and E. Koshchuk, E., 2004, Imaging scattering objects masked by specular reflections: 74th Annual International Meeting, SEG, Expanded Abstracts, 1131–1134.

Laughlin, K., P Garossino, and G. Partyka, 2002, Spectral decomp applied to 3D: AAPG Explorer, May issue.

Lazarevic, I., 2004, Factors impacting lateral velocity resolution: M.S. thesis, University of Houston.

Li, C.-F., and C. L. Liner, 2004, Wavelet-based analysis of singularities in seismic data: U. S. Patent 6 745 129.

Lin, I., K. J. Marfurt, and O. Johnson, 2003, Mapping 3-D multiattribute data into HLS color space: Applications to Vinton Dome, LA: 73rd Annual International Meeting, SEG, Expanded Abstracts,1728–1731.

Liner, C., C.-F. Li, A. Gersztenkorn, and J. Smythe, 2004, SPICE: A new general seismic attribute: 72nd Annual International Meeting, SEG, Expanded Abstracts, 433–436.

Lisle, R. J., 1994, Detection of zones of abnormal strains in structures using Gaussian curvature analysis: AAPG Bulletin, **78**, 1811–1819.

Liu, J., and K. J. Marfurt, 2005, Matching pursuit decomposition using Morlet wavelets: 75th Annual International Meeting, SEG, Expanded Abstracts, 786–790.

Luo, Y., S. al-Dossary, and M. Alfaraj, 2002, Edge-preserving smoothing and applications: The Leading Edge, **21**, 136–158.

Luo, Y., S. al-Dossary, and M. Marhoon, 2001, Generalized Hilbert transform and its application in geophysics: 71st Annual International Meeting, SEG, Expanded Abstracts, 1835–1838.

Luo, Y., S. al-Dossary, M. Marhoon, and M. Alfaraj, 2003, Generalized Hilbert transform and its application in geophysics: The Leading Edge, **22,** 198–202.

Maione, S. J., 2001, Discovery of ring faults associated with salt withdrawal basins, Early Cretaceous age, in the East Texas Basin: The Leading Edge, **20**, 818–829.

Marfurt, K., W. S. Duncan, P. Constance, 2002, Comparison of 3-D edge detection seismic attributes to Vinton Dome, Louisiana: 72nd Annual International Meeting, SEG, Expanded Abstracts, 577–580.

Marfurt, K. J., and R. L. Kirlin, 2000, 3-D broadband estimates of reflector dip and amplitude: Geophysics, **65**, 304–320.

———, 2001, Narrow-band spectral analysis and thin-bed tuning: Geophysics, **66**, 1274–1283.

Marfurt, K. J., R. L. Kirlin, S. H. Farmer, and M. S. Bahorich, 1998a, 3-D seismic attributes using a running window semblance-based algorithm: Geophysics, **63**, 1150–1165.

Marfurt, K. J., R. M. Scheet, J. A. Sharp, and M. G. Harper, 1998b, Suppression of the acquisition footprint for seismic sequence attribute mapping: Geophysics, **63**, 1024–1035.

Marfurt, K. J., V. Sudhakar, A. Gersztenkorn, K. D. Crawford, and S. E. Nissen, 1999, Coherency calculations in the presence of structural dip: Geophysics, **64**, 104–111.

Masaferro, J. L., R. Bourne, and J.-C. Jauffred, 2003, 3-D seismic imaging of carbonate reservoirs and structures: The Leading Edge, **22**, 18–25.

Meyer, D. E., E. L. Harvey, T. E. Bulloch, J. C. Voncannon, and T. M. Sheffield, 2001, Use of seismic attributes in 3-D geovolume interpretation: The Leading Edge, **20,** 1377–1380, 1400.

Nissen, S. E., T. R. Carr, and K. J. Marfurt, 2005, Using new 3-D seismic attributes to identify subtle fracture trends in mid-continent Mississippian carbonate reservoirs: AAPG Annual Meeting, Calgary, Alberta, Canada, Abstracts, #90039.

Nissen, S. E., N. L. Haskell, C. Steiner, and K. Coterill, 1999, Debris flow outrunner blocks on the Nigerian continental slope delineated by 3-D seismic coherency: The Leading Edge, **18**, 595–599.

Oyedele, O., 2005, 3-D high resolution seismic imaging of deep water systems, SE Green Canyon, Sigsbee Escarpment, Gulf of Mexico: M.S. thesis, University of Houston.

Partyka, G., 2001, Seismic thickness estimation: Three approaches, pros and cons: 71st Annual International Meeting, SEG, Expanded Abstracts, 503–506.

Partyka, G., J. Gridley, and J. A. Lopez, 1999, Interpretational applications of spectral decomposition in reservoir characterization: The Leading Edge, **18**, 353–360.

Peyton, L., R. Bottjer, and G. Partyka, 1998, Interpretation of incised valleys using new 3-D seismic techniques: A case history using spectral decomposition and coherency: The Leading Edge, **17**, 1294–1298.

Prior, D. B., B. D. Bornhold, and M. W. Johns, 1984, Depositional characteristics of a submarine debris flow: Journal of Geology, **92**, 707–727.

Reimer, J. D., and M. R. Teare, 1991, Reservoir development and resource emplacement in selected Paleozoic carbonates of northeastern British Columbia and northwestern Alberta: Canadian Society of Petroleum Geologists, Annual Meeting Abstracts, 116.

———, 1992, Thermochemical sulfate reduction and hydrothermal dolomitization (TSR-HTD): A diagenetic process that created and modified Middle Devonian gas reservoirs in northeastern British Columbia: Core-poster presentation, CSPG-AAPG meeting, Calgary, Alberta, Canada.

Rietveld, W. E., J. H. Kommedal, and K. J. Marfurt, 1999, The effect of prestack depth migration on 3-D seismic attributes: Geophysics, **64**, 1553–1561.

Roberts, A., 2001, Curvature attributes and their application to 3D interpreted horizons: First Break, **19**, 85–99.

Sagan, J., and B. Hart, 2004, Seismic and structural analysis of a Trenton–Black River hydrothermal dolomite reservoir: Search and Discovery Article #40129, http://www.searchanddiscovery.com/documents/2004/hart/index.htm (accessed June 22, 2005).

Saller, A., B. Ball, S. Robertson, B. McPherson, C. Wene, R. Nims, and J. Gogas, 2001, Reservoir characteristics of Devonian cherts and their control on oil recovery: Dollarhide field, west Texas: AAPG Bulletin, **85**, 35–50.

Saller, A. H., I. A. T. Noah, A. P. Ruzuar, and R. Schneider, 2004, Linked lowstand delta to basin-floor fan deposition, offshore Indonesia: An analog for deep-water reservoir systems: AAPG Bulletin, **88**, 21–46.

Serrano, I., C. H. Blumentritt, E. C. Sullivan, K. J. Marfurt, and M. Murphy, 2003, Seismic attribute delineation of lineaments and reservoir compartmentalization: An example from the Devonian Dollarhide field, Central Basin Platform, west Texas: 73rd Annual International Meeting, SEG, Expanded Abstracts, 1744–1747.

Sheriff, R. E., and L. Geldart, 1995, Exploration seismology: Cambridge University Press.

Short, N. M. Sr., and R. W. Blair Jr., 1986,Geomorphology from space: Goddard Space Center Data and Information Service Center, http://disc.sci.gsfc.nasa.gov/geomorphology (accessed June 22, 2005).

Sigismondi, E. M., and C. J. Soldo, 2003, Curvature attributes and seismic interpretation: Case studies from Argentina basins: The Leading Edge, **22**, 1122–1126.

Simon, Y. S., 2005, Stress and fracture characterization in a shale reservoir, north Texas, using correlation between new seismic attributes and well data: M.S. thesis, University of Houston.

Skirius, C., S. Nissen, N. Haskell, K. Marfurt, S. Hadley, D. Ternes, K. Michel, I. Reglar, D. DíAmico, F. Deliencourt, T. Romero, R. DíAngelo, and B. Brown, 1999, 3-D seismic attributes applied to carbonates: The Leading Edge, **18**, 384–393.

Smythe, J., A. Gersztenkorn, B. Radovich, C.-F. Li, and, C. Liner, 2004, SPICE: Layered Gulf of Mexico shelf framework from spectral imaging: The Leading Edge, **23**, 921–926.

Story, C., P. P. Peng, C. Heubeck, C. Sullivan, and L. J. Dong, 1999, An integrated study of the Liuhua 11-1 Field using an ultra high resolution 3D seismic dataset: South China Sea: 69th Annual International Meeting, SEG, Expanded Abstracts, 905–908.

Taner, M. T., 2000, Attributes revisited: http://www.rocksolidimages.com/pdf/attrib_revisited.htm (accessed June 22, 2005).

Taner, M. T., F. Koehler, and R. E. Sheriff, 1979, Complex seismic trace analysis: Geophysics, **44**, 1041–1063.

Veeken, P., M. Rauch-Davies, R. Gallardo, R. Villasenor, and D. Tebar, 2002, Model-driven inversion of the Fortuna National 3D seismic data in the Macuspana Basin, onshore Mexico: 72nd Annual International Meeting, SEG, Expanded Abstracts, 898–901.

Wessels, R., N. Haskell, and G. A. Partyka, 1996, Satellite imaging technology applied to multiple-attribute maps: Presentation to the Geophysical Society of Tulsa.

West, B., S. May, J. E. Eastwood, and C. Rossen, 2002, Interactive seismic facies classification using textural and neural networks: The Leading Edge, **21**, 1042–1049.

Widess, M. B., 1973, How thin is a thin bed?: Geophysics, **38**, 1176–1254.

Woodward, D. L., and P. J. Flynn, 2004, Finger surface as a biometric identifier: Biometric Consortium Conference, Arlington, Virginia.